Springer Collected Works in Mathematics

For further volumes:
http://www.springer.com/series/11104

Photo by T. Charles Erickson, Yale Photographic Services

Serge Lang
(ca. late 1980s)

Serge Lang

Collected Papers IV

1990–1996

Reprint of the 2000 Edition

 Springer

Serge Lang (1927–2005)
Department of Mathematics
Yale University
New Haven, CT
USA

ISSN 2194-9875
ISBN 978-1-4614-6138-8   (Softcover)
        978-0-387-98804-7   (Hardcover)
DOI 10.1007/978-1-4614-7121-9
Springer New York Heidelberg Dordrecht London

Library of Congress Control Number: 2012954381

Mathematics Subject Classification (1991): 11Gxx, 14Kxx, 11Fxx, 11Jxx, 14H

Contents

Bibliography (through 1999)   vii

[1990c] *Lectures on Nevanlinna Theory* 1

[1995a] Mordell's Review, Siegel's Letter to Mordell, Diophantine
Geometry, and 20th Century Mathematics 177

[1995b] Some History of the Shimura-Taniyama Conjecture 198

[1996a] La Conjecture de Bateman-Horn 213

[1996b] Comments on Chow's Works 217

[1996d] *Topics in Cohomology of Groups* 225

[1998] The Kirschner Article and HIV: Scientific and Journalistic
(Ir)responsibilities 457

[1999b] Response to the Steele Prize 470

Bibliography (through 1999)

(Boldface items are books or Lecture Notes.)

[1952a] On quasi algebraic closure, *Ann. of Math.* **55** No. 2 (1952) pp. 373–390.

[1952b] Hilbert's nullstellensatz in infinite dimensional space, *Proc. AMS* **3** No. 3 (1952) pp. 407–410.

[1952c] (with J. TATE) On Chevalley's proof of Luroth's theorem, *Proc. AMS* **3** No. 4 (1952) pp. 621–624.

[1953] The theory of real places, *Ann. of Math.* **57** No. 2 (1953) pp. 378–391.

[1954a] Some applications of the local uniformization theorem, *Am. J. Math.* **76** No. 2 (1954) pp. 362–374.

[1954b] (with A. WEIL) Number of points of varieties in finite fields, *Am. J. Math.* **76** No. 4 (1954) pp. 819–827.

[1955] Abelian varieties over finite fields, *Proc. NAS* **41** No. 3 (1955) pp. 174–176.

[1956a] Unramified class field theory over function fields in several variables, *Ann. of Math.* **64** No. 2 (1956) pp. 285–325.

[1956b] On the Lefschetz principle, *Ann. of Math.* **64** No. 2 (1956) pp. 326–327.

[1956c] L-series of a covering, *Proc. NAS* **42** No. 7 (1956) pp. 422–424.

[1956d] Sur les séries *L* d'une variété algébrique, *Bull. Soc. Math. France* **84** (1956) pp. 385–407.

[1956e] Algebraic groups over finite fields, *Am. J. Math.* **78** (1956) pp. 555–563.

[1957a] (with J.-P. SERRE) Sur les revêtements non ramifiés des variétés algébriques, *Am. J. Math.* **79**, No. 2 (1957) pp. 319–330.

[1957b] (with W.-L. CHOW) On the birational equivalence of curves under specialization, *Am. J. Math.* **79**, No. 3 (1957) pp. 649–652.

[1957c] Divisors and endomorphisms on an abelian variety, *Am. J. Math.* **79** No. 4 (1957) pp. 761–777.

[1957d] Families algébriques de Jacobiennes (d'après IGUSA), *Séminaire Bourbaki* No. 155, 1957/1958.

[1958a] Reciprocity and correspondences, *Am. J. Math.* **80** No. 2 (1957) pp. 431–440.

[1958b] (with J. TATE) Principal homogeneous spaces over abelian varieties, *Am. J. Math.* **80** No. 3 (1958) pp. 659–684.

[1958c] (with E. KOLCHIN), Algebraic groups and the Galois theory of differential fields, *Am. J. Math.* **80** No. 1 (1958) pp. 103–110.

[1958d] *Introduction to algebraic geometry*, Wiley-Interscience, 1958.

[1959a] (with A. NÉRON) Rational points of abelian varieties over function fields, *Am. J. Math.* **81** No. 1 (1959) pp. 95–118.

[1959b] Le théorème d'irreductibilité de Hilbert, *Séminaire Bourbaki* No. 201, 1959/1960.

[1959c] *Abelian varieties*, Wiley-Interscience, 1959; Springer-Verlag, 1983.

[1960a] (with E. KOLCHIN) Existence of invariant bases, *Proc. AMS* **11** No. 1 (1960) pp. 140–148.

[1960b] Integral points on curves, *Pub. IHES* No. **6** (1960) pp. 27–43.

[1960c] Some theorems and conjectures in diophantine equations, *Bull. AMS* **66** No. 4 (1960) pp. 240–249.

[1960d] On a theorem of Mahler, *Mathematika* **7** (1960) pp. 139–140.

[1960e] L'équivalence homotopique tangentielle (d'apres MAZUR), *Séminaire Bourbaki* No. 222, 1960/1961.

[1961]   Review: Elements de géométrie algébrique (A. Grothendieck). *Bull. AMS* **67** No. 3 (1961) pp. 239–246.

[1962a]   A transcendence measure for E-functions, *Mathematika* **9** (1962) pp. 157–161.

[1962b]   Transcendental points on group varieties, *Topology* **1** (1962) pp. 313–318.

[1962c]   Fonctions implicities et plongements Riemanniens, *Séminaire Bourbaki* 1961/1962, No. 237, May 1962.

[1962d] *Introduction to Differential Manifolds*, Addison Wesley, 1962.

[1962e] *Diophantine Geometry*, Wiley-Interscience, 1962.

[1963]   *Transzendente Zahlen*, Bonn Math. Schr. No. 21 (1963).

[1964a]   Diophantine approximations on toruses, *Am. J. Math.* **86** No. 3 (1964) pp. 521–533.

[1964b]   Les formes bilinéaires de Néron et Tate, *Séminaire Bourbaki* 1963/64 Fasc. 3 Exposé 274, Paris 1964.

[1964c] *First Course in Calculus*, Addison Wesley 1964; Fifth edition by Springer-Verlag, 1986.

[1964d] *Algebraic and Abelian Functions*, W.A. Benjamin Lecture Notes, 1964. See also [1982c].

[1964e] *Algebraic Numbers*, Addison Wesley, 1964; superceded by [1970d].

[1965a]   Report on diophantine approximations, *Bulletin Soc. Math. France* **93** (1965) pp. 177–192.

[1965b]   Division points on curves, *Annali Mat. pura ed applicata*, Serie IV **70** (1965) pp. 229–234.

[1965c]   Algebraic values of meromorphic functions, *Topology* **3** (1965) pp. 183–191.

[1965d]   Asymptotic approximations to quadratic irrationalities I, *Am. J. Math.* **87** No. 2 (1965) pp. 481–487.

[1965e]   Asymptotic approximations to quadratic irrationalities II, *Am. J. Math.* **87** No. 2 (1965) pp. 488–496.

[1965f]   (with W. ADAMS) Some computations in diophantine approximations, *J. reine angew. Math.* Band **220** Heft 3/4 (1965) pp. 163–173.

[1965g]   Corps de fonctions méromorphes sur une surface de Riemann (d'après ISS'SA), *Séminaire Bourbaki* No. 292, 1964/65.

[1965h] *Algebra*, Addison Wesley, 1965; second edition 1984; third edition 1993.

[1966a]   Algebraic values of meromorphic functions II, *Topology* **5** (1960) pp. 363–370.

[1966b]   Asymptotic diophantine approximations, *Proc. NAS* **55** No. 1 (1966) pp. 31–34.

[1966c] *Introduction to transcendental numbers*, Addison Wesley, 1966.

[1966d] *Introduction to Diophantine Approximations*, Addison Wesley 1966; see [1995d].

[1966e] *Rapport sur la cohomologie des groupes*. Benjamin, 1966.

[1967]   *Algebraic structures*, Addison Wesley 1967.

[1968]   *Analysis I,* Addison Wesley, 1968; (superceded by [1983c]).

[1969]   *Analysis II*, Addison Wesley, 1969; (superceded by [1993c]).

[1970a]   (with E. BOMBIERI) Analytic subgroups of group varieties, *Inventiones Math.* **11** (1970) pp. 1–14.

[1970b]   Review: L.J. Mordell's *Diophantine Equations, Bull. AMS* **76** (1970) pp. 1230–1234.

[1970c] *Introduction to Linear Algebra*, Addison Wesley 1970; see also [1986b].

[1970d] *Algebraic Number Theory*, Addison Wesley 1970; see also [1994c].

[1971a]   Transcendental numbers and diophantine approximations, *Bull. AMS* **77** No. 5 (1971) pp. 635–677.

[1971b]   On the zeta function of number fields, *Invent. Math.* **12** (1971) pp. 337–345.

[1971c]   The group of automorphisms of the modular function field, *Invent. Math.* **14** (1971) pp. 253–254.

[1971d] *Linear Algebra*, Addison Wesley 1971; see also [1987b].

[1971e] *Basic Mathematics*, Addison Wesley 1971; Springer-Verlag 1988.

[1972a]   Isogenous generic elliptic curves, *Amer. J. Math.* **94** (1972) pp. 661–674.

[1972b]   (with H. TROTTER) Continued fractions for some algebraic numbers, *J. reine angew. Math.* **255** (1972) pp. 112–134.

[1972c]   *Differential manifolds*, Addison Wesley, 1972.

[1972d]   *Introduction to Algebraic and Abelian Functions*, Benjamin-Addison Wesley, 1972; second edition see [1982c].

[1973a]   Frobenius automorphisms of modular function fields, *Amer. J. Math.* **95** (1973) pp. 165–173.

[1973b]   *Calculus of Several Variables*, Addison Wesley 1973; Third edition see [1987d].

[1973c]   *Elliptic functions*, Addison Wesley 1973; second edition see [1987d].

[1974a]   Higher dimensional diophantine problems, *Bull. AMS* **80** No. 5 (1974) pp. 779–787.

[1974b]   (with H. TROTTER) Addendum to "Continued fractions of some algebraic numbers," *J. reine angew. Math.* **267** (1974) pp. 219–220.

[1975a]   Diophantine approximations on abelian varieties with complex multiplication, *Advances Math.* **17** (1975) pp. 281–336.

[1975b]   Division points of elliptic curves and abelian functions over number fields, *Amer. J. Math.* **97** No. 1 (1975) pp. 124–132.

[1975c]   (with D. KUBERT) Units in the modular function field I, Diophantine Applications, *Math. Ann.* **218** (1975) pp. 67–96.

[1975d]   (with D. KUBERT) Units in the modular function field II, A full set of units, *Math. Ann.* **218** (1975) pp. 175–189.

[1975e]   (with D. KUBERT) Units in the modular function field III, Distribution relations, *Math. Ann.* **218** (1975) pp. 273–285.

[1975f]   La conjecture de Catalan d'après Tijdeman, *Séminaire Bourbaki* 1975/76 No. 29.

[1975g/85]   SL$_2$(**R**), Addison Wesley, 1975; Springer-Verlag corrected second printing, 1985.

[1976a]   (with J. COATES) Diophantine approximation on abelian varieties with complex multiplication, *Invent. Math.* **34** (1976) pp. 129–133.

[1976b]   (with D. KUBERT) Distribution on toroidal groups, *Math. Z.* **148** (1976) pp. 33–51.

[1976c]   (with D. KUBERT) Units in the modular function field, in *Modular Functions in One Variable* V, Springer Lecture Notes **601** (Bonn Conference) 1976, pp. 247–275.

[1976d] . (with H. TROTTER) *Frobenius distributions in* GL_2-*extensions*, Springer Lecture Notes **504**, Springer-Verlag 1976.

[1976e] *Introduction to Modular Forms*, Springer-Verlag, 1976.

[1977a]   (with D. KUBERT) Units in the modular function field IV, The Siegel functions are generators, *Math. Ann.* **227** (1997) pp. 223–242.

[1977b]   (with H. TROTTER) Primitive points on elliptic curves, *Bull. AMS* **83** No. 2 (1977) pp. 289–292.

[1977c] *Complex Analysis*, Addison Wesley; second Edition Springer-Verlag 1985; fourth edition Springer-Verlag 1999.

[1978a]   (with D. KUBERT) The p-primary component of the cuspidal divisor class group of the modular curve $X(p)$, *Math. Ann.* **234** (1978) pp. 25–44.

[1978b]   (with D. KUBERT) Units in the modular function field V, Iwasawa theory in the modular tower, *Math. Ann.* **237** (1978) pp. 97–104.

[1978c]   (with D. KUBERT) Stickelberger ideals, *Math. Ann.* **237** (1978) pp. 203–212.

[1978d]   (with D. KUBERT) The index of Stickelberger ideals of order 2 and cuspidal class numbers, *Math. Ann.* **237** (1978) pp. 213–232.

[1978e]   Relations de distributions et exemples classiques, *Séminaire Delange-Pisot-Poitou (Théorie des Nombres)*, 1978 No. 40 (6 pages).

[1978f] *Elliptic curves: Diophantine Analysis*, Springer-Verlag 1978.

[1978g] *Cyclotomic Fields* I, Springer-Verlag, 1978.

[1979a]   (with D. KUBERT) Cartan-Bernoulli numbers as values of L-series, *Math. Ann.* **240** (1979) pp. 21–26.

[1979b]   (with D. KUBERT) Independence of modular units on Tate curves, *Math. Ann.* **240** (1979) pp. 191–201.

[1979c]   (with D. KUBERT) Modular units inside cyclotomic units, *Bull. Soc. Math. France* **107** (1979) pp. 161–178.

[1980] *Cyclotomic Fields* II, Springer-Verlag 1980.

[1981a]   (with N. KATZ) Finiteness theorems in geometric classfield theory, *Enseignement mathématique* **27** (3–4) (1981) pp. 285–314.

[1981b]   (with DAN KUBERT) *Modular Units*, Springer-Verlag 1981.

[1982a]   Représentations localement algébriques dans les corps cyclotomiques, Séminaire de Théorie des Nombres 1982, Birkhauser, pp. 125–136.

[1982b]   Units and class groups in number theory and algebraic geometry, *Bull. AMS* **6** No. 3 (1982) pp. 253–316.

[1982c]   *Introduction to algebraic and abelian functions, Second Edition*, Springer-Verlag, 1982.

[1983a]   Conjectured diophantine estimates on elliptic curves, in Volume I of *Arithmetic and Geometry*, dedicated to Shafarevich, M. Artin and J. Tate editors, Birkhauser (1983) pp. 155–171.

[1983b]   *Fundamentals of Diophantine Geometry*, Springer-Verlag 1983.

[1983c]   *Undergraduate Analysis*, Springer-Verlag, 1983.

[1983d]   (with GENE MURROW) *Geometry: A High School Course*, Springer Verlag, 1983 Second edition 1988.

[1983e]   *Complex Multiplication*, Springer-Verlag 1983.

[1984a]   Vojta's conjecture, *Arbeitstagung Bonn 1984*, Springer Lecture Notes **1111** 1985, pp. 407–419.

[1984b]   Variétés hyperboliques et analyse diophantienne, *Séminaire de théorie des nombres*, 1984/85, pp. 177–186.

[1985a]   (with W. FULTON) *Riemann-Roch Algebra*, Springer-Verlag, 1985.

[1985b]   *The Beauty of Doing Mathematics*, Springer-Verlag, 1985 originally published as articles in the *Revue du Palais de la Découverte*, Paris, 1982–1984, specifically:
    Une activité vivante: faire des mathématiques, *Rev. P.D.* Vol. **11** No. **104** (1983) pp. 27–62
    Que fait un mathématicien pure et pourquoi?, *Rev. P.D.* Vol. **10** No. **94** (1982) pp. 19–44
    Faire des Maths: grands problèmes de géométrie et de l'espace, *Rev. P.D.* Vol. **12** No. **114** (1984) pp. 21–72.

[1985c]   *Math! Encounters with High School Students*, Springer-Verlag, 1985 (French edition *Serge Lang, des Jeunes et des Maths*, Belin, 1984).

[1986a]   Hyperbolic and diophantine analysis, *Bulletin AMS* **14** No. 2 (1986) pp. 159–205.

[1986b]   *Introduction to Linear Algebra*, Second Edition, Springer-Verlag 1986.

[1987a]   Diophantine problems in complex hyperbolic analysis, *Contemporary Mathematics AMS* **67** (1987) pp. 229–246.

[1987b]   *Linear Algebra*, Third Edition, Springer-Verlag 1987.

[1987c]   *Undergraduate Algebra*, Springer-Verlag 1987.

[1987d]   *Elliptic functions*, Second Edition, Springer-Verlag 1987.

[1987e]   *Introduction to complex hyperbolic spaces*, Springer-Verlag 1987.

[1988a]   The error term in Nevanlinna theory, *Duke Math. J.* **56** No. 1 (1988) pp. 193–218.

[1988b]   *Introduction to Arakelov Theory*, Springer-Verlag 1988.

[1990a]   The error term in Nevanlinna Theory II, *Bull. AMS* **22** No. 1 (1990) pp. 115–125.

[1990b]   Old and new conjectured diophantine inequalities. *Bull. AMS* **23** No. 1 (1990) pp. 37–75.

[1990c]   *Lectures on Nevanlinna theory*, in *Topics in Nevanlinna Theory*, Springer Lecture Notes **1433** (1990) pp. 1–107.

[1990d]   *Cyclotomic Fields I and II*, combined edition with an appendix by Karl Rubin, Springer-Verlag, 1990.

[1991]   *Number Theory III*, Survey of Diophantine Geometry, Encyclopedia of Mathematical Sciences, Springer-Verlag 1991.

[1993a]   (with J. JORGENSON) On Cramér's theorem for general Euler products with functional equations, *Math. Ann.* **297** (1993) pp. 383–416.

[1993b]   (with J. JORGENSON) *Basic analysis of regularized series and products*, Springer Lecture Notes **1564** (1993).

[1993c]   *Real and Functional Analysis*, Springer-Verlag, 1993.

[1993d]   *Algebra, Third Edition*, Addison Wesley 1993.

[1994a]   (with J. JORGENSON) Artin formalism and heat kernels, *J. reine angew. Math.* **447** (1994) pp. 165–200.

[1994b]   (with J. JORGENSON) *Explicit Formulas for regularized products and series*, in Springer Lecture Notes **1593** pp. 1–134.

[1994c]   *Algebraic Number Theory, Second Edition*, Springer-Verlag 1994.

[1995a]   Mordell's review, Siegel's letter to Mordell, diophantine geometry, and 20th century mathematics, *Notices AMS* March 1995 pp. 339–350.

[1995b]   Some history of the Shimura-Taniyama conjecture, *Notices AMS* November 1995 pp. 1301–1307.

[1995c]   *Differential and Riemannian Manifolds*, Springer-Verlag 1995.

[1995d]   *Introduction to Diophantine Approximations, new expanded edition*, Springer-Verlag 1995.

[1996a]   La conjecture de Bateman-Horn, *Gazette des mathématiciens* January 1996 No. 67 pp. 82–84.

[1996b]   Comments on Chow's works, *Notices AMS* **43** (1996) No. 10 pp. 1117–1124.

[1996c]   (with J. JORGENSON) Extension of analytic number theory and the theory of regularized harmonic series from Dirichlet series to Bessel series, *Math. Ann.* **306** (1996) pp. 75–124.

[1996d]   *Topics in Cohomology of groups*, Springer-Verlag, Springer Lecture Notes **1625**, 1996 (English translation and expansion of *Rapport sur la Cohomologie des Groupes*, Benjamin, 1966).

[1997]   *Survey of Diophantine Geometry*, Springer-Verlag 1997 (same as *Number Theory III*, with corrections and additions).

[1998]   The Kirschner article and HIV: Scientific and journalistic (ir)responsibilities.

[1999a]   (with J. JORGENSON) Hilbert-Asai Eisenstein series, regularized products and heat kernels, *Nagoya Math. J.* **153** (1999) pp. 155–188.

[1999b]   Response to the Steele Prize, *Notices AMS* **46** No. 4, April 1999 p. 458.

[1999c]   *Fundamentals of Differential Geometry*, Springer-Verlag 1999.

[1999d]   *Complex analysis*, fourth edition, Springer-Verlag 1999.

[1999e]   *Math Talks for Undergraduates*, Springer-Verlag 1999.

The Zurich Lectures

I was invited by Wustholz for a decade to give talks to students in Zurich. I express here my appreciation, also to Urs Stammbach for his translations, and for his efforts in producing and publishing the articles in *Elemente der Mathematik*.

Primzahlen, *Elem. Math.* **47** (1992) pp. 49–61

Die abc-vermutung, *Elem. Math.* **48** (1993) pp. 89–99

Approximationssätze der Analysis, *Elem. Math.* **49** (1994) pp. 92–103

Die Wärmeleitung auf dem Kreis und Thetafunktionen, *Elem. Math.* **51** (1996) pp. 17–27

Globaler Integration lokal integrierbarer Vektorfelder, *Elem. Math.* **52** (1997) pp. 1–11

Bruhat-Tits-Raüme, *Elem. Math.* **54** (1999) pp. 45–63

Articles on Scientific Responsibility

Circular A-21. A history of bureaucratic encroachment, J. Society of Research Administrators, 1984

Questions de responsabilité dans le journalisme scientifique, *Revue du Palais de la Découverte* Paris February 1991 pp. 17–46

Questions of scientific responsibility: The Baltimore case, *J. Ethics and Behavior* **3(1)** (1993) pp. 3–72

The Kirschner article and HIV: Scientific and journalistic (ir)responsibilities, Refused publication by the *Notices AMS*, dated 5 January 1998

Books on Scientific Responsibility

The Scheer Campaign, W.A. Benjamin, 1966

The File, Springer-Verlag, 1981

Challenges, Springer-Verlag, 1998

Lecture Notes in Mathematics

Edited by A. Dold, B. Eckmann and F. Takens

1433

Serge Lang   William Cherry

Topics in Nevanlinna Theory

Springer-Verlag
Berlin Heidelberg New York London
Paris Tokyo Hong Kong Barcelona

Authors

Serge Lang
William Cherry
Department of Mathematics, Yale University
Box 2155 Yale Station, New Haven, CT 06520, USA

Mathematics Subject Classification (1980): 30D35, 32A22, 32H30

ISBN 3-540-52785-0 Springer-Verlag Berlin Heidelberg New York
ISBN 0-387-52785-0 Springer-Verlag New York Berlin Heidelberg

© Springer-Verlag Berlin Heidelberg 1990
Printed in Germany

Printing and binding: Druckhaus Beltz, Hemsbach/Bergstr.
2146/3140-543210 – Printed on acid-free paper

PART ONE

LECTURES ON NEVANLINNA THEORY

by Serge Lang

CHAPTER I

NEVANLINNA THEORY IN ONE VARIABLE

1. The Poisson-Jensen formula and the Nevanlinna functions   12
2. The differential geometric definitions and Green-Jensen's   20
   formula
3. Some calculus lemmas   29
4. Ramification and second main theorem   34
5. An estimate for the height transform   42
6. Variations and applications, the lemma on the logarithmic   48
   derivative

Appendix by Zhuan Ye. On Nevanlinna's error term   53

CHAPTER II

EQUIDIMENSIONAL HIGHER DIMENSIONAL THEORY

1. The Chern and Ricci forms   57
2. Some forms on $\mathbf{C}^n$ and $\mathbf{P}^{n-1}(\mathbf{C})$ and the Green-Jensen formula   66
3. Stokes' theorem with certain singularities on $\mathbf{C}^n$ 70
4. The Nevanlinna functions and the first main theorem   78
5. The calculus lemma   85
6. The trace and determinant in the main theorem   87
7. A general second main theorem (Ahlfors-Wong method)   91
8. Variations and applications   102

1

PART TWO

NEVANLINNA THEORY OF COVERINGS

by William Cherry

CHAPTER III

NEVANLINNA THEORY FOR MEROMORPHIC FUNCTIONS ON COVERINGS OF C

1. Notation and preliminaries   113
2. First main theorem   121
3. Calculus lemmas   126
4. Ramification and the second main theorem   128
5. A general second main theorem   131

CHAPTER IV

EQUIDIMENSIONAL NEVANLINNA THEORY ON COVERINGS OF C^n

1. Notation and preliminaries   143
2. First main theorem   151
3. Calculus lemmas   154
4. The second main theorem without a divisor   156
5. A general second main theorem   158
6. A variation   167

References   169

INDEX   173

2

PART ONE

LECTURES ON
NEVANLINNA THEORY

by Serge Lang

INTRODUCTION

These are notes of lectures on Nevanlinna theory, in the classical case of meromorphic functions (Chapter I) and the generalization by Carlson-Griffiths to equidimensional holomorphic maps $f: \mathbf{C}^n \to X$ where X is a compact complex manifold (Chapter II). Until recently, no special attention was paid to the significance of the error term in Nevanlinna's main inequality, see for instance Shabat's book [**Sh**]. In [**La 8**] I pointed to the existence of a structure to this error term and conjectured what could be essentially the best possible form of this error term in general. I also emphasized the importance of determining the best possible error term for each of the classical functions. I shall give a more detailed discussion of these problems in the introduction to Chapter I. In this way, new areas are opened in complex analysis and complex differential geometry. I shall also describe the way I was inspired by Vojta's analogy between Nevanlinna Theory and the theory of heights in number theory.

P.M. Wong used a method of Ahlfors to prove my conjecture in dimension 1 [**Wo**]. In higher dimension, there was still a discrepancy between his result and the one in [**La 8**], neither of which contained the other. By an analysis of Wong's proof, I was able to make a certain technical improvement at one point which leads to the desired result, conjecturally best possible in general. Using Wong's approach, I was also able to give the same type of structure to the error term in Nevanlinna's theorem on the logarithmic derivative. As a result, it seemed to me useful to give a leisurely exposition which might lead people with no background in Nevanlinna theory to some of the basic problems which now remain about the error term. The existence of these problems and

5

7

the possibly rapid evolution of the subject in light of the new viewpoints made me wary of writing a book, but I hope these lecture notes will be helpful in the meantime, and will help speed up the development of the subject. They might very well be used as a continuation for a graduate course in complex analysis, also leading into complex differential geometry. Sections 1 and 2 of Chapter I provide a natural bridge, and Chapter I is especially well suited to be used in conjunction with a course in complex analysis, to give applications for the Poisson and Jensen formulas which are usually proved at the end of such a course.

I have not treated Cartan's theorem, giving a second main theorem for holomorphic maps $f: \mathbf{C} \to \mathbf{P}^n$, because I gave Cartan's proof in [**La 7**], in a self contained way, and it is still the shortest and clearest. However, at the end of Chapter II, I give one application of the techniques to one case of a map $f: \mathbf{C} \to Y$ into a possibly non compact complex manifold as an illustration of the techniques in this case. I also have not given the theory of derived curves, which introduced complications of multilinear algebra in Cartan and Ahlfors [**Ah**], and prevented seeing more clearly certain phenomena having to do with the error term, which form our main concern here. I also want to draw attention to Vojta's result for maps $f: \mathbf{C} \to \mathbf{P}^n$ into projective space. In Cartan, Ahlfors, and Schmidt's version (the number theoretic case), it is assumed that the image of f does not lie in any hyperplane. Vojta was able to weaken this assumption to the image of f not lying in a finite union of hyperplanes, which he determines explicitly as generalized diagonals [**Vo 2**]. This result gives substantial new insight into the structure of the "exceptional set" in the linear case. Ultimately, this and other advances will also have to be included into a more complete book account of the theory, as it is now developing.

Serge Lang

6

Acknowledgement

I want to thank Donna Belli for typing these lecture notes, and doing such a beautiful job of computer setting in a triumph of person over machine.

S.L.

By **increasing** we shall mean weakly increasing throughout, so an increasing function is allowed to be constant. **Positive** will mean strictly positive.

The open **disc** of radius R centered at the origin is denoted $\mathbf{D}(r)$. The **circle** of radius r centered at the origin is $\mathbf{S}(r)$. The closed disc is $\overline{\mathbf{D}}(r)$.

In $\mathbf{C}^n$, the **ball** and **sphere** of radius r are denoted

$$\mathbf{B}(r), \ \overline{\mathbf{B}}(r), \ \text{and} \ \mathbf{S}(r)$$

respectively.

Let F_1, F_2 be two positive functions defined for all real numbers $\geq r_0$. We write

$$F_1 \ll F_2$$

to mean that $F_1 = O(F_2)$. We shall write

$$F_1 \gg \ll F_2$$

to mean that $F_1 \ll F_2$ and $F_2 \ll F_1$.

8

CHAPTER I

NEVANLINNA THEORY IN ONE VARIABLE

In the first part of this chapter we essentially follow Nevanlinna, as in his book [Ne]. The main difference lies in the fact that we are careful about the error term in Nevanlinna's main theorem. That this error term has an interesting structure was first brought up in [La 8], in analogy with a similar conjecture in number theory. Although Osgood [Os] did notice a similarity between the 2 in the Nevanlinna defect and the 2 in Roth's theorem, Vojta gave a much deeper analysis of the situation, and compared the theory of heights in number theory or algebraic geometry with the Second Main Theorem of Nevanlinna theory.

In [La 2] and [La 3] I defined a **type** for a number α to be an increasing function ψ such that

$$-\log\left|\alpha - \frac{p}{q}\right| - 2\log q \leq \log \psi(q)$$

for all but a finite number of fractions p/q in lowest form, $q > 0$. The **height** $h(p/q)$ is defined as

$$\log \max(|p|, |q|).$$

If p/q is close to α then $\log q$ has the same order of magnitude as the height, so $\log q$ is essentially the height in the above inequality. A theorem of Khintchine states that almost all numbers have type $\leq \psi$ if

$$\sum \frac{1}{q\psi(q)} < \infty.$$

9

The idea is that algebraic numbers behave like almost all numbers, although it is not clear a priori if Khintchine's principle will apply without any further restriction on the function ψ. Roth's theorem can be formulated as saying that an algebraic number has type $\leq q^\varepsilon$ for every $\varepsilon > 0$, and in the sixties I conjectured that this could be improved to having type $\leq (\log q)^{1+\varepsilon}$ in line with Khintchine's principle. Cf. [La 1], [La 3], [La 4] especially.[1] Thus for instance we would have the improvement of Roth's inequality

$$\left|\alpha - \frac{p}{q}\right| \geq \frac{C(\alpha, \varepsilon)}{q^2 (\log q)^{1+\varepsilon}}$$

which could also be written

$$-\log\left|\alpha - \frac{p}{q}\right| - 2\log q \leq (1 + \varepsilon)\log\log q$$

for all but a finite number of fractions p/q. However, except for quadratic numbers that have bounded type, there is no known example of an algebraic number about which one knows that it is or is not of type $(\log q)^k$ for some number $k > 1$. It becomes a problem to determine the type for each algebraic number and for the classical numbers. For instance, it follows from Adams' work [Ad 1], [Ad 2] that e has type

$$\psi(q) = \frac{C \log q}{\log \log q}$$

with a suitable constant C, which is much better than the "probabilistic" type $(\log q)^{1+\varepsilon}$.

In light of Vojta's analysis, it occurred to me to transpose my conjecture about the "error term" in Roth's theorem to the context of Nevanlinna theory, in one and higher dimension. Transposing to the analytic context, it becomes a problem to determine the "type" of the

---

[1] Unknown to me until much later, similar conjectures were made by Bryuno [Br] and Richtmyer, Devaney and Metropolis [RDM], see [L-T 1] and [L-T 2].

10

classical meromorphic functions, i.e. the best possible error term in the second main theorem which describes the value distribution of the function. It is classical, and easy, for example, that e^z has bounded type, i.e. that the error term in the Second Main Theorem is $O(1)$. Two problems arise here:

• To determine for "almost all" functions (in a suitable sense) whether the type follows the pattern of Khintchine's convergence principle.

• To determine the specific type for each concrete classical function, using the specific special properties of each such function: $\wp, \theta, \Gamma, \zeta, J$, etc.

I am much indebted to Ye for an appendix exhibiting functions of type corresponding to a factor of $1 - \varepsilon$ in number theory. Until he gave these examples, I did not even know a function which did not have bounded type.

In [La 8], using the singular volume form of Carlson-Griffiths or a variation of it, I was not able to prove my conjecture exactly, with the correct factor of $1 + \varepsilon$ (I got only 3/2 instead of 1).

P.M. Wong brought back to life a method which occurs in Ahlfors' original 1941 paper, and by this method he established my conjecture with the $1 + \varepsilon$. I pointed out to him that his method would also prove the desired result with an arbitrary type function satisfying only the convergence of the integral similar to the Khintchine principle. The role of the convergence principle becomes very clear in that method, which is given in the second part of this chapter. The method had also been tried improperly by Chern in the early sixties, and we shall have more to say on the technical aspects when we come to the actual theorems in §4. Ahlfors' method was obscured for a long time by other technical aspects of his paper, and I think Wong made a substantial contribution by showing how it could be applied successfully. I should also note that H. Wu also proved the conjecture with $1 + \varepsilon$ (unpublished) by the "averaging method" of Ahlfors. But the method used by Wong

11

lent itself better to give the generalized version with the function ψ.

Developing fully the two problems mentioned above would create a whole new area of complex analysis, digging into properties of meromorphic functions in general, and of the classical functions in particular, which up to now have been disregarded.

I, §1. THE POISSON-JENSEN FORMULA AND THE NEVANLINNA FUNCTIONS

By a meromorphic function we mean a meromorphic function on the whole plane, so its zeros and poles form a discrete set. A meromorphic function on a closed set (e.g. the closed disc $\overline{\mathbf{D}}(R)$) is by definition meromorphic on some open neighborhood of the set.

Theorem 1.1. (Poisson formula) *Let f be holomorphic on the closed disc $\overline{\mathbf{D}}(R)$. Let z be inside the disc, and write $z = re^{i\varphi}$. Then*

$$f(z) = \int_0^{2\pi} f(Re^{i\theta}) \operatorname{Re} \frac{Re^{i\theta} + z}{Re^{i\theta} - z} \frac{d\theta}{2\pi}$$

$$= \int_0^{2\pi} f(Re^{i\theta}) \frac{R^2 - r^2}{R^2 - 2R\cos(\theta - \varphi) + r^2} \frac{d\theta}{2\pi}$$

$$= \int_0^{2\pi} \operatorname{Re} f(Re^{i\theta}) \frac{Re^{i\theta} + z}{Re^{i\theta} - z} \frac{d\theta}{2\pi} + iK \quad \textit{for some real constant } K.$$

Proof: By Cauchy's theorem,

$$f(0) = \frac{1}{2\pi i} \int_{S_R} \frac{f(\zeta)}{\zeta} d\zeta = \int_0^{2\pi} f(Re^{i\theta}) \frac{d\theta}{2\pi}.$$

Let g be the automorphism of $\overline{\mathbf{D}}(R)$ which interchanges 0 and z. Then

$$f(z) = f \circ g(0).$$

12

We apply the above formula to $f \circ g$. We then change variables, and use the fact that $g \circ g = \mathrm{id}$, $\zeta = g(w)$, $d\zeta = g'(w)dw$. For $R = 1$,

$$g(w) = \frac{z - w}{1 - w\overline{z}}.$$

The desired formula drops out as in the first equation. The identity

$$\mathrm{Re}\, \frac{Re^{i\theta} + z}{Re^{i\theta} - z} = \frac{R^2 - r^2}{R^2 - 2R\cos(\theta - \varphi) + r^2}$$

is immediate by direct computation. The third equation comes from the fact that f and the integral on the right hand side of the third equation are both analytic in z, and have the same real part, so differ by a pure imaginary constant. This concludes the proof.

For $a \in \mathbf{D}(R)$ define

$$G_R(z, a) = G_{R,a}(z) = \frac{R^2 - \overline{a}z}{R(z - a)}.$$

Then $G_{R,a}$ has precisely one pole on $\overline{\mathbf{D}}(R)$ and no zeros. We have

$$|G_{R,a}(z)| = 1 \text{ for } |z| = R.$$

Theorem 1.2. (Poisson-Jensen formula) *Let f be meromorphic non constant on $\mathbf{D}(R)$. Then for any simply connected open subset of $\mathbf{D}(R)$ not containing the zeros or poles of f, there is a real constant K such that for z in the open set we have*

$$\log f(z) = \int_0^{2\pi} \log|f(Re^{i\theta})| \frac{Re^{i\theta} + z}{Re^{i\theta} - z} \frac{d\theta}{2\pi}$$

$$- \sum_{a \in \mathbf{D}(R)} (\mathrm{ord}_a f) \log G_R(z, a) + iK.$$

The constant K depends on a fixed determination of the logs.

13

Proof: Suppose first that f has no zeros or poles on $S(R)$. Let

$$h(z) = f(z) \prod \left(\frac{R^2 - \bar{a}z}{R(z-a)} \right)^{\text{ord}_a f}$$

where the product is taken for $a \in D(R)$. Then h has no zero or pole on $\overline{D}(R)$, and so $\log h$ is defined as a holomorphic function to which we can apply Theorem 1.1 to get the present formula. Then we use the fact that the log of a product is the sum of the logs plus a pure imaginary constant, on a simply connected open set, to conclude the proof in the present case.

Suppose next that f may have zeros and poles on $S(R)$. Note that

$$\theta \longmapsto \log |f(Re^{i\theta})|$$

is absolutely integrable, because where there are singularities, they are like $\log|x|$ in a neighborhood of the origin $x = 0$ in elementary calculus. Let R_n be a sequence of radii having R as a limit. For R_n sufficiently close to R, the zeros and poles of f inside the disc of radius R_n are the same as the zeros and poles of f inside the disc of radius R, except for the zeros and poles lying on the circle $S(R)$. The left hand side of the formula is independent of R_n. Let

$$\varphi_n(\theta) = \log |f(R_n e^{i\theta})| \quad \text{and} \quad \varphi(\theta) = \log |f(Re^{i\theta})|.$$

Then φ_n converges to φ. Outside small intervals around the zeros or poles of f on $S(R)$, the convergence is uniform. Near the zeros and poles of f on the circle, the contribution of the integrals over small θ-intervals is small. Hence

$$\int_0^{2\pi} \log |f(R_n e^{i\theta})| \frac{Re^{i\theta} + z}{Re^{i\theta} - z} \frac{d\theta}{2\pi} \quad \text{converges to} \quad \int_0^{2\pi} \log |f(Re^{i\theta})| \frac{Re^{i\theta} + z}{Re^{i\theta} - z} \frac{d\theta}{2\pi}.$$

thus proving the formula in general.

14

From the Poisson-Jensen formula, we deduce a slightly simpler relationship for the real parts, namely:

For all $z \in \mathbf{D}(R)$ which are not zeros or poles of f, we have

$$\log|f(z)| = \int_0^{2\pi} \log|f(Re^{i\theta})| \operatorname{Re} \frac{Re^{i\theta} + z}{Re^{i\theta} - z} \frac{d\theta}{2\pi}$$
$$- \sum_{a \in \mathbf{D}(R)} (\operatorname{ord}_a f) \log|G_R(z,a)|.$$

If 0 is not a zero or pole of f, then

$$\log|f(0)| = \int_0^{2\pi} \log|f(Re^{i\theta})| \frac{d\theta}{2\pi} - \sum_{\substack{a \in \mathbf{D}(r) \\ a \neq 0}} (\operatorname{ord}_a f) \log\left|\frac{R}{a}\right|.$$

In general, let $f(z) = c_f z^e + \cdots$ where $c_f \neq 0$ is the leading coefficient. Then

$$\log|c_f| = \int_0^{2\pi} \log|f(Re^{i\theta})| \frac{d\theta}{2\pi} - \sum_{a \neq 0} (\operatorname{ord}_a f) \log|R/a| - e \log R.$$

This last formula follows by applying the previous formula to the function $f(z)/z^e$.

Following Nevanlinna, we define the **counting functions**

$$n_f(0, R) = \text{number of zeros of } f \text{ in } \overline{\mathbf{D}}(R)$$
$$n_f(\infty, R) = \text{number of poles of } f \text{ in } \overline{\mathbf{D}}(R)$$
$$= n_f(R).$$

Thus $n_f(0, R) = n_{1/f}(\infty, R) = n_{1/f}(R)$. We also define:

$$N_f(0, R) = \sum_{\substack{a \in \mathbf{D}(R) \\ a \neq 0, f(a) = 0}} (\operatorname{ord}_a f) \log\left|\frac{R}{a}\right| + (\operatorname{ord}_0 f) \log R$$

$$N_f(\infty, R) = N_{1/f}(0, R).$$

15

We may rewrite Jensen's formula with the above notation:

Theorem 1.3 *Let f be meromorphic on $\overline{D}(R)$. Then*

$$\int_0^{2\pi} \log |f(Re^{i\theta})| \frac{d\theta}{2\pi} = \log |c_f| + N_f(0, R) - N_f(\infty, R).$$

If $a \in C$ we define

$$n_f(a, r) = n_{f-a}(0, r) \text{ and } N_f(a, r) = N_{f-a}(r).$$

Proposition 1.4 *Let $n_f(r)$ and $N_f(r)$ denote $n_f(0, r)$ and $N_f(0, r)$. Then*

$$N_f(R) = \int_0^R [n_f(t) - n_f(0)] \frac{dt}{t} + n_f(0) \log R.$$

Proof: The function

$$t \mapsto n_f(t) - n_f(0)$$

is a step function which is equal to 0 for t sufficiently close to 0. If we decompose the interval $[0, R]$ into subintervals whose end points are the jumps in the absolute values of the zeros of f, integrate over each such interval where the integrand is constant, and take the sum, then the formula of the proposition drops out.

Remark: If $R > 1$ we have $N_f(a, R) \geq 0$. If f has no zero at the origin, then $N_f(0, R) \geq 0$ for all $R > 0$. The only possibly non positive contribution to N_f is the term with $n_f(0, R) \log R$ with $0 < R < 1$. For simplicity, we shall often assume that f has no zero or pole at the origin, to get rid of this term.

For $\alpha > 0$ we define

$$\log^+ \alpha = \max(0, \log \alpha).$$

16

18

Proposition 1.5 *Let $b \in \mathbf{C}$. Then*

$$\int_0^{2\pi} \log |b - e^{i\theta}| \frac{d\theta}{2\pi} = \log^+ |b|.$$

This is an immediate consequence of Theorem 1.3, but can also be proved directly using the mean value property of harmonic functions. Indeed, if $|b| > 1$ then $\log |b - z|$ for $|z| < 1 + \epsilon$ is harmonic, and $\log^+ |b| = \log |b|$, so the formula is true by the mean value property for harmonic functions. If $|b| < 1$, then subtracting $\log |b|$ from the left hand side, replacing θ by $-\theta$, and using the first part of the proof shows that the integral comes out to be 0, which is $\log^+ |b|$ again. Finally for $|b| = 1$, the formula follows by continuity of each side in b, and the absolute convergence of the integral on the left. This proves the proposition.

Remark: Proposition 1.5 shows that the expression $\log^+ |b|$ is in some sense natural.

As usual, we let the projective line $\mathbf{P}^1$ consist of $\mathbf{C}$ and a point ∞. Points of $\mathbf{P}^1$ can be identified with equivalence classes of pairs (w_1, w_2) with complex numbers w_1, w_2 not both 0, and

$$(w_1, w_2) \sim (cw_1, cw_2) \quad \text{if} \quad c \neq 0.$$

Each point of $\mathbf{P}^1$ has an affine representative $(w, 1)$ or $\infty = (1, 0)$ with $w \in \mathbf{C}$. A meromorphic function f can be identified with a holomorphic map into $\mathbf{P}^1$. Namely we write $f = f_1/f_0$ where f_1, f_0 are entire functions without common zeros, which can always be done by the Weierstrass factorization theorem. Then we obtain a holomorphic map into $\mathbf{P}^1$ given in terms of coordinates by

$$z \mapsto (f_0(z), f_1(z)).$$

17

19

We define a "distance" on $\mathbf{P}^1$ between two points w, w' to be

$$\|w, w'\|^2 = \frac{|w - w'|^2}{(1 + |w|^2)(1 + |w'|^2)}.$$

The formula makes sense if $w, w' \in \mathbf{C}$. If one but not both of $w, w' = \infty$, then we let

$$\|w, \infty\|^2 = \|\frac{1}{w}, 0\|^2 = \frac{1}{(1 + |w|^2)}.$$

We let $\|0, \infty\| = 1$. This distance satisfies the triangle inequality, as one can show by identifying it with what is sometimes called the "chordal distance" under stereographic projection, but we don't go into this here. Observe that if $w \neq w'$ then

$$0 < \|w, w'\| \leq 1, \text{ and } < 1 \text{ except for antipodal points.}$$

Furthermore, the function $(w, w') \mapsto \|w, w'\|^2$ is a C^∞ (even real analytic) function on $\mathbf{P}^1 \times \mathbf{P}^1$.

Let f be a non-constant meromorphic function and let $a \in \mathbf{P}^1$. We define the **mean proximity** function to be

$$m_f(a, r) = \int_0^{2\pi} - \log \|f(re^{i\theta}), a\| \frac{d\theta}{2\pi}.$$

The function $z \mapsto - \log \|f(z), a\|$ is a convenient normalization for the more general notion of a **Weil function**. Given $a \in \mathbf{P}^1$, we define a **Weil function associated with** a to be a continuous map

$$\lambda_a : \mathbf{P}^1 - \{a\} \longrightarrow \mathbf{R}$$

having the property that in some open neighborhood of a there exists a continuous function α such that

$$\lambda_a(z) = - \log |z - a| + \alpha(z).$$

18

20

The difference between two Weil functions is a continuous function on $\mathbf{P}^1$, which is bounded. Another way of normalizing a Weil function is to take

$$\lambda_a(z) = \log^+ \frac{1}{|z - a|} \qquad \text{if} \quad a \neq \infty$$

$$\lambda_a(z) = \log^+ |z| \qquad \text{if} \quad a = \infty.$$

However, the previous normalization is more natural and more symmetric in many respects. Given any Weil function, one can define a corresponding proximity function by

$$m_f(\lambda_a, r) = \int_0^{2\pi} \lambda_a(f(re^{i\theta})) \frac{d\theta}{2\pi}.$$

It differs from $m_f(a, r)$ by a bounded function. Our normalization of m_f insures that

$$m_f(a, r) \geq 0.$$

For simplicity we shall assume throughout that

$$f(0) \neq 0, \infty \text{ and } f'(0) \neq 0.$$

We define the **height** to be the function $T_f \colon \mathbf{R}_{>0} \to \mathbf{R}$ given by

$$T_f(\infty, r) = T_f(r) = m_f(\infty, r) + N_f(\infty, r) + \log \|f(0), \infty\|,$$

and more generally if $f(0) \neq a$

$$T_{f,a}(r) = m_f(a, r) + N_f(a, r) + \log \|f(0), a\|.$$

Theorem 1.6 (First Main Theorem). *The function $T_f(a, r)$ is independent of $a \in \mathbf{P}^1$, provided $f(0) \neq a$.*

Proof: Applying Jensen's formula to $f - a$, and $a \neq \infty$, we get

$$-\int_0^{2\pi} \log |f(re^{i\theta}) - a| \frac{d\theta}{2\pi} = \log |f(0) - a| - N_{f-a}(0, r) + N_{f-a}(\infty, r).$$

19

But the poles of $f - a$ are the same as the poles of f, so $N_{f-a}(\infty, r) = N_f(\infty, r)$. On the other hand, by definition of the symbols,

$$T_{f,a}(r) = -\int_0^{2\pi} \log \|f(re^{i\theta}), a\| \frac{d\theta}{2\pi} + N_{f,a}(r) + \log \|f(0), a\|$$

$$= -\int_0^{2\pi} \log |f(re^{i\theta}) - a| \frac{d\theta}{2\pi} + \frac{1}{2} \int_0^{2\pi} \log(1 + |f(re^{i\theta})|^2) \frac{d\theta}{2\pi}$$

$$+ \frac{1}{2} \log(1 + |a|^2) + N_{f-a}(0, r) + \log \|f(0), a\|.$$

Using the definitions and the above Jensen formula, it is then immediate to verify that the right hand side is equal to $T_f(\infty, r)$, thus concluding the proof.

In light of the theorem, one simply writes $T_f(r)$ for the Nevanlinna height. Observe that the only non positive contribution to $T_f(r)$ is the constant term involving $f(0)$. Note that as a function of $a \in \mathbf{P}^1$, this term is absolutely integrable on $\mathbf{P}^1$ with respect to any C^∞ form.

I, §2. THE DIFFERENTIAL GEOMETRIC DEFINITIONS AND GREEN-JENSEN'S FORMULA

Let $f: \mathbf{C} \to \mathbf{P}^1$ be a holomorphic map, which we identify with a meromorphic function. We define the **Fubini-Study** form ω on $\mathbf{P}^1$ to be the form given in terms of an affine coordinate w by

$$\omega = \frac{\sqrt{-1}}{2\pi} \frac{dw \wedge d\bar{w}}{(1 + |w|^2)^2}.$$

We let the **euclidean form** Φ on $\mathbf{C}$ be the form

$$\Phi = \frac{\sqrt{-1}}{2\pi} dz \wedge d\bar{z} = 2r dr \frac{d\theta}{2\pi}.$$

20

Then

$$\text{letting} \quad \gamma_f = \frac{|f'|^2}{(1 + |f|^2)^2}, \quad \text{we have} \quad f^*\omega = \gamma_f \Phi.$$

Note that γ_f is C^∞ (actually real analytic). By the Weierstrass factorization theorem, we can write $f = f_1/f_0$ where f_1, f_0 are entire functions without common zero. We let

$$W = W(f_0, f_1) = f_0 f_1' - f_0' f_1$$

be the **Wronskian**. Then we can write

$$\gamma_f = \frac{|W|^2}{\left(|f_0|^2 + |f_1|^2\right)^2}.$$

Note that $|f_0|^2 + |f_1|^2$ is a positive C^∞ function.

Let ∂ and $\bar{\partial}$ be the usual differential operators, so for instance for a C^∞ function α we have

$$\partial \alpha = \frac{\partial \alpha}{\partial z} dz \quad \text{and} \quad \bar{\partial} \alpha = \frac{\partial \alpha}{\partial \bar{z}} d\bar{z}.$$

We let as usual

$$d = \partial + \bar{\partial} \quad \text{and} \quad d^c = \frac{1}{2\pi} \frac{\partial - \bar{\partial}}{2i}.$$

Then

$$dd^c = \frac{\sqrt{-1}}{2\pi} \partial \bar{\partial}.$$

Let g be a meromorphic function. Then outside the zeros and poles of g, we have

$$\partial \bar{\partial} g = 0$$

because of the Cauchy-Riemann equation $\bar{\partial} g = 0$. Also

$$dd^c \log |g|^2 = 0$$

21

outside the zeros and poles of g, because $\log(g\bar{g}) = \log|g|^2$ is defined and equal to $\log g + \log \bar{g}$ in any simply connected open set, up to an additive constant, and again applying $\partial\bar{\partial}$ kills this expression by Cauchy-Riemann. We emphasize that when we write a differential operator applied to a function with singularities (zeros and poles) as above, we mean the operator applied to the function on the complement of the singularities. An appropriate notation will be developed later when we need to consider the singularities explicitly.

We note that in polar coordinates (r, θ) the operator d^c has the form

$$d^c\alpha = \frac{1}{2}r\frac{\partial\alpha}{\partial r}\frac{d\theta}{2\pi} - \frac{1}{4\pi}\frac{1}{r}\frac{\partial\alpha}{\partial\theta}dr.$$

The term with dr disappears when we restrict the function to a circle centered at the origin.

Proposition 2.1 *We have (outside the singularities)*

$$\gamma_f\Phi = dd^c\log(1+|f|^2) = dd^c\log(|f_0|^2 + |f_1|^2)$$
$$= -\frac{1}{2}dd^c\log\gamma_f$$

Proof: Using the fact that d, d^c are derivations, one has

$$dd^c\log(1+u) = \frac{dd^cu}{1+u} - \frac{du \wedge d^cu}{(1+u)^2}.$$

But, $u^2 dd^c\log u = u\,dd^cu - du \wedge d^cu$. Applying this to $u = |f|^2$ and $dd^c\log|f|^2 = 0$ will give the formula. We use the fact that

$$dd^c\log|f_0|^2 = 0$$

for the second equality sign.

We shall apply Stokes' theorem in the following context.

Proposition 2.2. *Let α be a C^2 function except for a discrete set of singularities. Suppose that α has no singularity on the circle $S(t)$.*

22

Let Z be the set of singularities in $\mathbf{D}(t)$, and let $S(Z, \varepsilon)$ be the (formal) sum of small circles around these singularities. Then

$$
(1) \qquad \int_{\mathbf{D}(t)} dd^c\alpha = \int_{\mathbf{S}(t)} d^c\alpha - \lim_{\varepsilon \to 0} \int_{S(Z,\varepsilon)} d^c\alpha.
$$

If α has no singularities, i.e. Z is empty, then

$$
(2) \qquad \int_{\mathbf{D}(t)} dd^c\alpha = \int_{\mathbf{S}(t)} d^c\alpha.
$$

On the other hand, if $\alpha = \log |g|^2$ where g is holomorphic, then

$$
(3) \qquad \lim_{\varepsilon \to 0} \int_{S(Z,\varepsilon)} d^c \log |g|^2 = n_g(0, t).
$$

Proof: The first formula is simply the Stokes-Green formula in the plane. For the third formula, we are reduced to proving it for a circle around a single zero or pole, so without loss of generality, let us assume that the point is the origin. We then have to prove

$$
\lim_{\varepsilon \to 0} \int_{\mathbf{S}(\varepsilon)} d^c \log |g|^2 = \mathrm{ord}_0(g).
$$

Let $k = \mathrm{ord}_0(g)$. We can write $g\bar{g} = r^{2k} h(r, \theta)$ where h is C^∞ and > 0. The operator $d^c \log$ transforms multiplication to addition, so we are reduced to replacing $|g|^2$ by r^{2k}. In this case, using the expression for d^c in polar coordinates concludes the proof.

In the next result, we shall need to differentiate under an integral sign, and we also need continuity with respect to parameters. There is a basic set of conditions under which these properties hold, carried out in detail for instance in my *Real Analysis*. We recall these here in our context.

23

25

Let α be a C^2 function on $\mathbf{C}$ except for a discrete set of points, and assume α continuous at 0. We consider the following conditions on α, which will insure continuity and differentiability (**CD**):

CD 1. *For each r, the function $\theta \mapsto \alpha(re^{i\theta})$ is absolutely integrable.*

CD 2. *Given $r_0 > 0$ there exists an open interval around r_0 and an L^1 function α_1 of θ alone such that*

$$|\alpha(re^{i\theta})| \le \alpha_1(\theta) \quad \text{for all} \quad r \quad \text{in the interval.}$$

CD 3. *Given $r_0 > 0$ there exists an open interval around r_0 and an L^1 function α_2 of θ alone such that*

$$|\frac{\partial}{\partial r}\alpha(re^{i\theta})| \le \alpha_2(\theta) \text{ for all } r \text{ in the interval.}$$

Proposition 2.3. Green-Jensen Formula *Let α be a C^2 function except at a discrete set of points, and continuous at 0. Assume that α also satisfies the three **CD** conditions. Then*

$$r \mapsto \int_0^{2\pi} \alpha(re^{i\theta})\frac{d\theta}{2\pi}$$

is continuous for $r \ge 0$, and

$$\int_0^r \frac{dt}{t} \int_{\mathbf{D}(t)} dd^c\alpha + \int_0^r \frac{dt}{t} \lim_{\varepsilon \to 0} \int_{\mathbf{S}(Z,\varepsilon)(t)} d^c\alpha = \frac{1}{2}\int_0^{2\pi}\alpha(re^{i\theta})\frac{d\theta}{2\pi} - \frac{1}{2}\alpha(0).$$

Proof: The proof is immediate by integrating the previous relation, and taking into account the polar expression for d^c. We can take $\partial/\partial t$ outside the integral sign, namely

$$\int_0^r \frac{dt}{t} \int_{\mathbf{S}(t)} d^c\alpha = \int_0^r \frac{dt}{t} \int_0^{2\pi} \frac{1}{2}t\frac{\partial}{\partial t}\alpha(te^{i\theta})\frac{d\theta}{2\pi} - \frac{1}{2}\int_0^r \frac{\partial}{\partial t}\left(\int_0^{2\pi} \alpha(te^{i\theta})\frac{d\theta}{2\pi}\right)dt$$

24

so the proposition falls out by Stokes' formula and the standard criteria for differentiating under the integral sign.

Remark: We are avoiding here the language of distribution, which actually would not shorten what we are doing. In that language, if we denote by $[\alpha]$ the distribution associated with the function α, then the **Green-Jensen formula** may be stated in the form

$$\int_0^r \frac{dt}{t} \int_{\mathbf{D}(t)} dd^c[\alpha] = \frac{1}{2} \int_0^{2\pi} \alpha(re^{i\theta}) \frac{d\theta}{2\pi} - \frac{1}{2}\alpha(0).$$

For our purposes, we simple **define** $dd^c[\alpha]$ to be the functional such that

$$\int \frac{dt}{t} \int_{\mathbf{D}(t)} dd^c[\alpha] = \int_0^r \frac{dt}{t} \int_{\mathbf{D}(t)} dd^c\alpha + \mathrm{Sing}_\alpha(r),$$

where

$$\mathrm{Sing}_\alpha(r) = \int_0^r \frac{dt}{t} \lim_{\varepsilon \to 0} \int_{\mathbf{S}(Z,\varepsilon)(t)} d^c\alpha.$$

We call the first integral with $dd^c\alpha$ the **regular part** of $dd^c[\alpha]$, and we call $\mathrm{Sing}_\alpha(r)$ the **singular part**. In practice, the conditions under which we apply the formula are varied, and require a determination of the singular part, which sometimes will contribute to the formula in an essential way, and sometimes will be 0. Each time one must determine just what it is, so using "distributions" in a abstract nonsense sort of way does not help. The singular part may be 0 for different reasons. First, if the function α is C^∞, or even C^2, then the singular part is 0 because the limit as $\varepsilon \to 0$ is 0. But the singularities of α may be weak enough so that the limit is still 0, and examples of this phenomenon will be given below. On the other hand, when the singularities are like poles, then they contribute to the expression in an essential way, as in Proposition 2.2.

25

We shall give important examples of the Green-Jensen formula. First comes an alternative definition of the height.

Theorem 2.4 (Ahlfors-Shimizu). *Let f be meromorphic. Then*

$$T_f(r) = \int_0^r \frac{dt}{t} \int_{D(t)} \gamma_f \Phi.$$

Proof: In Proposition 2.3, let

$$\alpha = \log(1 + |f|^2).$$

Let $f = f_1/f_0$ where f_1, f_0 are holomorphic without common zeros. Let z_j be the zeros of f_0, that is the poles of f in the disc $D(t)$. By Proposition 2.2,

$$\lim_{\varepsilon \to 0} \int_{S(z,\varepsilon)} d^c \alpha = \operatorname{ord}_{z_j}(f_0),$$

because

$$d^c \log(1 + |f|^2) = d^c \log(|f_0|^2 + |f_1|^2) - d^c \log |f_0|^2,$$

and the smooth part contributes 0 to the limit. Thus the singular part contributes the number of poles of f in the disc, and its integral against dt/t contributes $N_f(\infty, r)$. On the other hand, the other part coming from the circle of radius r is precisely

$$m_f(\infty, r) + \log \|f(0), \infty\|,$$

thus proving the theorem.

Integrals of the form

$$\int_0^r \frac{dt}{t} \int_{D(t)} \alpha \Phi$$

will be encountered in a fundamental way, and will be given a name in the next section.

26

As an application of some of the above computations, we give formulations of special cases in terms of the Nevanlinna functions.

Corollary 2.5 *For f meromorphic non-constant, we have:*

$$T_f(r) = \int_0^r \frac{dt}{t} \int_{\mathbf{D}(t)} dd^c \log(1 + |f|^2)$$

$$N(f, a, r) = \int_0^r \frac{dt}{t} \int_{\mathbf{D}(t)} dd^c [\log |f - a|^2]$$

$$T_f(r) - N(f, a, r) = -\int_0^r \frac{dt}{t} \int_{\mathbf{D}(t)} dd^c [\log \|f, a\|^2].$$

Proof: The first formula for $T_f(r)$ comes from Proposition 2.1 and Theorem 2.4. The formula for $N(f, a, r)$ is a special case of Proposition 2.2(3). The last formula for the difference comes from the definition of $\|f, a\|^2$ and the fact that the operator $dd^c \log$ transforms multiplication into addition.

Next we give an example such that the function has singularities, but these singularities are sufficiently mild so that the singular term in the Green-Jensen formula vanishes.

Proposition 2.6 *Let $0 < \lambda$ and let $h = \|f, a\|^{2\lambda}$. Then*

$$\int_0^r \frac{dt}{t} \int_{\mathbf{D}(t)} dd^c \log(1 + h)$$

$$= \frac{1}{2} \int_0^{2\pi} \log(1 + h(re^{i\theta})) \frac{d\theta}{2\pi} - \frac{1}{2} \log(1 + h(0)).$$

Proof: The function h is C^∞ except possibly at the zeros of $f - a$ or the poles of f if $a = \infty$. If no zero or pole lies on $S(t)$, then we shall

27

prove that

$$\int_{\mathbf{D}(t)} dd^c \log(1+h) = \int_{\mathbf{S}(t)} d^c \log(1+h).$$

Integrating this against dt/t from 0 to r just as in the smooth case of Theorem 2.2 then proves the proposition.

For simplicity, we assume that f is holomorphic, and we have to show that the term coming from the singularities in Stokes' theorem give zero contribution.

Let $\beta = \log(1 + h)$, so β is C^∞ except at the set of zeros of $f - a$, which we denote by Z. We let

$$S_\varepsilon(Z)(r) = \text{circles of radius } \varepsilon \text{ around each point of } Z$$

$$\text{lying inside } \mathbf{D}(r).$$

If $f - a$ has no zero on the circle $\mathbf{S}(t)$, then Stokes' theorem shows that

$$\int_{\mathbf{S}(t)} d^c \beta = \int_{\mathbf{D}(t)} dd^c \beta + \lim_{\varepsilon \to 0} \int_{S_\varepsilon(Z)(t)} d^c \beta.$$

For our special β, we claim that the limit on the right is 0. We are reduced to proving this limit in the neighborhood of each zero of $f - a$. Then in such a neighborhood, let z be a complex coordinate. Locally we are reduced to proving that if g is holomorphic at the origin, and

$$\beta = \log(1 + |g|^{2\lambda} \alpha_1)$$

where α_1 is > 0 and C^∞, then

$$\lim_{\varepsilon \to 0} \int_{r=\varepsilon} r \frac{\partial}{\partial r} \log(1 + |g|^{2\lambda} \alpha_1) \frac{d\theta}{2\pi} = 0.$$

28

30

We can rewrite $|g|^{2\lambda}\alpha_1 = r^{2m\lambda}\alpha(r,\theta)$ where α is > 0 and C^{∞}, $m > 0$. Then

$$r\frac{\partial}{\partial r}\log(1 + r^{2m\lambda}\alpha(r,\theta)) = \frac{r2m\lambda r^{2m\lambda-1}\alpha(r,\theta) + r^{2m\lambda+1}\partial\alpha/\partial r}{1 + r^{2m\lambda}\alpha(r,\theta)}.$$

Setting $r = \varepsilon$ and letting $\varepsilon \to 0$ immediately shows that the desired limit is 0, thus proving the proposition.

I, §3. SOME CALCULUS LEMMAS

The expression of Theorem 2.4 will now be viewed as a transform, and some of its properties will depend only on calculus.

Lemma 3.1. *Let F be a positive increasing function defined for $r > 0$ with piecewise continuous derivative. Suppose there exists r_1 such that $F(r_1) \geq e$, say. Let ψ be a positive increasing function such that*

$$\int_e^{\infty} \frac{1}{u\psi(u)}du = b_0(\psi)$$

is finite. Then we have the inequality

$$F'(r) \leq F(r)\psi(F(r))$$

for $r \geq r_1$ outside a set of measure $\leq b_0(\psi)$.

Proof: The measure of the set of numbers $r \geq r_1$ such that

$$F'(r) \geq F(r)\psi(F(r))$$

is bounded by

$$\int_{r_1}^{\infty} \frac{F'(r)}{F(r)\psi(F(r))}dr \leq \int_e^{\infty} \frac{1}{u\psi(u)}du = b_0(\psi),$$

29

which proves the lemma.

Note: The calculus lemma was stated and proved in that generality by Nevanlinna. However, Nevanlinna and subsequent contributors to the field instead of keeping the function arbitrary, subject only to the convergent integral, specialized this function, and thereby prevented seeing clearly its formal role in the theory, analogous to the Khintchine convergence principle in diophantine approximations. I suggested keeping an arbitrary ψ systematically and defined the error term function a priori as follows:

Let F be a positive increasing function of class C^1 such that $r \mapsto rF'(r)$ is positive increasing. Let ψ be positive increasing, and let r, c be positive numbers. We define the **error term function**

$$S(F, c, \psi, r) = \log\{F(r)\psi(F(r))\psi(cr\, F(r)\psi(F(r)))\}$$
$$= \log F(r) + \log \psi(F(r)) + \log \psi(cr\, F(r)\psi(F(r))).$$

Note that S is monotone in all its four variables.

In practice, we may take ψ to be a slowly growing function, so that for instance

$$\psi(u) = (\log u)^{1+\epsilon} \text{ with } \epsilon > 0.$$

Then the measure of the exceptional set in Lemma 3.1 is bounded by $1/\epsilon$. Furthermore, with such ψ we see that the error term is of type

$$S = \log F(r) + \log\log \text{ terms.}$$

Lemma 3.2. *Let F be a function of class C^2 defined on $(0, \infty)$. Assume that both $F(r)$ and $rF'(r)$ are positive increasing functions of r, and that there exists $r_1 \geq 1$ such that*

$$F(r_1) \geq e.$$

30

Let $b_1(F)$ be the smallest number $b_1 \geq 1$ such that

$$b_1 r F'(r) \geq e \quad \text{for} \quad r \geq 1.$$

(Such a number trivially exists.) Then for all $r \geq r_1$ outside a set of measure $\leq 2b_0(\psi)$, and all $b_1 \geq b_1(F)$ we have

$$\frac{1}{r}\frac{d}{dr}\left(r\frac{dF}{dr}\right) \leq F(r)\psi(F(r))\psi(b_1 r F(r)\psi(F(r))),$$

and so

$$\log \frac{1}{r}\frac{d}{dr}\left(r\frac{dF}{dr}\right) \leq S(F, b_1, \psi, r).$$

Proof: We apply Lemma 3.1 twice, first to $b_1 r F'(r)$ and then to $F(r)$ to get the desired inequality with $b_1 = b_1(F)$. We can then use any other $b_1 \geq b_1(F)$ by the monotonicity of S.

If we choose $b_1 = b_1(F)$ in Lemma 3.2, then we sometimes omit b_1 from the notation, and write simply $S(F, \psi, r)$. But in some cases, we must view b_1 as a separate variable which we bound in its own right uniformly for a family of functions F.

Similarly, we let $r_1(F)$ be the smallest number $r_1 \geq 1$ such that

$$F(r_1) \geq e.$$

Note that if $F \leq G$ and $F' \leq G'$ then $r_1(F) \geq r_1(G)$ and $b_1(F) \geq b_1(G)$.

In the sequel we assume throughout that ψ satisfies the conditions as in Lemma 3.1, so ψ is positive increasing, and with the convergent integral.

Smoothness properties of ψ will be irrelevant. Just to fix ideas, one might assume ψ continuous or piecewise continuous, just to make the integral have a naive sense.

31

We shall now describe how to construct a function F to which we apply Lemmas 3.1 and 3.2.

We let α be a function on $\mathbf{C}$ such that:

(a) α is continuous and > 0 except at a discrete set of points;

(b) for each $r > 0$, the integral

$$\int_0^{2\pi} \alpha(r,\theta)\frac{d\theta}{2\pi}$$

is absolutely convergent, and gives a continuous function of r.

We recall the **euclidean form**

$$\Phi = \frac{\sqrt{-1}}{2\pi}dz \wedge d\bar{z} = 2rdrd\theta/2\pi,$$

and define the **height transform** F_α of α by

$$F_\alpha(r) = \int_0^r \frac{dt}{t} \int_{\mathbf{D}(t)} \alpha\Phi \qquad \text{for } r > 0.$$

Note that F_α is differentiable and has positive derivative, so is strictly increasing. We shall need to assume here a condition which is satisfied in practice, namely:

(c) There is some number $r_1 \geq 1$ such that $F_\alpha(r_1) \geq e$.

Lemma 3.3 *The function F_α is of class C^2 and we have*

$$\frac{1}{r}\frac{\partial}{\partial r}\left(r\frac{\partial}{\partial r}F_\alpha\right) = 2\int_0^{2\pi}\alpha(r,\theta)\frac{d\theta}{2\pi}.$$

Proof: First,

$$r\frac{\partial F}{\partial r} = 2\int_0^r\int_0^{2\pi}\alpha(t,\theta)tdt\frac{d\theta}{2\pi}.$$

32

34

Differentiating once more yields the desired formula.

We note that $rF'_\alpha(r)$ is an increasing function of r, and is > 0 for $r > 0$ in light of the assumption (a) on α. In particular, there exists $b_1 > 0$ such that

$$b_1 r F'_\alpha(r) \geq e \text{ for } r \geq 1,$$

and so the assumptions of Lemma 3.2 are satisfied.

The function $r \mapsto S(F, \psi, r)$ will be viewed as an error term, and depends on F and ψ. If $F = F_\alpha$ then this function S depends on α and ψ. We shall fix ψ throughout the applications, but we shall apply the inequality to various choices of α.

Putting Lemma 3.2 and Lemma 3.3 together we get:

Lemma 3.4. *Given α satisfying* (a), (b), (c), *we have*

$$\log \int_0^{2\pi} \alpha(r, \theta) \frac{d\theta}{2\pi} \leq S(F_\alpha, \psi, r)$$

for all $r \geq r_1(F_\alpha)$ outside a set of measure $\leq 2b_0(\psi)$.

Lemma 3.4 will be applied by taking a log in and out of an integral. We recall that result here for the convenience of the reader.

Lemma 3.5 *Let X be a measured space with positive measure μ and total measure 1. Let α be a real valued function ≥ 0 such that $\log \alpha$ is integrable. Then*

$$\int_X \log \alpha \, d\mu \leq \log \int_X \alpha d\mu.$$

Proof: Let

$$\int_X \alpha \, d\mu = c.$$

33

35

If $c = 0$, then α is 0 almost everywhere, $\log \alpha = -\infty$ almost everywhere, and the result is trivial. Suppose $c > 0$. We have to show

$$\int\limits_X \log \alpha \, d\mu \le \int\limits_X (\log c) d\mu,$$

or equivalently,

$$\int\limits_X \log(\alpha/c) d\mu \le 0.$$

Write $\alpha/c = 1 + x$ and use $\log(1 + x) \le x$ for $x > -1$. The desired inequality drops out from the definitions.

I, §4. RAMIFICATION AND SECOND MAIN THEOREM

We shall now apply §3 to the special case of a meromorphic function f, viewed as a holomorphic map of $\mathbf{C}$ into $\mathbf{P}^1$. Recall that

$$f^* \omega = \gamma_f \Phi.$$

Then from the definitions, the height transform of γ_f is precisely T_f, that is

$$F_{\gamma_f}(r) = T_f(\infty, r) = T_f(r).$$

We now define the ramification. Suppose $f(z_0) = a$. With $e \ge 1$, write

$$f(z) = c_0 + c_e(z - z_0)^e + \text{ higher terms with } c_e \ne 0 \text{ if } a \ne \infty$$
$$f(z) = c_e(z - z_0)^{-e} + \text{ higher terms with } c_e \ne 0 \text{ if } a = \infty.$$

We call e the **ramification index**, and we let $e - 1$ be the **order of the ramification divisor** of f at z_0. If $f = f_1/f_0$ where f_0, f_1 are entire without common zero, then it is immediately verified that

$$e - 1 = \text{order of } f_0 f_1' - f_0' f_1 \text{ at } z_0 = \text{order of } W(f_0, f_1).$$

34

Thus the zeros of the Wronskian, with multiplicities, define the ramification divisor of f. We define

$$n_{f,\mathrm{Ram}}(r) = n_W(0,r) \quad \text{and} \quad N_{f,\mathrm{Ram}}(r) = N_W(0,r).$$

Theorem 4.1. *Let f be a meromorphic function, non-constant, such that $f(0) \neq 0, \infty$ and $f'(0) \neq 0$. Then*

$$N_{f,\mathrm{Ram}}(r) - 2T_f(r) \leq \frac{1}{2}S(T_f, \psi, r) - \frac{1}{2}\log \gamma_f(0)$$

for $r \geq r_1(T_f)$ outside a set of measure $\leq 2b_0(\psi)$.

Proof: In Proposition 2.3 we take

$$\alpha = \log \gamma_f$$

where γ_f is the C^∞ function given by

$$\gamma_f = \frac{|f'|^2}{(1+|f|^2)^2} = \frac{|W|^2}{(|f_0|^2 + |f_1|^2)^2}.$$

Then

$$dd^c[\log \gamma_f] = dd^c[\log |W|^2] - 2dd^c \log(|f_0|^2 + |f_1|^2).$$

By Proposition 2.1 and the Ahlfors-Shimizu expression for the height in Proposition 2.4, and by Proposition 2.3 which gives us the singular part for the term with W, we find

$$N_{f,\mathrm{Ram}}(r) - 2T_f(r) = \int_0^r \frac{dt}{t} \int_{D(t)} dd^c[\log \gamma_f]$$

$$= \frac{1}{2} \int_0^{2\pi} \log \gamma_f(re^{i\theta}) \frac{d\theta}{2\pi} - \frac{1}{2} \log \gamma_f(0)$$

$$[\text{by Lemma 3.5}] \leq \frac{1}{2} \log \int_0^{2\pi} \gamma_f(re^{i\theta}) \frac{d\theta}{2\pi} - \frac{1}{2} \log \gamma_f(0).$$

35

Since $F_{\gamma_f} = T_f$ we can apply Lemma 3.4 to conclude the proof.

Note: In the future, it will be convenient to give a name to the constant appearing on the right hand side, so we let

$$b_2(f) = -\frac{1}{2} \log \gamma_f(0).$$

Note: The argument for the proof of Theorem 4.1 goes back to Nevanlinna, who used the error term $O(\log rT_f(r))$, as did subsequent authors. The emphasis on trying to find a "best possible" error term stems from [**La 8**], and getting the factor $1/2$ in front of the error term stems from [**La 8**]. The error term as given here is due to Lang.

We shall extend Theorem 4.1 to the case when f can approximate a finite set of points, by measuring this closeness.

Theorem 4.2. *Let $q \geq 1$. Let $a_1, \ldots, a_q$ be a finite set of distinct points of $\mathbf{P}^1$, and let $f: \mathbf{C} \to \mathbf{P}^1$ be a non-constant holomorphic map such that $f(0) \neq 0, \infty, a_j$ for all j and $f'(0) \neq 0$. Let r_1 be a number ≥ 1 such that $T_f(r_1) \geq e$. There are constants $b = b(f, a_1, \ldots, a_q)$ and B_q (depending on q), such that for all $r \geq r_1$ outside a set of measure $\leq 2b_0(\psi)$ and all $b_1 \geq b_1(T_f)$ we have*

$$(q-2)T_f(r) - \sum N(f, a_j, r) + N_{f, \mathrm{Ram}}(r) \leq \frac{1}{2} S(B_q T_f^2, b_1, \psi, r) + b.$$

We can take $B_q = 12q^2 + q^3 \log 4$, for instance.

Before proving the theorem, we make some remarks on the error term. We pick ψ to grow slowly. Then 2 cancels $1/2$, and the dominant term in the error term is

$$\log T_f(r),$$

while the remaining terms have lower orders of magnitude, which can be made arbitrarily small, compatible with the convergence of the integral for ψ.

36

The rest of this section is devoted to the proof.

For $w, a \in \mathbf{C}$ recall that

$$\|w, a\|^2 = \frac{|w - a|^2}{(1 + |w|^2)(1 + |a|^2)} \leq 1,$$

and we make the similar definition if w or $a = \infty$, to get essentially a distance between two points on $\mathbf{P}^1$.

Remark. (Product into sum) *Let:*

$$s = \frac{1}{3} \min_{i \neq j} \|a_i, a_j\| \quad and \quad b_3 = b_3(a_1, ..., a_q) = \frac{1}{s^{2(q-1)}}$$

Then for all $w \in \mathbf{P}^1$ and all λ with $0 < \lambda < 1$ we have

$$\prod_j \|w, a_j\|^{-2(1-\lambda)} \leq b_3 \sum_j \|w, a_j\|^{-2(1-\lambda)}.$$

Indeed, for our choice of s, for all $w \in \mathbf{P}^1$ there exists at most one index j_0 such that $\|w, a_{j_0}\| \leq s$. Note that b_3 is independent of λ, again because the maximum value is reached when $\lambda = 0$.

Let Λ be a decreasing function of r with $0 < \Lambda < 1$. We allow Λ being constant. Following Ahlfors, Stoll [St], and Wong [Wo], we define the **Ahlfors-Wong function**

$$\gamma_\Lambda = \prod_j \|f, a_j\|^{-2(1-\Lambda)} \gamma_f.$$

We view a function of z as a function of (r, θ) so $\Lambda(z) = \Lambda(r)$. We have

$$-\log \gamma_\Lambda(0) \leq -\log \gamma_f(0) = b_2(f).$$

This is immediate from the fact that $\|f, a_j\| \leq 1$.

We define

$$\alpha_\Lambda = \sum_j \|f, a_j\|^{-2(1-\Lambda)} \gamma_f.$$

37

Then by the remark, we have the inequality

Lemma 4.3. $\gamma_\Lambda \leq b_3 \alpha_\Lambda$.

We now come to the main part of the proof.

Our first step is to estimate the integral which comes up in the Green-Jensen formula.

Proposition 4.4. *Let*

$$\Lambda(r) = \begin{cases} 1/qT_f(r) & \text{for} \quad r \geq r_1 \\ \text{constant} & \text{for} \quad r \leq r_1 \end{cases}.$$

Then

$$\frac{1}{2} \log \int_0^{2\pi} \gamma_\Lambda(re^{i\theta}) \frac{d\theta}{2\pi} \leq \frac{1}{2} S(B_q T_f^2, b_1, \psi, r) + \frac{1}{2} \log b_3.$$

for $r \geq r_1$ outside a set of measure $\leq 2b_0(\psi)$.

Let us postpone the proof of Proposition 4.4, and see how the proposition implies the theorem.

Note that for each a_j, we have

$$0 \leq \|f, a_j\|^2 = \frac{|f - a_j|^2}{(1 + |f|^2)(1 + |a_j|^2)}$$

$$= \frac{|f_1 - af_0|^2}{(|f_0|^2 + |f_1|^2)(1 + |a_j|^2)}.$$

The function $f_1 - af_0$ is entire and we shall count its zeros, which are precisely the points where $f = a$.

Suppose now that λ is constant. Note that for g holomorphic,

$$\log |g|^{2(1-\lambda)} = (1 - \lambda) \log |g|^2.$$

When g is meromorphic we can apply the Green-Jensen formula to $|g|^{2(1-\lambda)}$ by the homomorphic property of the log, and to a product of

38

40

such factors. By Corollary 2.5, we have

$$(1 - \lambda)T_f(r) - (1 - \lambda)N(f, a_j, r) = -\int_0^r \frac{dt}{t} \int_{\mathbf{D}(t)} dd^c[\log \|f, a_j\|^{2(1-\lambda)}].$$

Combining this with the terms which arise from Theorem 4.1, and using the Green-Jensen formula, we get:

$$q(1 - \lambda)T_f(r) - (1 - \lambda)\sum N(f, a_j, r) + N_{f,\mathrm{Ram}}(r) - 2T_f(r)$$

$$= \int_0^r \frac{dt}{t} \int_{\mathbf{D}(t)} dd^c[\log \gamma_\lambda]$$

$$= \frac{1}{2}\int_0^{2\pi} \log \gamma_\lambda(re^{i\theta})\frac{d\theta}{2\pi} - \frac{1}{2}\log \gamma_\lambda(0).$$

From here on, we do not need to assume Λ constant, and we obtain the further inequality for arbitrary Λ:

$$\frac{1}{2}\int_0^{2\pi} \log \gamma_\Lambda(re^{i\theta})\frac{d\theta}{2\pi} - \frac{1}{2}\log \gamma_\Lambda(0) \leq \frac{1}{2}\log \int_0^{2\pi}\gamma_\Lambda(re^{i\theta})\frac{d\theta}{2\pi} - \frac{1}{2}\log \gamma_f(0).$$

Let Λ be as in Proposition 4.4, that is $\Lambda = 1/qT_f$ for $r \geq r_1$. Then first we use that proposition to estimate the last integral; and second, putting $\lambda = \Lambda(r)$ we see that the bad term $-q\Lambda(r)T_f(r)$ now becomes equal to -1. Furthermore, the factor $(1 - \lambda)$ before $\sum N(f, a_j, r)$ can be replaced by 1 since this sum occurs with a minus sign in front. Hence we have proved Theorem 4.2, with the constant

$$\boxed{b(f, a_1, \dots, a_q) = \frac{1}{2}\log b_3 - \frac{1}{2}\log \gamma_f(0) + 1.}$$

Proposition 4.4 will be proved in the next section.

Remarks: With an error term $O(\log r + \log T_f(r))$ (outside an exceptional set), the theorem is due to Nevanlinna, and with the good

39

41

error term, it is due to P.M. Wong [**Wo**], except for the use of the general function ψ which I suggested. In [**La 8**] I did not see how to prove my conjecture that the error term should be $(1+\varepsilon)\log T_f(r)$. As in Stoll [**St**], formula 11.7, Wong used a method of Ahlfors, namely the variable function $\Lambda(r)$. Ahlfors did not pay attention to constant factors, and used $\Lambda(r) = 1/T(r)$. Wong uses $\Lambda(r) = 1/qT(r)$, thereby getting rid of extraneous terms by this method, so that the error term is independent of $a_1, \ldots, a_q$ except in the additive constant $b(f, a_1, \ldots, a_q)$.

Ahlfors' method had been partly used by Chern [**Ch**]. However, Chern improperly reproduced this method by taking only a constant exponent λ, so that his argument is erroneous. When he takes the limit as $\lambda \to 0$, the exceptional set depends on λ, and may a priori enlarge so that it covers all the real numbers, and so Chern's argument, apparently shortcutting parts of Ahlfors, fails. The trick of Lemma 4.3, used by Wong to estimate products by sums, also comes from Ahlfors, and is a key step for getting rid of extraneous terms. I did not know this trick in [**La 8**].

Ahlfors' paper mixed the combinatorial tricks of handling the function Λ and the products into sums argument with an entirely different problem, which was to describe the differential geometry of derived curves in the context of Nevanlinna theory. Thus confusion and some ignorance about this particular combinatorial aspect of his proof was rather widespread for a long time, since nobody noticed the difficulty in Chern's paper until I pointed it out in [**La 8**]. For instance, Chern's incorrect presentation of Ahlfors' proof is reproduced in [**Gr**] p. 81.

By taking the limit illegitimately, Chern actually obtained an error term of the form

$$(\frac{1}{2} + \varepsilon) \log T_f(r),$$

just as in Theorem 4.1, estimating the ramification divisor. My guess that this is false in general, was confirmed by examples of Ye which

40

will be found in the appendix. Two problems are involved here:

(a) In general, is $(1 + \varepsilon) \log T_f(r)$ essentially the best possible error term?

(b) For each specific classical function, what is the best possible error term?

I expect that there exist examples such that $\log T_f(r)$ (asymptotically, of course) cannot be an error term (i.e. without the epsilon) possibly by constructing Weierstrass products. More generally, as in the metric Khintchine theory of the real numbers, is the error term with the functions ψ having convergent integral the best possible error term for "almost all" holomorphic functions, in a suitable sense of "almost all"? It is also a theorem of Khintchine that if φ is a positive function such that the series

$$\sum_{q=1}^{\infty} \varphi(q)$$

diverges, then for almost all real numbers α, there exist infinitely many solutions to the inequality

$$\left| \alpha - \frac{p}{q} \right| < \frac{\varphi(q)}{q}.$$

For a proof see Khintchine's book [**Kh**]. What is the analogue of this theorem in Nevanlinna theory? Using the type function, I have shown how to give an asymptotic estimate for the number of solutions of such an inequality with an error term involving the type function, [**La 2**] and [**La 3**] Chapter II, §3. What is the analogue of these asymptotics in Nevanlinna theory?

It is known and easy to show that the error term in Nevanlinna theory for the function e^z is $O(1)$. In number theory it is easy to show that the quadratic numbers have bounded type. It is conjectured that the only algebraic real numbers of bounded type are quadratic. Is there a similar phenomenon in Nevanlinna theory? Also there are several

41

characterizations for numbers of bounded type, see [**La 3**] Chapter II, §2. What are the analogues in nevanlinna theory?

For the classical functions, one would have to use their special properties, individually for $\wp, \theta, \Gamma, \zeta, J$ and their variations, to determine their "type", i.e. their best possible error term in each special case. Thus I would define the **type** of a function f to be a function ψ such that the error term has the form

$$\log \psi(T_f) + O(1).$$

Finally, we observe that without much change one could formulate and prove a second main theorem when the image space is a compact Riemann surface. This will be a special case of the theorem in Chapter II. Of course, there exist holomorphic maps $f \colon \mathbf{C} \to X$ into such a surface only in case of genus 0 or 1. However, the error term is given in such a way that it applies to a map of a disc $f \colon \mathbf{D}(R) \to X$, and can be used in general to give a bound on the radius of the disc since the constants entering into the error term are given entirely as invariants of X, in the case of higher genus. This is in line with the classical Landau-Schottky theorem. The same remark applies to all the main theorems since the error term is given in explicit form each time.

I, §5. AN ESTIMATE FOR THE HEIGHT TRANSFORM

We shall prove Proposition 4.4. At first, we can work with an arbitrary Λ, not necessarily the one we selected specifically in that proposition. By Lemma 4.3 and the calculus Lemma 3.4, we find

$$\frac{1}{2} \log \int_0^{2\pi} \gamma_\Lambda(r, \theta) \frac{d\theta}{2\pi} \le \frac{1}{2} \log \int_0^{2\pi} \alpha_\Lambda(r, \theta) \frac{d\theta}{2\pi} + \frac{1}{2} \log b_3$$

$$\le \frac{1}{2} S(F_{\alpha_\Lambda}, \psi, r) + \frac{1}{2} \log b_3$$

42

44

for $r \geq r_1(F_{\alpha_\Lambda})$ outside a set of measure $\leq 2b_0(\psi)$.

But since $\alpha_\Lambda \geq \gamma_f$ we get $F_{\alpha_\Lambda} \geq T_f$, whence

$$r_1(F_{\alpha_\Lambda}) \leq r_1(T_f) = r_1, \qquad \text{where } T_f(r_1) \geq e$$

and we can use $r \geq r_1$ independently of Λ, as in Theorem 2.1. Furthermore the same inequality $\alpha_\Lambda \geq \gamma_f$ implies $F'_{\alpha_\Lambda} \geq T'_f$, and therefore

$$b_1(F_{\alpha_\Lambda}) \leq b_1(T_f) = b_1$$

again independently of Λ. Then we have the inequality

$$S(F_{\alpha_\Lambda}, \psi, r) \leq S(F_{\alpha_\Lambda}, b_1, \psi, r) \text{ for } r \geq r_1.$$

To estimate $S(F_{\alpha_\Lambda}, \psi, r)$ completely in terms of T_f, we shall prove:

Lemma 5.1. *We have*

$$F_{\alpha_\Lambda} \leq \frac{q \log 4}{\Lambda^2} + \frac{12qT_f}{\Lambda}.$$

This lemma amounts to a curvature computation, and will be proved in the next section. Now letting $\Lambda = 1/qT_f$ concludes the proof of Proposition 4.4.

Since α_Λ is additive in the a_j, it will suffice to deal with a single a_j. We restate the result to be proved above.

Proposition 5.2. *Let $a \in \mathbf{P}^1$. Let Λ be a decreasing function of r with $0 < \Lambda(r) < 1$. Define*

$$\alpha_\Lambda = \|f, a\|^{-2(1-\Lambda)} \gamma_f.$$

Then

$$F_{\alpha_\Lambda} \leq \frac{\log 4}{\Lambda^2} + \frac{12}{\Lambda} T_f(r).$$

43

45

For the proof I follow P.M. Wong [**Wo**]. We shall need two lemmas.

Lemma 5.3. *Let λ be a real number, $0 < \lambda < 1$. Then*

$$\lambda^2 \|f, a\|^{-2(1-\lambda)} \gamma_f \Phi \leq 4 dd^c \log(1 + \|f, a\|^{2\lambda}) + 12\lambda\gamma_f \Phi.$$

Assuming this lemma for the moment, and putting

$$\alpha_\lambda = \|f, a\|^{-2(1-\lambda)} \gamma_f,$$

we get

$$F_{\alpha_\lambda}(r) = \int_0^r \frac{dt}{t} \int_{D(t)} \|f, a\|^{-2(1-\lambda)} \gamma_f \Phi$$

$$\leq \frac{4}{\lambda^2} \int_0^r \frac{dt}{t} \int_{D(t)} dd^c \log(1 + \|f, a\|^{2\lambda}) + \frac{12}{\lambda} T_f(r).$$

By Proposition 2.6, the Green-Jensen formula is valid in the present instance in the form we need it, that is:

Let f be meromorphic. Then

$$\int_0^r \frac{dt}{t} dd^c \log(1 + \|f, a\|^{2\lambda})$$

$$= \frac{1}{2} \int_0^{2\pi} \log(1 + \|f(re^{i\theta}), a\|^{2\lambda}) \frac{d\theta}{2\pi} - \frac{1}{2} \log(1 + \|f(0), a\|^{2\lambda}).$$

For any w we have $\|w, a\| \leq 1$. It follows that

$$F_{\alpha_\lambda}(r) \leq \frac{\log 4}{\lambda^2} + \frac{12}{\lambda} T_f(r).$$

44

46

Since Λ is a decreasing function and $0 < \Lambda(r) < 1$, for a given value of r we have

$$\|f(\zeta), a\|^{-2(1-\Lambda(|\zeta|))} \leq \|f(\zeta), a\|^{-2(1-\Lambda(r))}$$

for all $|\zeta| \leq r$. Take $\lambda = \Lambda(r)$. Then

$$F_{\alpha_\Lambda}(r) \leq F_{\alpha_\lambda}(r)$$

thus concluding the proof of Proposition 5.2.

We now come to the proof of Lemma 5.3. We shall need:

Lemma 5.4. *Let λ be a real number > 0. Then*

$$dd^c\|f, a\|^{2\lambda} = \{\lambda^2\|f, a\|^{-2(1-\lambda)} - \lambda(\lambda+1)\|f, a\|^{2\lambda}\}\gamma_f\Phi.$$

This is done by brute force. To organize such a computation, one may wish to use the following additional formulas, which are also useful in another context. First, using the definition of $\|f, a\|^2$ as a product, and the homomorphic property of the operator $dd^c\log$, transforming multiplication into addition, and killing $|f - a|^2$, we see that

5.4.1 $$dd^c\log\|f, a\|^{2\lambda} = -\lambda\gamma_f\Phi.$$

Second we have

5.4.2 $$d\|f, a\|^2 \wedge d^c\|f, a\|^2 = \|f, a\|^2(1 - \|f, a\|^2)\gamma_f\Phi.$$

To prove this formula, first note that up to a suitable constant, we may use ∂ and $\bar{\partial}$ instead of d and d^c on the left hand side. We use the definition

$$\|f, a\|^2 = \frac{(f - a)(\bar{f} - \bar{a})}{(1 + f\bar{f})(1 + a\bar{a})}.$$

45

47

Then we take $\partial \|f, a\|^2$ by using the rule for the derivative of a quotient, keeping in mind that $\partial \bar{f} = 0$. With respect to the operator ∂, note that the factor $(\bar{f} - \bar{a})$ behaves like a constant. We get:

$$\partial \|f, a\|^2 = \frac{\bar{f} - \bar{a}}{(1 + a\bar{a})} \frac{1}{(1 + f\bar{f})^2} [(1 + f\bar{f})\partial f - (f - a)\bar{f}\partial f]$$

$$= \frac{\bar{f} - \bar{a}}{(1 + a\bar{a})} \frac{1}{(1 + f\bar{f})^2} (1 + a\bar{f})\partial f.$$

Putting a complex conjugate over both sides yields $\bar{\partial} \|f, a\|^2$. Wedging yields

$$\partial \|f, a\|^2 \wedge \bar{\partial} \|f, a\|^2 = \frac{|f - a|^2}{(1 + a\bar{a})^2(1 + f\bar{f})^4} (1 + a\bar{f} + \bar{a}f + a\bar{a}f\bar{f})\partial f \wedge \bar{\partial}\bar{f}.$$

Using the definition of $\|f, a\|^2$ and substituting in the right hand side of 5.4.2 one finds the corresponding expression which proves 5.4.2.

Third, for any function u we have formulas which are verified directly from the fact that d and d^c are derivations:

dd^c1 $\qquad u^2 dd^c \log u = u dd^c u - du \wedge d^c u$

dd^c2 $\qquad dd^c \log(1 + u) = \dfrac{dd^c u}{1 + u} - \dfrac{du \wedge d^c u}{(1 + u)^2}$

dd^c3 $\qquad\qquad\quad = \dfrac{dd^c u}{(1 + u)^2} + \dfrac{u^2 dd^c \log u}{(1 + u)^2}$ [using **dd^c1**]

dd^c4 $\qquad\qquad\quad = \dfrac{1}{u(1 + u)^2} du \wedge d^c u + \dfrac{u}{1 + u} dd^c \log u.$

Now to prove Lemma 5.4, let $u = \|f, a\|^2$. Then the left hand side in Lemma 5.4 is $dd^c u^\lambda$, and we have

$$dd^c u^\lambda = d(\lambda u^{\lambda-1} d^c u) = \lambda(\lambda - 1)u^{\lambda-2} du \wedge d^c u + \lambda u^{\lambda-1} dd^c u.$$

Note that 5.4.2 gives $du \wedge d^c u$ in terms of $\gamma_f \Phi$, and formula **dd^c1** gives

$$dd^c u = u dd^c \log u + u^{-1} du \wedge d^c u.$$

46

48

Therefore again we can use 5.4.2 combined with 5.4.1 to get $dd^c u$ in terms of $\gamma_f \Phi$. If one combines all the factors of $\gamma_f \Phi$, one finds precisely the factor on the right hand side of Lemma 5.4.

We now conclude the proof of Lemma 5.3, that is for $0 < \lambda < 1$ we prove the inequality

$$dd^c \log(1 + \|f, a\|^{2\lambda}) \geq \frac{\lambda^2}{4} \|f, a\|^{-2(1-\lambda)} \gamma_f \Phi - 3\lambda \gamma_f \Phi.$$

Let $u = \|f, a\|^{2\lambda}$. Then using Lemma 5.4 and 5.4.1 we get

$$dd^c \log(1 + \|f, a\|^{2\lambda}) = \frac{dd^c \|f, a\|^{2\lambda}}{(1 + \|f, a\|^{2\lambda})^2} + \frac{\|f, a\|^{4\lambda} dd^c \log \|f, a\|^{2\lambda}}{(1 + \|f, a\|^{2\lambda})^2}$$

$$\geq \{\frac{1}{4}\lambda^2 \|f, a\|^{-2(1-\lambda)} - \lambda(\lambda + 1)\|f, a\|^{2\lambda}\}\gamma_f \Phi - \lambda \gamma_f \Phi$$

using the fact that $\|f, a\| \leq 1$. Replacing $\lambda + 1$ by 2 and combining the last two terms concludes the proof of Lemma 5.3, and therefore also of Proposition 5.2.

Note: The result here in its precise form is due to P.M. Wong [Wo], and replaces a curvature computation. A weaker version appears in Stoll [St], as an "Ahlfors Estimate", Theorem 10.3. The introduction of curvature in Nevanlinna theory is due to F. Nevanlinna (cf. [Ne] Chapter IX, §4), but the way it is carried out here, using the singular form with variable Λ is closer to Ahlfors [Ah]. Indeed, the occurrence of T_f^2 in the error term is already in Ahlfors, pp. 26-27 in a similar curvature computation, although as usual, Ahlfors does not keep track of constants. It is the presence of T_f^2 in Theorem 4.2 and Proposition 4.4 combined with the factor $1/2$ in front which gave rise to the proof of my conjecture that the error term should be of the form $(1 + \varepsilon) \log T_f(r)$. As long as one was not looking for the best possible error term, no great significance was attached to this particular structure of the proof, and the significance of T_f^2 went unnoticed.

47

I, §6. VARIATIONS AND APPLICATIONS. THE LEMMA ON THE LOGARITHMIC DERIVATIVE

We start by giving a proof of Nevanlinna's lemma on the logarithmic derivative. The basic differential geometric pattern was given by Nevanlinna [Ne] p. 259. I did not quite get the desired error term in [La 8], but I am now able to get it by using Wong's idea, with the Ahlfors-Wong function Λ.

We write m_f for $m_{f,\infty}$ and similarly for T_f.

Theorem 6.1 (Lemma on the logarithmic derivative). *Let f be a non-constant meromorphic function such that $f(0) \neq 0, \infty$ and $f'(0) \neq 0$. Then*

$$m_{f'/f}(r) \leq \frac{1}{2} S(B_2 T_f^2, \psi, b_1, r) - \log \frac{|f'(0)|^2}{(1+|f(0)|^2)^2} + 1 + \frac{1}{2} \log b_3(0, \infty)$$

for $r \geq r_1$ outside a set of measure $\leq 2b_0(\psi)$.

Proof: Let $\Lambda = 1/2T_f$ for $r \geq r_1$ and constant for $r \leq r_1$, as in Proposition 4.4. We let $D = (0) + (\infty)$, so $q = 2$. Then directly from the definitions,

$$\gamma_\Lambda = |f'/f|^2 h^\Lambda \qquad \text{where} \qquad h = \|f, 0\|^2 \|f, \infty\|^2.$$

Let

$$u = |f'/f|^2 \qquad \text{and} \qquad v = h^\Lambda, \quad \text{so} \quad 0 \leq v \leq 1.$$

48

Denote by $S(r)$ the circle of radius r, and let $\sigma = d\theta/2\pi$. Then

$$m_{f'/f}(r) = \frac{1}{2} \int_{S(r)} (\log^+ u)\sigma$$

$$= \frac{1}{2} \int_{S(r)} (\log^+ u + \log v)\sigma - \frac{1}{2} \int_{S(r)} (\log v)\sigma$$

$$= \frac{1}{2} \int_{S(r)} (\log e^{\log^+ u + \log v})\sigma - \frac{1}{2}\Lambda(r) \int_{S(r)} (\log h)\sigma$$

$$\leq \frac{1}{2} \log \int_{S(r)} uv\sigma + 1 + \Lambda(r) m_{f,(0)+(\infty)}(r).$$

Indeed, for this last inequality, we pull the first log out of the integral. Then we use that if $0 \leq v \leq 1$ then

$$e^{\log^+ u + \log v} \leq uv + 1 = \gamma_\Lambda + 1,$$

which gives us the first term by Proposition 4.4. As to the second term,

$$m_{f,(0)+(\infty)} \leq 2T_f - \log \frac{|f'(0)|^2}{(1 + |f(0)|^2)^2}.$$

Hence our choice of $\Lambda(r)$ gives the desired bound.

The lemma on the logarithmic derivative was essentially a variation on the Second Main Theorem. Next, we give an actual application of this theorem due to Nevanlinna, and bounding the number of totally ramified values.

Given a point $a \in \mathbf{P}^1$ we consider those elements $z \in \mathbf{C}$ such that $f(z) = a$. We say that f is **totally ramified over** a if for every such element z we have

$$\operatorname{ord}_z(f - a) > 1.$$

Theorem 6.2. *Let f be a non-constant meromorphic function. Then f has at most four totally ramified values.*

49

Proof: Let $a_1, \ldots, a_q$ be distinct totally ramified values. We have to prove that $q \leq 4$. Write

$$n_f(a, r) = n_f^*(a, r) + n_{f,\mathrm{Ram}}(a, r)$$

where $n_f^*(a, r)$ is the number of $z \in \mathbf{D}(r)$ such that $f(z) = a$, counted with multiplicity 1, and $n_{f,\mathrm{Ram}}(a, r)$ is the number counted with the ramification index, i.e. if f is ramified of order e at z_0 then z_0 is counted with multiplicity $e-1$, then we get the corresponding counting functions

$$N_f^*(a, r) + \int_0^r n_f^*(a, t) \frac{dt}{t} + n_f^*(a, 0) \log r,$$

so

$$N_f(a, r) = N_f^*(a, r) + N_{f,\mathrm{Ram}}(a, r).$$

By the Second Main Theorem,

$$\sum_{j=1}^q m_f(a_j, r) + N_{f,\mathrm{Ram}}(r) - 2T_f(r) \leq o_{\mathrm{exc}}(T_f(r)).$$

Here I use an extension o_{exc} of the usual symbol o. Namely, I define

$$\alpha = o_{\mathrm{exc}}(\beta)$$

to mean that there exists a set E of finite measure such that

$$\lim_{\substack{r \to \infty \\ r \notin E}} \frac{\alpha(r)}{\beta(r)} = 0.$$

Using our expression for $N_{f,\mathrm{Ram}}$, we find by the First Main Theorem:

$$(q - 2)T_f(r) \leq \sum_{j=1}^q N_f^*(a_j, r) + o_{\mathrm{exc}}(T_f(r))$$

$$\leq \sum_{j=1}^q \frac{1}{2} N_f(a_j, r) + o_{\mathrm{exc}}(T_f(r))$$

$$\leq \frac{q}{2} T_f(r) + o_{\mathrm{exc}}(T_f(r)).$$

50

Hence

$$\frac{q}{2}T_f(r) \leq 2T_f(r) + o_{\text{exc}}(T_f(r)),$$

which proves $q \leq 4$, and concludes the proof of the theorem.

Note that the Weierstrass $\wp$-function has exactly 4 totally ramified values, so 4 is best possible in general.

51

Appendix. On Nevanlinna's Error Term
by Zhuan Ye

This note is a part of an ongoing more detailed study. In Theorem 4.2 of Chapter I of these notes, Serge Lang presents the explicit estimate of Pit-Mann Wong [**Wo**] conjectured by Lang that

(1)
$$\sum_{\nu=1}^{q} m(r,a_\nu) - 2T(r,f) + N_{\text{Ram}}(r,f) \le \log T(r,f) + \text{ lower-order terms}$$

outside an exceptional set as $r \to \infty$. We show here that this estimate is essentially best possible, in the sense that "1" on the right side of (1) cannot be replaced by a smaller number. We have the following

Theorem. *Given $\varepsilon > 0$, there exists an entire function f of finite order and a finite set $\{a_\nu\}$ such that*

(2)
$$\sum m(r,a_\nu) - 2T(r,f) + N_{\text{Ram}}(r,f) > (1-\epsilon)\log T(r,f),$$

for all large r.

Proof: We consider the function

$$f(z) = \int_0^z e^{-t^q}\,dt$$

where $q \ge 2$ is a integer, $a_\nu = e^{2\pi i \nu/q}\int_0^\infty e^{-t^q}\,dt$ and examine the computation of [Ne, VI §2.3].

If $|\arg z - 2\pi\nu q^{-1}| \le \frac{1}{2}\pi q^{-1}$, we see that

$$f(z) - a_\nu = -\int_z^\infty e^{-t^q}\,dt = -\frac{e^{-z^q}}{qz^{q-1}} + \frac{q-1}{q}\int_z^\infty \frac{e^{-t^q}}{t^q}\,dt$$

$$= -\frac{e^{-z^q}}{qz^{q-1}}(1+o(1)), \qquad (1 \le \nu \le q).$$

53

Thus, when $|\arg z - 2\pi\nu q^{-1}| \le \frac{1}{2}\pi q^{-1}$, we find that

$$\log |f(re^{i\theta}) - a_\nu| = -(r^q \cos q\theta + (q-1)\log r + O(1)).$$

Since the a_ν are distinct, this shows that

$$m(r, a_\nu, f) = \frac{1}{\pi q}r^q + \frac{q-1}{2q}\log r + O(1), \qquad (1 \le \nu \le q).$$

The computation of $T(r, f)$ is similar. In

$$|\arg z - (2\nu - 1)\pi q^{-1}| < \frac{1}{2}\pi q^{-1},$$

we find that

$$f(z) = \frac{e^{-z^q}}{qz^{q-1}}(1 + o(1)) + O(1),$$

and hence, on summing over these q sectors,

$$T(r, f) = m(r, f) = \frac{1}{\pi}r^q - \frac{q-1}{2}\log r + O(1).$$

Thus, $\log T(r, f) = (q-1)\log r + O(1)$. In particular, $f(z)$ has order q, mean type. Obviously $N_{\mathrm{Ram}}(r, f) \equiv 0$, and hence

$$\sum_{\nu=1}^{q} m(r, f, a_\nu) + m(r, f) - 2T(r, f) + N_{\mathrm{Ram}}(r, f)$$

$$= (q-1)\log r + O(1) = \frac{q-1}{q}\log T(r, f) + O(1).$$

If we choose q so large that $q\epsilon > 1$, we have shown (2).

This example may be typical. For example, if we consider $g(z) = e^{z^p}$, slightly easier computations show that $m(r, g, 0) = m(r, g) = T(r, g) = \frac{1}{\pi}r^p$, so that

$$\log T(r, g) = p \log r + O(1).$$

54

Finally, the only ramification is at the origin, where $f(z)$ has a zero of multiplicity $p - 1$. Thus

$$m(r, g, 0) + m(r, g) + N_{\text{Ram}}(r, g) - 2T(r, g)$$
$$\equiv N_{\text{Ram}}(r, g) = (p - 1)\log r + O(1)$$
$$= \frac{p - 1}{p}\log T(r, g) + o(1).$$

By choosing p large, we again obtain (2).

In view of the ubiquity of such examples, it seems interesting to find the exact bound in (1).

Finally, I thank D. Drasin for presenting this problem to me.

55

CHAPTER II
EQUIDIMENSIONAL HIGHER DIMENSIONAL THEORY

Here following Carlson-Griffiths [C-G] we consider the higher dimensional situation which is closest to that of dimension one, namely a mapping

$$f: \mathbf{C}^n \to X$$

where X is a compact complex manifold of dimension n. Then the theorems and proofs are completely parallel, with no additional difficulty except that which comes from the extra formal manipulation of higher dimensions. But the theory is in good shape, in that once one has understood the basic principles and mechanisms which must be introduced to deal with the higher dimensional case, then the translation from dimension one becomes essentially automatic. These mechanisms are essentially fundamental to complex analysis and complex differential geometry, especially concerning holomorphic curvature, and are of interest independently of Nevanlinna theory.

Carlson-Griffiths used a singular volume form more or less generalizing such use by F. Nevanlinna. Here we follow mostly Ahlfors and P.M. Wong [Wo], who uses a more efficient singular form on $\mathbf{C}^n$ instead of X.

II, §1. THE CHERN AND RICCI FORMS

We begin by fixing some notation and terminology.

We recall the notion of a divisor, both on X and on $\mathbf{C}^n$, so let Y be a complex manifold, not necessarily compact. Consider pairs (U, φ)

consisting of an open set U and a meromorphic function φ. We say that two pairs (U, φ) and (V, ψ) are **equivalent** if $\varphi\psi^{-1}$ is holomorphic invertible on $U \bigcap V$, so $\varphi^{-1}\psi$ is also holomorphic. By a (Cartier) **divisor** on Y we mean a maximal family of equivalent pairs $\{(U_i, \varphi_i)\}$ such that the open sets U_i cover Y. If one is given a family of such pairs which is not maximal, but such that the U_i cover Y then we say that this family **represents** the divisor. If (U, φ) is a pair in the family, we say that this pair **represents** the divisor on U, or that φ represents the divisor on U. Thus a divisor is defined by local conditions. If all φ_i are holomorphic, we say that the divisor is **effective**. On the whole we assume that the reader is acquainted with elementary properties of divisors, e.g. that they form a group, the existence of pull backs, etc. It is a basic fact that on $\mathbf{C}^n$, given a divisor Z, there always exists a meromorphic function g on $\mathbf{C}^n$ which represents the divisor globally, namely $(\mathbf{C}^n, g)$ is a representative pair for the divisor. Furthermore, on $\mathbf{C}^n$ every meromorphic function is the quotient of two holomorphic functions. See Gunning-Rossi [**GuR**] theorems III-N4, III-K5 and III-K6. Of course such properties cannot be true on compact manifolds: they have to do with so-called Stein manifolds. However if

$$f : \mathbf{C}^n \to X$$

is holomorphic and D is a divisor on X such that $f(\mathbf{C}^n)$ is not contained in D, then the pull back $f^{-1}(D)$ is a divisor on $\mathbf{C}^n$, which can therefore be represented by a meromorphic function on $\mathbf{C}^n$.

Given a divisor defined on an open set U by a function φ. Let O be the ring of holomorphic functions at a point of U. Then O has unique factorization, and we can factorize

$$\varphi = \varphi_1^{m_1} \cdots \varphi_r^{m_r}$$

where $\varphi_1, \ldots, \varphi_r$ are irreducible elements. If we replace φ by the product

$$\varphi_1 \cdots \varphi_r$$

58

and do this for every pair (U, φ), then we obtain another divisor, called the **reduced divisor**. If $r = 1$ in the above expression, and $m_1 = 1$, then we say that the divisor is **irreducible locally**. Globally, every divisor on a compact manifold can be written as a sum of irreducible ones.

Let X be a complex manifold of dimension n, and let L be a holomorphic line bundle over X. As we deal only with holomorphic bundles, unless otherwise specified, we shall omit the word holomorphic to qualify them. Let $\{U_i\}$ be an open covering of X such that $L \mid U_i$ has a trivialization (holomorphic, according to our convention)

$$\varphi_i: \ L \mid U_i \to U_i \times \mathbf{C}.$$

Then

$$\varphi_{ij} = \varphi_i \circ \varphi_j^{-1}: \ (U_i \cap U_j) \times \mathbf{C} \to (U_i \cap U_j) \times \mathbf{C}$$

is an isomorphism, given by a holomorphic map

$$g_{ij}: U_i \cap U_j \to \mathbf{C}^* = GL_1(\mathbf{C})$$

such that

$$\varphi_{ij}(x, z) = (x, g_{ij}(x)z).$$

Let s be a holomorphic section of L over X. Then s is represented by a holomorphic map

$$s_i: U_i \to \mathbf{C},$$

satisfying

$$s_i = g_{ij} s_j.$$

Suppose a covering family $\{(U_i, \varphi_{ij})\}$ represents L as above. Suppose given for each i a function (smooth)

$$\rho_i: U_i \to \mathbf{R}_{>0}$$

59

such that on $U_i \cap U_j$ we have

$$\rho_i = |g_{ij}|^2 \rho_j.$$

Then we say that the family of triplets $\{(U_i, \varphi_{ij}, \rho_i)\}$ **represents a metric**. It is clear how to define compatible families, or compatible triples in this context, and the metric itself is an equivalence class of such covering triples, or is the maximal family of compatible triples. We could also write a representative family as $\{(U_i, \varphi_i, \rho_i)\}$ using the isomorphisms φ_i instead of the transition functions φ_{ij}.

If s is a section of L over some open set U_i containing a point P, then we define

$$|s(P)|^2 = \frac{|s_i(P)|^2}{\rho_i(P)}.$$

The value on the right-hand side is independent of the choice of U_i, as one sees at once from the transformation law.

Instead of using indices i, if L is trivial over an open set U, so

$$L_U \approx U \times \mathbf{C},$$

we write $s_U \colon U \to \mathbf{C}$ for the map representing a section $s \colon U \to L$ over U, and then we also write

$$|s|^2 = |s_U|^2 / \rho_U.$$

A metric as above will be denoted by ρ, for instance. Since at each point L is one-dimensional, we see that the metric is determined by a hermitian product in a trivial way. In particular, we obtain line bundles by considering the tangent or cotangent bundle of a one-dimensional complex manifold, usually called a **Riemann surface**.

We shall now associate some differential forms to a metric on a line bundle. First we review some terminology.

60

Let $z_1, \ldots, z_n$ be holomorphic coordinates for X over U. As usual, we have the operators ∂ and $\bar{\partial}$, where say for a function $f(z)$,

$$\partial f(z) = \sum_{k=1}^n \frac{\partial f}{\partial z_k} dz_k \quad \text{and} \quad \bar{\partial} f(z) = \sum_{k=1}^n \frac{\partial f}{\partial \bar{z}_k} d\bar{z}_k.$$

Then

$$d = \partial + \bar{\partial}.$$

The operators $\partial, \bar{\partial}$, and d extend to forms of arbitrary degree as usual, for instance

$$\bar{\partial}(f_{IJ}(z, \bar{z}) dz_I \wedge d\bar{z}_J) = \sum_{k=1}^n \frac{\partial f}{\partial \bar{z}_k} d\bar{z}_k \wedge dz_I \wedge d\bar{z}_J,$$

where $dz_I = dz_{i_1} \wedge \cdots \wedge dz_{i_p}$ and $d\bar{z}_J = d\bar{z}_{j_1} \wedge \cdots \wedge d\bar{z}_{j_q}$. A sum of terms

$$\omega = \sum_{\substack{|I|=p \\ |J|=q}} f_{IJ}(z, \bar{z}) dz_I \wedge d\bar{z}_J$$

is called a form of **type** (p, q). The numbers p, q do not depend on the choice of holomorphic coordinates, because if g is holomorphic, then $\bar{\partial} g = 0$ by the Cauchy-Riemann equations.

We define the operator

$$d^c = \frac{1}{4\pi \sqrt{-1}} (\partial - \bar{\partial}).$$

The advantage of such an operator is that it is a real operator. If ω is a form such that $\omega = \bar{\omega}$, then $d^c \omega$ also satisfies this property. Note that

$$dd^c = \frac{\sqrt{-1}}{2\pi} \partial \bar{\partial} = \frac{1}{2\pi \sqrt{-1}} \bar{\partial} \partial.$$

In dimension 1, and polar coordinates (r, θ) we have

$$\boxed{d^c = \frac{1}{2} r \frac{\partial}{\partial r} \otimes \frac{d\theta}{2\pi} - \frac{1}{4\pi} \frac{1}{r} \frac{\partial}{\partial \theta} \otimes dr.}$$

61

Restricting to the circle the term with dr vanishes, so

$$d^c \text{ restricted to the circle is } \frac{1}{2}r\frac{\partial}{\partial r}\frac{d\theta}{2\pi}.$$

Given a metric ρ on L we define the **Chern form** of the metric to be the unique form $c_1(\rho)$ such that on an open set U, in terms of the trivialization as above, we have

$$\boxed{c_1(\rho) \mid U = -dd^c \log |s|^2 = dd^c \log \rho_U}$$

for any holomorphic section s. The right-hand side is independent of the choice of holomorphic coordinates on U, because for any non-zero holomorphic function g (giving rise to a change of charts) we have

$$\partial\bar\partial \, \log |g|^2 = \partial\bar\partial \, \log(g\bar g) = 0.$$

so $dd^c \log(g\bar g) = 0$.

We say that a $(1,1)$-form

$$\omega = \frac{\sqrt{-1}}{2\pi} \sum h_{ij}(z)dz_i \wedge d\bar z_j$$

is **positive** and we write $\omega > 0$, if the matrix $h = (h_{ij})$ is hermitian positive definite for all values of z. This condition is independent of the choice of holomorphic coordinates $z_1, \ldots, z_n$.

A metric ρ is called **positive** if $c_1(\rho)$ is positive.

Example. Projective space. Let $X = \mathbf{P}^n$ be projective n-space. Let $T_0, \ldots, T_n$ be the homogeneous variables, and let U_i be the open set of points such that $T_i \neq 0$. We let

$$z_0^{(i)} = T_0/T_i, \ldots, z_n^{(i)} = T_n/T_i \quad \text{so} \quad z_i^{(i)} = 1.$$

Then $z_j^{(i)}$ with $j \neq i$ are complex coordinates on U_i. Then there is a line bundle L, called the **hyperplane line bundle**, whose transition functions are given by

$$g_{ij} = T_j/T_i.$$

62

Its sheaf of sections is usually denoted by $\mathcal{O}(1)$. Say for $i = 0$, we write simply $z = (z_1, \ldots, z_n)$ where $z_j = T_j/T_0$. The **standard metric** ρ on L is defined on U_0 by the function

$$\rho(z) = 1 + \sum_{\nu=1}^{n} z_\nu \bar{z}_\nu,$$

and similarly for U_i. Then its Chern form is computed using rules from freshman calculus for the derivative of a product and quotient, to give

$$c_1(\rho) = \frac{\sqrt{-1}}{2\pi} \partial \bar{\partial} \ \log \rho(z)$$

$$= \frac{\sqrt{-1}}{2\pi} \frac{1}{\rho(z)^2} \left(\sum_{i,j=1}^{n} h_{ij} dz_i \wedge d\bar{z}_j \right),$$

where $h = (h_{ij})$ is the matrix $h = \rho(z)I - (\bar{z}_i z_j)$. The metric on the cotangent bundle defined by this Chern form is called the **Fubini-Study metric**.

Proposition 1.1 . *The Fubini-Study metric is positive.*

Proof: We have to show that h is positive definite (it is obviously hermitian). For any complex vector $C = {}^t(c_1, \ldots, c_n)$ we expand ${}^t\bar{C} h C$ and the Schwartz inequality immediately shows that for $C \neq 0$ we have

$${}^t\bar{C} h C > 0.$$

This proves the proposition.

The example will play no role for the rest of this chapter, but becomes useful later.

We now pass to volume forms. In $\mathbf{C}^n$ we have what we call the **euclidean form**, expressed in terms of coordinates z by

$$\boxed{\Phi(z) = \prod_{i=1}^{n} \frac{\sqrt{-1}}{2\pi} dz_i \wedge d\bar{z}_i.}$$

63

Except for the factor involving π and the power of 2, it is just

$$dx_1 \wedge dy_1 \wedge \cdots \wedge dx_n \wedge dy_n.$$

The product sign is to be interpreted as the alternating product, but 2-forms commute with all forms, so it is harmless to write it as the usual product sign to emphasize this commutativity.

By a **volume form** on X, we mean a form of type (n, n), which locally in terms of complex coordinates can be written as

$$\Psi(z) = h(z)\Phi(z),$$

where h is C^∞ and $h(z) > 0$ for all z. This is invariant under a change of complex coordinates, since the factor coming out in such a change is of the form $g(z)\overline{g(z)}$, where $g(z)$ is holomorphic invertible.

Thus a volume form is a metric on the **canonical bundle**

$$K_X = \overset{\text{max}}{\bigwedge} T^\vee(X),$$

which is the $\max(n\,\text{th})$ exterior power of the cotangent bundle.

We define the **Ricci form** of Ψ to be the Chern form of this metric, so $\text{Ric}(\Psi)$ is the *real* $(1,1)$-form given by

$$\text{Ric}(\Psi) = c_1(\rho) = dd^c \log h(z) \quad \text{in terms of coordinates } z.$$

Remarks: *If C is a constant then*

$$\text{Ric}(C\Psi) = \text{Ric}(\Psi).$$

If u is a positive smooth function, then

$$\text{Ric}(u\Psi) = \text{Ric}(\Psi) + dd^c \log u.$$

Both assertions are trivial from the definition.

64

A 2-form commutes with all forms. By the n-th power

$$\text{Ric}(\Psi)^n$$

we mean the n-th exterior power. Then $\text{Ric}(\Psi)^n$ is an (n,n)-form, and in particular a max degree form on X. Since Ψ is a volume form, there is a unique function G on X such that

$$\frac{1}{n!}\text{Ric}(\Psi)^n = G\Psi.$$

We may also write symbolically

$$G = \frac{1}{n!}\text{Ric}(\Psi)^n/\Psi.$$

Note that G is a real-valued function. We call G the **Griffiths function** associated with the original volume form Ψ. We denote it by

$$G_\Psi \quad \text{or} \quad G(\Psi).$$

Special case *Let dim $X = 1$, and let z be a complex coordinate. Then the Griffiths function is given by*

$$G_\Psi(z) = \frac{1}{h(z)}\frac{\partial^2 \log h(z)}{\partial z \partial \bar{z}}.$$

Proof: Immediate from the definitions.

The function $-G$ in dimension 1 is classically called the **Gauss curvature**.

Remark: In the above definitions, we did not need to assume X compact. If X is compact, we shall also deal with volume forms Ψ which are defined only on the complement of a divisor in X. Then the same definitions will apply to such forms. In particular, we have the Ricci form and the Griffiths function defined outside the singularities.

65

II, §2. SOME FORMS ON $\mathbf{C}^n$ AND $\mathbf{P}^{n-1}(\mathbf{C})$ AND THE GREEN-JENSEN FORMULA

We shall deal constantly with the following differential forms on $\mathbf{C}^n$, letting $z = (z_1, \ldots, z_n)$ be complex coordinates:

$$\omega(z) = dd^c \log \|z\|^2$$
$$\varphi(z) = dd^c \|z\|^2$$
$$\sigma(z) = d^c \log \|z\|^2 \wedge \omega^{n-1}$$

The form φ will appear a little later, at first we deal with ω and σ. We shall in fact deal with the restriction of σ to the sphere $S(r)$ for $r > 0$.

Remark: Let

$$\pi \colon \mathbf{C}^n - \{0\} \to \mathbf{P}^{n-1}$$

be the natural map representing a point of projective space by homogeneous coordinates. Let τ be the Fubini-study metric, and let $\omega_{\mathbf{P}} = c_1(\tau)$ be the Fubini-Study form. Then

$$\omega = \pi^* \omega_{\mathbf{P}}.$$

This is immediate from the definitions. The fibering $\pi \colon \mathbf{C}^n - \{0\} \to \mathbf{P}^{n-1}$ restricts to a fibering on each sphere

$$\pi \colon S(r) \to \mathbf{P}^{n-1}$$

which we use for the rest of this section.

We can write the original formula for $\omega_{\mathbf{P}}$ on $U_0 = \mathbf{C}^n$ in the form

$$\omega_{\mathbf{P}} = \frac{\sqrt{-1}}{2\pi} \frac{1}{1 + \|z\|^2} \left(<dz, dz> - \frac{<dz, z> \wedge <z, dz>}{1 + \|z\|^2} \right),$$

with the notation of the hermitian product $<dz, dz> = \sum dz_k \wedge d\bar{z}_k$ for instance. The exterior powers (other than the first) of $<dz, z>$

66

and $< z, dz >$ are 0. Hence

$$\left(\frac{2\pi}{\sqrt{-1}}\right)^n (1 + \|z\|^2)^n \omega_{\mathbf{P}}^n$$
$$= \left(< dz, dz >^n - \frac{n < dz, dz >^{n-1} \wedge < dz, z > \wedge < z, dz >}{1 + \|z\|^2}\right).$$

But

$$< dz, dz >^n = n! \prod dz_k \wedge d\bar{z}_k \qquad \text{and}$$
$$< dz, dz >^{n-1} = (n-1)! \sum dz_1 \wedge d\bar{z}_1 \wedge \cdots \wedge \widehat{dz_k \wedge d\bar{z}_k} \wedge \cdots \wedge dz_n \wedge d\bar{z}_n,$$

so we get the formula

$$\omega_{\mathbf{P}}^n(z) = \frac{n!}{(1 + \|z\|^2)^{n+1}} \Phi(z).$$

We shall use **Fubini's theorem** in a global version. In general, let

$$\pi : X \to Y$$

be a fibering of real manifolds, meaning a C^∞ map of manifolds, locally isomorphic to a product. Let $q = \dim Y$ and $\dim X = p + q$, so p is the dimension of a fiber. Let η be a p-form on X and let ω_Y be a q-form on Y. Then

$$\int_X \pi^* \omega_Y \wedge \eta = \int_{y \in Y} \left(\int_{\pi^{-1}(y)} \eta \right) \omega_Y(y),$$

under conditions of absolute convergence. For instance, if η has compact support, then the formula is valid. The proof reduces to a local statement by partitions of unity, and locally, the relation is merely Fubini's theorem that a double integral is equal to a repeated integral.

Proposition 2.1. $\displaystyle \int_{S(r)} \sigma = 1 = \int_{\mathbf{P}^{n-1}} \omega_{\mathbf{P}}^{n-1}.$

67

Proof: We consider the fibering

$$\pi : S(r) \to \mathbf{P}^{n-1}.$$

Let I_n be the integral on the left of the proposition, and let J_{n-1} be the integral on the right. For $\pi : S(r) \to \mathbf{P}^{n-1}$ we get

$$I_n = \int_{S(r)} \sigma = \int_{y \in \mathbf{P}^{n-1}} \left(\int_{\pi^{-1}(y)} d^c \log \|z\|^2 \right) \omega_{\mathbf{P}}^{n-1}(y)$$

$$= \int_{\pi^{-1}(y)} d^c \log \|z\|^2 \cdot J_{n-1}$$

$$= J_{n-1},$$

because the first factor is independent of $y \in \mathbf{P}^{n-1}$, and one computes directly in dimension 1, using the formula for d^c given in polar coordinates by

$$d^c = \frac{1}{2} r \frac{\partial}{\partial r} \otimes \frac{d\theta}{2\pi} \qquad \text{on the circle.}$$

Now to compute J_n, since the complement of U_0 is a hyperplane, having measure 0, we have

$$J_n = \int_{\mathbf{P}^n} \omega_{\mathbf{P}}^n = \int_{\mathbf{C}^n} \omega_{\mathbf{P}}^n = n! \int_{\mathbf{C}^n} \frac{1}{(1 + \|z\|^2)^{n+1}} \prod_{k=1}^{n} \frac{\sqrt{-1}}{2\pi} dz_k \wedge d\bar{z}_k.$$

This is now easily evaluated by induction, leaving $z_1, \ldots, z_{n-1}$ fixed and integrating with respect to z_n, using polar coordinates $z_n = re^{i\theta}$. One finds that $J_n = J_{n-1}$, whence $J_n = 1$ by a direct computation on J_1. This concludes the proof.

The next proposition generalizes the formula for d^c in higher dimension.

Proposition 2.2. *For smooth functions α, we have*

$$d^c \alpha \wedge \omega^{n-1} | S(t) = \frac{1}{2} t \frac{\partial \alpha}{\partial t} \sigma$$

68

70

Proof: Apply the left-hand side as a distribution, i.e., as a functional on smooth functions with compact support. Again use the fibration

$$\pi: \mathbf{S}(r) \to \mathbf{P}^{n-1}.$$

Suppose β is C^∞ on $\mathbf{S}(t)$. It suffices to verify

$$\int\limits_{y \in \mathbf{P}^{n-1}} \Big(\int\limits_{\pi^{-1}(y)} \beta d^c \alpha \Big) \omega_Y^{n-1}(y) = \int\limits_{y \in \mathbf{P}^{n-1}} \frac{r}{2} \Big(\int\limits_{\pi^{-1}(y)} \beta \frac{\partial \alpha}{\partial r} \frac{d\theta}{2\pi} \Big) \omega_Y^{n-1}(y),$$

which is true because $\pi^{-1}(y)$ is a circle, and on the circle we already know the formula for d^c. Thus the formula follows in general.

Observe that the formula extends to more general functions β by general integration theory.

We now want to extend the theorems in one variable having to do with various integrations and the height transform. We shall consider functions whose singularities are on divisors of $\mathbf{C}^n$. They will be of three types:

I. C^∞ functions, for which no problem will arise in applying Stokes' theorem, or differentiating under the integral sign.

II. Functions which have singularities on a divisor of $\mathbf{C}^n$, and which are locally of the form

$$\alpha = \log |g|^2$$

where g is holomorphic.

III. Functions locally of the form $\log(1 + (|g|^2 h)^\lambda)$ with $0 < \lambda \le 1$, where h is > 0 and C^∞ and g is holomorphic.

We shall also consider linear combinations of these functions with constant coefficients.

69

We could call the space generated by such functions the vector space of **admissible functions**. I have not found yet a very easy characterization of functions for which the formal arguments which we shall apply are valid. For our purposes at the moment, we shall suspend the definition of an appropriate space for which the formal arguments which follow are valid.

Theorem 2.3 *For any admissible function α which is C^∞ near the origin, we have*

$$\int_0^r \frac{dt}{t} \int_{S(t)} d^c\alpha \wedge \omega^{n-1} = \frac{1}{2} \int_{S(r)} \alpha\sigma - \frac{1}{2}\alpha(0).$$

Proof: By Proposition 2.2, and after permuting a derivative with the integral, we have

$$\int_0^r \frac{dt}{t} \int_{S(t)} d^c\alpha \wedge \omega^{n-1} = \int_0^r \frac{dt}{t} \frac{1}{2} t \frac{\partial}{\partial t} \int_{S(t)} \alpha\sigma$$

$$= \frac{1}{2} \int_{S(r)} \alpha\sigma - \frac{1}{2}\alpha(0)$$

thus proving the theorem.

II, §3. STOKES' THEOREM WITH CERTAIN SINGULARITIES ON $\mathbf{C}^n$

Although we ultimately study holomorphic mappings $f: \mathbf{C}^n \to X$ into a compact complex manifold (in fact, algebraic), these are studied via pull-backs to $\mathbf{C}^n$, and basic relations for f are reduced to certain relations on $\mathbf{C}^n$. These relations involve Stokes' theorem, and we shall now develop them directly on $\mathbf{C}^n$.

A version of Stokes' theorem with singularities is given in [La 5] and [La 6]. The idea is that if the singularities are of dimension one

70

less than the dimension of the set of regular points on the boundary of the manifold in question, then Stokes' formula is valid. The integral of a form over an analytic space is by definition the integral over the set of its regular points. Since this set may not be compact, it is of course necessary to know the absolute convergence of all integrals involved. The proof of Stokes' formula with singularities is obtained as follows. In that reference, an a priori definition of a **negligible** set on the boundary is given, such that if the singularities are negligible then Stokes' formula applies. One multiplies the given form by a C^∞ function that is 0 near the singularities and 1 slightly farther out from the singularities. One then takes a limit over such functions, depending on a parameter which tends to 0. One has to estimate the derivatives that come into the proof when taking d of the form. All details are given in the above references to carry out this idea.

Throughout this section, we let Z be a divisor on $\mathbf{C}^n$ and $0 \notin Z$. We let L_Z be a line bundle with a meromorphic section s with divisor

$$(s) = Z.$$

Certain relations will involve Z linearly, and consequently to prove such relations we may assume that Z is effective, and so s is holomorphic. We suppose ρ is a hermitian metric on L_Z.

In practice, we shall be given a holomorphic map $f: \mathbf{C}^n \to X$ into a complex manifold, and the divisor Z will be the pull back of a divisor on X. Since X is assumed a manifold, this implies that Z is defined as a Cartier divisor, and the metric on L_Z is defined as the pull back of a metric on a line bundle on X.

So suppose Z is effective. One way to define a tubular neighborhood of $Z \cap \mathbf{B}(r)$ is to take the set

$$V(Z, \varepsilon)(r) = \{z \in \mathbf{B}(r) \text{ such that } |s(z)|_\rho^2 < \varepsilon\}.$$

71

Locally in the neighborhood of a point, s can be represented by a holomorphic function g, and $|s|_\rho^2 = |g|^2 h$ where h is positive and C^∞. Then locally, $V(Z, \varepsilon)$ consists of those $z \in \mathbf{B}(r)$ such that $|g(z)| < \varepsilon/h(z)$. For all but a discrete set of sufficiently small ε, this neighborhood has a regular boundry. We let its boundry be

$$S(Z, \varepsilon)(r) = \{z \in \mathbf{B}(r) \text{ such that } |s(z)|_\rho^2 = \varepsilon\}.$$

Letting r_1 be a number slightly bigger than r, the boundry of $\mathbf{B}(r) - V(Z, \varepsilon)(r)$ consists of

$$\mathbf{S}(r) - V(Z, \varepsilon)(r_1) \quad \text{and} \quad S(Z, \varepsilon)(r) \quad \text{(suitably oriented)}.$$

Note that the tubular neighborhood could also be defined using a reduced section s_0 defining the reduced divisor Z_0 rather than Z itself. In the following application of Stoke's theorem, all we are using is a family of tubular neighborhoods $S(Z, \varepsilon)$, depending on a parameter ε, such that this family shrinks to Z as ε approaches 0.

Theorem 3.1. *Let α be a C^∞ function on $\mathbf{C}^n - Z$, and let β be a C^∞ form, of type $(n-1, n-1)$ except on a negligible set of points on the boundary of $\mathbf{S}(r) - Z$. Assuming that all subsequent integrals are absolutely convergent, we have*

$$\int_{\mathbf{B}(r)} (dd^c\alpha \wedge \beta - \alpha \wedge dd^c\beta) = \int_{\mathbf{S}(r)} (d^c\alpha \wedge \beta - \alpha \wedge d^c\beta)$$

$$- \lim_{\varepsilon \to 0} \int_{S(Z,\varepsilon)(r)} (d^c\alpha \wedge \beta - \alpha \wedge d^c\beta).$$

Proof: We have

$$d(d^c\alpha \wedge \beta) = dd^c\alpha \wedge \beta - d^c\alpha \wedge d\beta$$
$$d(\alpha \wedge d^c\beta) = d\alpha \wedge d^c\beta + \alpha \wedge dd^c\beta.$$

72

Since β is of type $(n-1, n-1)$, we get $\partial\alpha \wedge \partial\beta = 0$ and $\bar{\partial}\alpha \wedge \bar{\partial}\beta = 0$, so

$$d^c\alpha \wedge d\beta = d^c\beta \wedge d\alpha$$

Applying Stokes and subtracting the two expressions above yields the theorem.

Note that we did not assume β closed, so that in the proof we can apply the technique of multiplying β by a function with appropriate small support in the neighborhood of a negligible singularity. However, in applications, we shall deal only with the case when β is closed, and more specially when $\beta = \omega^{n-1}$. Indeed, since $d\omega = 0$, we also get

$$d\omega^{n-1} = 0.$$

We thus specialize Stokes' theorem to the special case which we shall use:

If α is an admissible function, then for all but a discrete set of r:

$$\boxed{\int_{B(r)} dd^c\alpha \wedge \omega^{n-1} = \int_{S(r)} d^c\alpha \wedge \omega^{n-1} - \lim_{\varepsilon \to 0} \int_{S(Z,\varepsilon)(r)} d^c\alpha \wedge \omega^{n-1}}$$

and similarly with φ^{n-1} instead of ω^{n-1}.

In order to evaluate the limit on the right, we shall need a specific computation which determines the limit in each case. Here again, although we apply the computation to the case when $\beta = \omega^{n-1}$, we must carry out the statement with a general β, since we wish to multiply β by suitable functions in the proof.

The limit on the right will be called the **singular term** in Stokes' theorem.

We now compute the limit on the right for each case of admissible functions.

73

Lemma 3.2. *If α is C^∞ then*

$$\lim_{\varepsilon \to 0} \int_{S(Z,\varepsilon)(r)} d^c\alpha \wedge \beta = 0.$$

Proof: In this case, by Stoke's theorem,

$$\int_{S(Z,\varepsilon)(r)} d^c\alpha \wedge \beta = \int_{V(Z,\varepsilon)(r)} dd^c\alpha \wedge \beta - d^c\alpha \wedge d\beta,$$

and since the volume of $V(Z,\varepsilon)$ shrinks to 0 as $\varepsilon \to 0$, it follows that the right hand side approaces 0, whence the lemma follows.

The next result is classical, and is the standard case involving singularities which are not negligible. First we make a preliminary remark. Over a ball $\mathbf{B}(r_1)$ for r_1 slightly bigger than r, we can write the divisor as a Weil divisor

$$Z = \sum m_i Z_i$$

where the components Z_i are reduced and irreducible, and m_i are positive integers. Then an integral over Z is defined as a sum of integrals

$$\int_Z = \sum m_i \int_{Z_i},$$

and each integral over Z_i is defined as an integral over the regular points of Z_i. Note that the intersections of two or more components has complex codimension at least 2, so real codimension at least 4, whence this intersection is negligible from the point of view of Stoke's theorem.

Lemma 3.3 (Poincaré). *Let β be as in the theorem, and s holomorphic, defining the divisor Z. Let $S(Z,\varepsilon)$ be defined by a reduced section s_0 of L_Z as mentioned above. Then*

$$\lim_{\varepsilon \to 0} \int_{S(Z,\varepsilon)(r)} d^c \log |s|_\rho^2 \wedge \beta = \int_{Z(r)} \beta.$$

74

Proof: Using a partition of unity, and the absolute convergence of the integrals which is assumed, we reduce the proof of the formula to the case when β has support at a regular point of Z, so that in a neighborhood of that point, s can be represented by a holomorphic function g, and s_0 can be represented by a complex coordinate function w_1, the first of a system of complex coordinates $w_1, \ldots, w_n$. The neighborhood can be chosen small enough that we can write

$$g(w) = w_1^m g_1(w)$$

where g_1 does not vanish at $w_1 = 0$ (defining Z locally). Then

$$|s|_\rho^2 = |w_1|^{2m} h(w)$$

where h is positive C^∞. In this case, putting $w_1 = ue^{i\theta}$ we have on the circle $u = \varepsilon$:

$$d^c \log |w_1|^2 = u \frac{\partial}{\partial u} \log u \frac{d\theta}{2\pi} = \frac{d\theta}{2\pi},$$

and $\log h$ is locally C^∞, so we can apply Lemma 3.2 to the term coming from the $d^c \log h$ contribution. As to the other term, we get

$$\lim_{\varepsilon \to 0} \int_{S(Z,\varepsilon)} d^c \alpha \wedge \beta = \lim_{\varepsilon \to 0} \int_{|w_1| = \varepsilon} d^c \log |w_1|^{2m} \wedge \beta$$

$$= m \int_{w_1 = 0} \beta = \int_{Z(r)} \beta,$$

which proves Poincaré's lemma.

The next case occurs when singularities are weak enough so that the singular term in Stokes' theorem vanishes.

Lemma 3.4. *Let α be of type III. Then the singular term in Stokes' formula vanishes; that is for any C^∞ form β of type $(n-1, n-1)$*

75

77

we have

$$\lim_{\varepsilon \to 0} \int_{S(Z,\varepsilon)(r)} d^c\alpha \wedge \beta = 0.$$

Proof: Again by using a partition of unity on β and the absolute convergence of the integrals involved, we are reduced to a local evaluation, when β has support in a small neighborhood of a point of Z, and α has a representation

$$\alpha = \log(1 + u^\lambda) \quad \text{where} \quad u = |g|^2 h,$$

where g is holomorphic, and h is positive C^∞. One could again argue as before, but I shall us an argument which Stoll uses [**St**], namely for $\delta > 0$, the function

$$\alpha_\delta = \log\left(1 + (\delta + u)^\lambda\right)$$

is C^∞ and we can apply Lemma 3.2. But

$$d^c\alpha_\delta = \frac{\lambda(\delta + u)^{\lambda-1} d^c u}{1 + (\delta + u)^\lambda}.$$

One can then take the limit under the integral sign as $\delta \to 0$ to conclude the proof.

Remark: Instead of β in the lemma, we may use ω^{n-1}. For one thing, the arguments still work for a form with negligible singularity at the origin. But also we assumed that $0 \notin Z$, so multiplying ω^{n-1} with a C^∞ function which is equal to 1 outside a small neighborhood of the origin does not change the validity of the formula to be proved, and reduces it to the case of the C^∞ form β as stated.

If we combine Lemma 3.4 with Stokes' theorem and Theorem 2.3, then we have proved:

76

Theorem 3.5. *Suppose $\alpha = \log \gamma$ is of type I or III, and is C^∞ near the origin. Then*

$$\int_0^r \frac{dt}{t} \int_{\mathbf{B}(t)} dd^c \log \gamma \wedge \omega^{n-1} = \frac{1}{2} \int_{\mathbf{S}(r)} (\log \gamma)\sigma - \frac{1}{2} \log \gamma(0).$$

For functions of mixed type I and II, we do get a singular contribution as follows.

Theorem 3.6. *Let Z be a divisor on $\mathbf{C}^n$ and let s be a meromorphic section of L_Z such that $(s) = Z$. Suppose $0 \notin Z$. Let ρ be a hermitian metric on L_Z and let $\gamma = |s|_\rho^2$. Then*

$$\int_0^r \frac{dt}{t} \int_{\mathbf{B}(t)} dd^c \log \gamma \wedge \omega^{n-1} + \int_0^r \frac{dt}{t} \int_{Z(t)} \omega^{n-1}$$
$$= \frac{1}{2} \int_{\mathbf{S}(r)} (\log \gamma)\sigma - \frac{1}{2} \log \gamma(0).$$

Proof: This is immediate from Lemmas 3.2 and 3.3, also using Theorem 2.3.

We shall use Theorems 3.5 and 3.6 in products and inverses of functions as in those theorems, and we use the homomorphic property of

$$\gamma \mapsto dd^c \log \gamma$$

to get relations in the next section.

77

II, §4. THE NEVALINNA FUNCTIONS AND THE FIRST MAIN THEOREM

In this section we introduce the definitions of the three basic Nevanlinna functions, and translate Theorems 3.5 and 3.6 into the new terminology. We are here in the equidimensional case, which stemmed from Carlson-Griffiths [C-G]. We shall make additional historical remarks later. We start by making the definitions on $\mathbf{C}^n$, and then get the theorems for any non-degenerate holomorphic map

$$f: \mathbf{C}^n \to X$$

into a compact complex manifold of dimension n by pull back.

The Nevanlinna functions on $\mathbf{C}^n$

Let Z be a divisor on $\mathbf{C}^n$, such that $Z = (g)$ where g is a meromorphic function, $0 \notin Z$. We define the **pre counting function** and **counting function** by

$$\mathbf{n}_Z(t) = \int_{Z(t)} \omega^{n-1} \quad \text{and} \quad N_Z(r) = \int_0^r \mathbf{n}_Z(t)\frac{dt}{t}$$

For a positive function γ sufficiently smooth such that the integrals are absolutely convergent, we define the **pre proximity function** and **proximity function** by

$$\mathbf{m}_\gamma^0(t) = \int_{\mathbf{S}(t)} d^c \log \gamma \wedge \omega^{n-1} \quad \text{and} \quad m_\gamma^0(r) = \int_0^r \mathbf{m}_\gamma^0(r)\frac{dt}{t}$$

If $\log \gamma$ is admissible, then by Theorem 2.3,

$$m_\gamma^0(r) = \frac{1}{2} \int_{\mathbf{S}(r)} (\log \gamma)\sigma - \frac{1}{2}\log \gamma(0).$$

78

80

Suppose that we define a metric ρ on the line bundle L_Z having the meromorphic section s represented by g, by the formula

$$|s|_\rho^2 = |g|^2 h$$

where h is a positive C^∞ function. We then take $\gamma = |s|_\rho^{-2}$ and we use the notation

$$m_{Z,\rho}^0(r) = \int_0^r \frac{dt}{t} \int_{S(t)} - d^c \log |s|_\rho^2 \wedge \omega^{n-1}.$$

Thirdly, we define the **pre height** and **height functions** associated to a $(1,1)$ form η by

$$t_\eta(t) = \int_{B(t)} \eta \wedge \omega^{n-1} \qquad \text{and} \qquad T_\eta(r) = \int_0^r \frac{dt}{t} \int_{B(t)} \eta \wedge \omega^{n-1}.$$

This definition will be applied to the special case when $\eta = dd^c \log \gamma$ for suitable functions γ, in which case we also use the notation

$$\text{Ric } \gamma = dd^c \log \gamma \qquad \text{and so} \qquad T_{\text{Ric } \gamma}(r) = \int_0^r \frac{dt}{t} \int_{B(t)} dd^c \log \gamma \wedge \omega^{n-1}.$$

Thus we use the abbreviation Ric γ instead of $\text{Ric}(\gamma\Phi)$, according to our definition of the Ricci form in §1. We may also call T_η the **height transform** of η. For the height, recall that we defined the Chern form of a metric ρ by

$$c_1(\rho) = -dd^c \log h.$$

Then we also use the notation

$$T_\rho(r) = \int_0^r \frac{dt}{t} \int_{B(t)} c_1(\rho) \wedge \omega^{n-1}.$$

79

Observe that each one of the proximity and height functions are homomorphic in γ, in other words

$$\gamma \mapsto m_\gamma^0 \qquad \text{and} \qquad \gamma \mapsto T_{\text{Ric } \gamma}$$

sends products into sums. The counting function is homomorphic in Z, that is

$$Z \mapsto N_Z$$

is a homomorphism. For the height,

$$\rho \mapsto T_\rho$$

is a homomorphism, where on the left we take the tensor product of metrics on the tensor product of line bundles.

We may now reformulate Theorem 3.6 in terms of the terminology we have just defined.

Theorem 4.1 *Let Z be a divisor on $\mathbf{C}^n$ and let s be a meromorphic section s of L_Z. Let ρ be a metric on L_Z. Suppose $0 \notin Z$. Then*

$$\mathbf{t}_\rho = \mathbf{n}_Z + \mathbf{m}_{Z,\rho}^0, \qquad \text{and} \qquad T_\rho = N_Z + m_{Z,\rho}^0.$$

The Nevanlinna functions for a mapping $f\colon \mathbf{C}^n \to X$

Consider now a holomorphic map

$$f\colon \mathbf{C}^n \to X$$

which is non-degenerate, i.e., is a local isomorphism somewhere, so its image is not contained in any divisor of the compact complex manifold X. If D is a divisor on X, s a meromorphic section of L_D with $(s) = D$, and ρ is a metric on L_D then we can define the three Nevanlinna functions by pull back.

80

82

Pre height and height

$$t_{f,\rho}(r) = \int_{B(r)} f^*c_1(\rho) \wedge \omega^{n-1} \quad \text{and} \quad T_{f,\rho}(r) = \int_0^r t_{f,\rho}(t)\frac{dt}{t}$$

If η is a $(1,1)$ form on X, we also use the same notation

$$t_{f,\eta}(r) = \int_{B(r)} f^*\eta \wedge \omega^{n-1} \quad \text{and} \quad T_{f,\eta} = \int_0^r t_{f,\eta}(t)\frac{dt}{t}.$$

Pre proximity and proximity functions

$$m_{f,D,\rho}^0(r) = \int_{S(r)} - d^c \log |s \circ f|_\rho^2 \wedge \omega^{n-1} \quad \text{and}$$

$$m_{f,D,\rho}^0(r) = \int_0^r m_{f,D,\rho}^0(t)\frac{dt}{t}.$$

Then

$$m_{f,D,\rho}^0(r) = \frac{1}{2}\int_{S(r)} (\log |s \circ f|_\rho^{-2})\sigma + \frac{1}{2}\log |s \circ f(0)|_\rho^2.$$

Counting functions

$$n_{f,D}(r) = \int_{f^*D(r)} \omega^{n-1} \quad \text{and} \quad N_{f,D}(r) = \int_0^r n_{f,D}(t)\frac{dt}{t}.$$

Proposition 4.2. *Let ρ, ρ' be two metrics on the line bundle L on X. Then*

$$T_{f,\rho'} = T_{f,\rho} + O(1).$$

81

In other words, two height functions associated with a line bundle differ by a bounded function.

Proof: There is a positive C^∞ function $\gamma > 0$ such that $\rho' = \gamma\rho$. Then $c_1(\rho)$ and $c_1(\rho')$ differ by the form $dd^c \log \gamma$, and $\log \gamma$ is bounded because it is a continuous function on the compact space X. Pulling back to $\mathbf{C}^n$, and using Stokes' theorem in its simplest form when the singular divisor is 0, we get

$$T_{f,\rho} - T_{f,\rho'} = \int_0^r \frac{dt}{t} \int_{\mathbf{B}(t)} dd^c \log \gamma \wedge \omega^{n-1} = \frac{1}{2} \int_{\mathbf{S}(r)} (\log \gamma)\sigma - \frac{1}{2} \log \gamma(0).$$

Since $\log \gamma$ is bounded, the right hand side is bounded, thus proving the proposition.

Proposition 4.3. *Let η, η' be $(1,1)$ forms, and assume η positive. Then*

$$T_{f,\eta'} = O(T_{f,\eta}).$$

In particular, if η, η' are both positive, then

$$T_{f,\eta'} \gg \ll T_{f,\eta}.$$

Proof: Since X is compact, there exist a constant k such that

$$\eta' \leq k\eta$$

so the proposition is obvious.

Remark: The proposition means that the heights associated with positive $(1,1)$ forms are of the same order of magnitude. In some applications we shall take their logs, and so we get

$$\log T_{f,\eta'} = \log T_{f,\eta} + O(1).$$

82

In particular, if we are interested only in the functions mod $O(1)$ (multiplicatively or additively), we may simply write T_f or $\log T_f$ respectively, without any further subscript. In the one-dimensional theory, one usually writes

$$T_f = T_{f,\infty}$$

if f is a holomorphic map into the projective line, but one is only interested in the order of magnitude of T_f in certain estimates.

Theorem 4.4. *For any metric ρ on L_D we have*

$$T_{f,\rho} = N_{f,D} + m^0_{f,D,\rho}.$$

Proof: This is simply a restatement of Theorem 4.1, taking the pull back into account. Of course, the theorem also holds at the pre level, that

$$\mathbf{t}_{f,\rho} = \mathbf{n}_{f,D} + \mathbf{m}^0_{f,D,\rho}$$

as functions defined for all but a discrete set of values r.

Let Ω be a volume form on X. Let us write

$$f^*\Omega = \gamma_f \Phi, \quad \text{with} \quad \gamma_f = \gamma_{f,\Omega},$$

where Φ is the euclidean volume form on $\mathbf{C}^n$ as defined in §1. Then locally,

$$\gamma_f = |\Delta|^2 h$$

where Δ is holomorphic, $h > 0$ is C^∞, and locally, (Δ) is the **ramification divisor**

$$(\Delta) = \mathrm{Ram}_f$$

defined locally by the Jacobian determinant of f in terms of complex coordinates. The volume form may be viewed as defining a metric κ on the canonical bundle, and we also define the **Ricci form**

$$\mathrm{Ric}\,\Omega = c_1(\kappa).$$

83

85

The pull back of the Chern form of this metric is given outside Ram_f by

$$f^*\text{Ric}\,\Omega = f^*c_1(\kappa) = dd^c \log \gamma_f = dd^c \log h.$$

Let us define the **height associated with the volume form** Ω to be

$$T_{f,\text{Ric}\,\Omega} = T_{f,\kappa} = \int_0^r \frac{dt}{t} \int_{\mathbf{B}(t)} dd^c \log \gamma_f \wedge \omega^{n-1}.$$

As a special case of Theorem 4.1, actually Theorem 3.6 plus notation, we then get:

Theorem 4.5.

$$T_{f,\kappa}(r) + N_{f,\text{Ram}}(r) = \frac{1}{2} \int_{\mathbf{S}(r)} (\log \gamma_f)\sigma - \frac{1}{2} \log \gamma_f(0).$$

If D denotes a divisor on X, we denote by $T_{f,D}$ the height function $T_{f,\rho}$ for any metric ρ on L_D. By Proposition 4.2, $T_{f,D}$ is well defined mod $O(1)$, i.e. mod bounded functions. Thus we could write $T_{f,K}$ instead of $T_{f,\kappa}$ if K is a canonical divisor, with the understanding that the relation of Theorem 4.4 then holds with an added term $O(1)$. In most applications, one is looking only for relations mod $O(1)$.

So far in this section, we have had formal relationships derived from an application of Stokes' theorem and a computation of the singular term. In the next section, we estimate terms like those occurring on the right hand side in Theorem 4.4, and activate another technique.

84

II, §5. THE CALCULUS LEMMA

The calculus lemma will be applied in the same way as before, but with dt/t replaced by dt/t^{2n-1}. We use Lemma 3.1 of Chapter I with the function ψ unchanged, however.

Lemma 5.1. *Let F be a function of one variable $r > 0$ such that the first derivative exists, and F' is piecewise of class C^1. Suppose that both $F(r)$ and $r^{2n-1}F'(r)$ are positive increasing functions of r, and that there exists r_1 such that $F(r_1) \geq e$ for $r \geq r_1$. Let $b_1 \geq 1$ be the smallest number such that*

$$b_1 r^{2n-1} F'(r) \geq e \quad \text{for all} \quad r \geq 1.$$

Then

$$\frac{1}{r^{2n-1}} \frac{d}{dr}\left(r^{2n-1}\frac{dF}{dr}\right) \leq F(r)\psi(F(r))\psi(r^{2n-1}b_1 F(r)\psi(F(r)))$$

for all $r \geq r_1$ outside a set of measure $\leq 2b_0(\psi)$.

The proof is again by a double application of Lemma 3.1 of Chapter I. For our purposes now, we define the **error term**

$$S(F, b_1, \psi, r) = \log\{F(r)\psi(F(r))\psi(r^{2n-1}b_1 F(r)\psi(F(r)))\},$$

i.e. the log of the right hand side in the above inequality for the repeated derivative.

We shall obtain a function F to which we apply the inequality as follows.

Suppose α is a function which is positive and continuous except on a divisor of $\mathbf{C}^n$. We shall need that α is continuous at 0 for convenience as usual, but also that

$$t \mapsto \int_{S(t)} \alpha\sigma$$

85

is piecewise continuous. I leave to a future write up what I hope will be a neat characterization of such α, but I don't know how to do it now.

For such α, we define the **height transform**

$$F_\alpha(r) = \int_0^r \frac{dt}{t^{2n-1}} \int_{B(t)} \alpha\Phi, \quad \text{where} \quad \Phi = \prod \frac{\sqrt{-1}}{2\pi} dz_i \wedge d\bar{z}_i.$$

Then first F_α is differentiable, with derivative

$$F'_\alpha(r) = \frac{1}{r^{2n-1}} \int_{B(r)} \alpha\Phi;$$

and using spherical coordinates in higher dimension, writing $\alpha(t, u)$ where u is the variable on the sphere of radius 1, we have

$$r^{2n-1} F'_\alpha(r) = \frac{2}{(n-1)!} \int_0^r dt \int_{S(1)} \alpha(t, u)t^{2n-1}\sigma(u).$$

Hence

$$\frac{d}{dr}\left(r^{2n-1} \frac{dF_\alpha}{dr}\right) = \frac{2}{(n-1)!} r^{2n-1} \int_{S(r)} \alpha\sigma.$$

Hence finally,

$$\frac{1}{r^{2n-1}} \frac{d}{dr}(r^{2n-1} F'_\alpha(r)) = \frac{2}{(n-1)!} \int_{S(r)} \alpha\sigma.$$

Thus F_α satisfies the hypotheses of Lemma 5.1, and by that lemma, we find:

Proposition 5.2. *We have the inequality*

$$\log \int_{S(r)} \alpha\sigma \leq S(F_\alpha, b_1, \psi, r) + \log \frac{(n-1)!}{2}$$

for all $r \geq r_1(F_\alpha)$ outside a set of measure $\leq 2b_0(\psi)$.

86

II, §6. THE TRACE AND DETERMINANT IN THE MAIN THEOREM

So far, the theory in higher dimension has essentially been completely parallel to the one dimensional case. We now meet an additional phenomenon, where the higher dimension plays a role which was not noticeable before.

Let η be a $(1,1)$ form on an open set of $\mathbf{C}^n$. We can write

$$\eta = \sum \eta_{ij} \frac{\sqrt{-1}}{2\pi} dz_i \wedge d\bar{z}_j.$$

We define the **trace** and **determinant** of η by

$$\mathrm{tr}(\eta) = \sum \eta_{ii} \quad \text{and} \quad \det \eta = \det(\eta_{ij}).$$

Trivially from the definitions, we get the coordinate free expression

$$\boxed{(\det \eta)\Phi = \frac{1}{n!}\eta^n.}$$

To study the trace further, we introduce a new form, namely

$$\varphi(z) = dd^c \|z\|^2 = \sum \frac{\sqrt{-1}}{2\pi} dz_i \wedge d\bar{z}_i.$$

Then

$$\varphi^{n-1} = (n-1)! \sum \left(\frac{\sqrt{-1}}{2\pi}\right)^{n-1} dz_1 \wedge d\bar{z}_1 \wedge \cdots \wedge \widehat{dz_j \wedge d\bar{z}_j} \wedge \cdots \wedge dz_n \wedge d\bar{z}_n.$$

Therefore we get a neat expression for the trace:

$$\boxed{\eta \wedge \varphi^{n-1} = (n-1)!\,\mathrm{tr}(\eta)\Phi.}$$

Next, we show how the height can be defined in terms of φ instead of the form ω which we used previously.

87

First remark that *restricted to the sphere* $\mathbf{S}(r)$,

$$d^c\|z\|^2 = r^2 d^c \log \|z\|^2 \quad \text{and} \quad dd^c\|z\|^2 = r^2 dd^c \log \|z\|^2.$$

Both relations are immediate.

Proposition 6.1. *If η is a closed C^∞ $(1,1)$ form, then*

$$r^{2n-2} \int_{\mathbf{B}(r)} \eta \wedge \omega^{n-1} = \int_{\mathbf{B}(r)} \eta \wedge \varphi^{n-1}.$$

The same relation holds if $\eta = dd^c\alpha$, where α is C^∞ except for singularities on a divisor such that the singular term in Stokes' theorem vanishes, as in Lemma 3.4.

Proof: If $n = 1$ the relation is trivial. Let $n \geq 2$. In the C^∞ case, we have:

$$r^{2n-2} \int_{\mathbf{B}(r)} \eta \wedge \omega^{n-1} = r^{2n-2} \int_{\mathbf{B}(r)} d(\eta \wedge d^c \log \|z\|^2 \wedge \omega^{n-2})$$

$$= r^{2n-2} \int_{\mathbf{S}(r)} \eta \wedge d^c \log \|z\|^2 \wedge \omega^{n-2}$$

$$= \int_{\mathbf{S}(r)} \eta \wedge d^c\|z\|^2 \wedge \varphi^{n-2}$$

$$= \int_{\mathbf{B}(r)} \eta \wedge \varphi^{n-1}.$$

This proves the proposition when η is C^∞. Now suppose $\eta = dd^c\alpha$, where α is such that the singular term vanishes in Stokes' theorem.

88

Then we have directly

$$r^{2n-2} \int_{\mathbf{B}(r)} dd^c\alpha \wedge \omega^{n-1} = r^{2n-2} \int_{\mathbf{S}(r)} d^c\alpha \wedge \omega^{n-1}$$

$$= \int_{\mathbf{S}(r)} d^c\alpha \wedge \varphi^{n-1}$$

$$= \int_{\mathbf{B}(r)} \eta \wedge \varphi^{n-1}$$

thus concluding the proof.

Our first application is to the C^∞ case, to give another formula for the height in certain cases.

Let as before X be a compact complex manifold, and let η be a closed positive $(1,1)$-form. We let

$$\Omega = \frac{1}{n!}\eta^n$$

so Ω is a volume form. We let $f\colon \mathbf{C}^n \to X$ be a non-degenerate map, and

$$f^*\Omega = \gamma_f \Phi \qquad \text{so} \qquad \gamma_f = \det(f^*\eta).$$

The next proposition shows how the height is given by a height transform.

Proposition 6.2. *Let* $\tau_f = \operatorname{tr} f^*\eta$. *Then*

$$T_{f,\eta} = (n-1)!F_{\tau_f}.$$

Proof: From the definitions and Proposition 6.1, we find:

89

$$T_{f,\eta}(r) = \int_0^r \frac{dt}{t} \int_{\mathbf{B}(t)} f^* \eta \wedge \omega^{n-1}$$

$$= \int_0^r \frac{dt}{t^{2n-1}} \int_{\mathbf{B}(t)} f^* \eta \wedge \varphi^{n-1}$$

$$= (n-1)! \int_0^r \frac{dt}{t^{2n-1}} \int_{\mathbf{B}(t)} (\operatorname{tr} f^* \eta) \Phi = (n-1)! F_{\tau_f}(r)$$

as was to be shown.

We can now pick up the relation of Theorem 4.4 and combine it with the calculus lemma and the expression of the height as a height transform, to get the Second Main Theorem when the divisor D is 0, so as in the one dimensional case, this amounts to a theorem about the ramification divisor.

Remark: *If η is hermitian semipositive, then*

$$(\det \eta)^{1/n} \leq \frac{1}{n} \operatorname{tr} \eta.$$

This is immediate, since the relation is one of linear algebra at each point, and so the relation follows immediately after diagonalizing the matrix of the form.

Theorem 6.3. *Let η be a closed, positive $(1,1)$-form, with $\Omega = \eta^n/n!$, and let $T_{f,\kappa}$ be the height associated with the volume form Ω, so $T_{f,\kappa} = T_{f,\operatorname{Ric}\Omega}$. Then*

$$T_{f,\kappa}(r) + N_{f,\operatorname{Ram}}(r) \leq \frac{n}{2} S(T_{f,\eta}, b_1, \psi, r) - \frac{1}{2} \log \gamma_f(0)$$

for all $r \geq r_1$ outside a set of measure $\leq 2b_0(\psi)$, where

$$r_1 = r_1 \left(\frac{1}{(n-1)!} T_{f,\eta} \right) \qquad and \qquad b_1 = b_1 \left(\frac{1}{(n-1)!} T_{f,\eta} \right).$$

90

92

Proof: By the remark,

$$\gamma_f^{1/n} = (\det f^* \eta)^{1/n} \leq \frac{1}{n} \operatorname{tr} f^* \eta.$$

We then estimate the right hand side in Theorem 4.5. We get:

$$\frac{1}{2} \int_{S(r)} (\log \gamma_f) \sigma = \frac{n}{2} \int_{S(r)} (\log \gamma_f^{1/n}) \sigma \leq \frac{n}{2} \log \int_{S(r)} \gamma_f^{1/n} \sigma$$

$$\leq \frac{n}{2} \log \int_{S(r)} \tau_f \sigma$$

$$[\text{by Proposition 5.2}] \quad \leq \frac{n}{2} S(F_{\tau_f}, b_1, \psi, r) + \log(n-1)!$$

$$[\text{by Proposition 6.2}] \quad \leq \frac{n}{2} S(T_{f,\eta}, b_1, \psi, r)$$

for $r \geq r_1$ outside the exceptional set, thus proving the theorem.

Remark: By Proposition 4.3, we know that two height functions associated with positive $(1,1)$ forms are of the same order of magnitude. Since the error term $S(T_{f,\eta}, \psi, r)$ grows slowly, essentially like a log, we may replace $T_{f,\eta}$ by any constant multiple of $T_{f,\eta}$, at the cost of adding $O(1)$ at the end of the estimate.

The theorem with the usual error term $O(\log r + \log T_{f,\eta}(r))$ is due to Carlson-Griffiths. The improvement of the error term with the precise factor $n/2$ and the arbitrary estimating function ψ is due to Lang [**La 8**], see [**Wo**] who followed my suggestion. As before, it is not known if the factor $n/2$ is best possible in general.

II, §7. A GENERAL SECOND MAIN THEOREM (AHLFORS-WONG METHOD)

As in the one dimensional case, we formulate and prove a Second Main Theorem. The method stems partly originally from F. Nevanlinna, who was the first to use a singular volume form in the one-

dimensional case. Then Carlson-Griffiths carried out the higher dimensional case [C-G], while P.M. Wong [Wo] used a different singular
volume form making more precise a method of Ahlfors. We shall make
additional historical comments at the end.

Let X as before be a compact complex manifold. Let D be a divisor
on X. We say that D has **simple normal crossings** if

$$D = \sum_j D_j$$

where each D_j is irreducible, non-singular, and at each point of X there
exist complex coordinates $z_1, \ldots, z_n$ such that D in a neighborhood of
this point is defined by

$$z_1 \cdots z_k = 0 \qquad \text{with} \quad k \leq n.$$

When $n = 1$, then the property of D having simple normal crossings
is equivalent to the property that D consists of distinct points, taken
with multiplicity 1. The maximal value of k which can occur will be
called the **complexity** of D. Throughout we consider a non-degenerate
holomorphic map

$$f \colon \mathbf{C}^n \to X.$$

For the rest of this section we let:

$D = \sum D_j$ have simple normal crossings of complexity k;

$L_j = L_{D_j}$ be the line bundle associated with D_j, with a metric ρ_j;

$\Omega =$ volume form on X, defining a metric κ on the canonical
    bundle L_K so $c_1(\kappa) = \text{Ric}\,\Omega$;

$\gamma_f =$ the function such that $f^*\Omega = \gamma_f \Phi$;

$\eta =$ a closed, positive $(1,1)$-form such that $c_1(\rho_j) \leq \eta$ for all j, and
    $\Omega \leq \eta^n/n!$.

92

Theorem 7.1. *Let f, D, ρ_j, κ, and η be as above. Suppose that $f(0) \notin D$ and $0 \notin \mathrm{Ram}_f$. Then*

$$T_{f,\kappa}(r) + \sum T_{f,\rho_j}(r) - N_{f,D}(r) + N_{f,\mathrm{Ram}}(r)$$
$$\leq \frac{n}{2} S(BT_{f,\eta}^{1+k/n}, b_1, \psi, r) - \frac{1}{2} \log \gamma_f(0) + 1$$

for all $r \geq r_1$ outside a set of measure $\leq 2b_0(\psi)$, and some constant $B = B(\Omega, D, \eta)$. The constants B, b_1, and r_1 will be determined explicitly below.

The general shape of the theorem stems from Carlson-Griffiths, but the good error term stems from Wong except for three improvements which I made on Wong's method: the use of the general function ψ; the exponent $1 + k/n$ instead of 2 (in this way, one recovers the essential estimate of [**La 8**]); and third, Wong takes the irreducible components of D in one linear system, but this is not necessary, and the statement of the theorem as well as its proof are easier without this extraneous assumption.

Note that the exponent $1 + k/n$ is also valid for $k = 0$, in which case we recover Theorem 6.3. This exponent therefore interpolates in the error term for the singularities of D. If D does not have normal crossings, it would be interesting to give an error term showing the dependence on the complexity of the singularities of D in this more general case.

We shall give Wong's proof, suitably adjusted. The proof is a refinement of Ahlfors' method [**Ah**] pp. 22-27, and also makes use of the curvature ideas of Carlson-Griffiths [**C-G**], but somewhat more efficiently.

As in the one-dimensional case, we observe that we do not necessarily need to work with a map of $\mathbf{C}^n$ into X. We could work with a map $f \colon \mathbf{B}(R) \to X$, and the final estimate gives an implicit bound on R in case the canonical class K is ample. In this case, there do not exist

93

non-degenerate maps of $\mathbf{C}^n$ into X, but a bound on the radius R still provides a theorem.

For the rest of this section we let:

s_j = holomorphic section of L_j such that $(s_j) = D_j$; after multiplying s_j by a small constant if necessary, we may assume without loss of generality that

$$|s_j|_{\rho_j} \leq 1/e, \quad \text{say}$$

$|s_j \circ f|_j^2 = |s_j \circ f|_{\rho_j}^2$

Λ = positive decreasing function of r with $0 < \Lambda < 1$.

As in Wong [**Wo**] and Stoll [**St**], we define the **Ahlfors-Wong singular form** with constant λ to be

$$\Omega(D)_\lambda = \frac{\Omega}{\Pi |s_j|_j^{2(1-\lambda)}}.$$

If Λ is not constant, then we shun taking $dd^c \log$. But with $\Lambda = \lambda$ equal to a constant, then we shall take $dd^c \log$, and the only thing that happens is that the factor $(1-\lambda)$ comes out in front. We let

$$\Omega(D)_{f,\Lambda} = \prod_j |s_j \circ f|_j^{-2(1-\Lambda)} f^* \Omega = \gamma_\Lambda \Phi$$

where

$$\gamma_\Lambda = \gamma_{f,D,\Lambda} = \frac{\gamma_f}{\Pi |s_j \circ f|_j^{2(1-\Lambda)}}$$

Then in particular, from the assumption $|s_j|_j \leq 1/e \leq 1$, we have

$$\gamma_f \leq \gamma_\Lambda.$$

Note that with non-constant Λ, $\Omega(D)_{f,\Lambda}$ is not the pullback of a form on X. As in dimension 1, we shall prove the fundamental estimate:

94

96

Proposition 7.2. *Let* $r_1 = r_1(F_{\gamma_j^{1/n}})$, *and let*

$$\Lambda(r) = \begin{cases} 1/qT_{f,\eta}(r) & \text{for } r \geq r_1 \\ \text{constant} & \text{for } r \leq r_1. \end{cases}$$

Let k be the complexity of D. Then for some constants B, b_1 we have

$$\frac{n}{2} \log \int_{S(r)} \gamma_\Lambda^{1/n} \sigma \leq \frac{n}{2} S(BT_{f,\eta}^{1+k/n}, b_1, \psi, r)$$

for $r \geq r_1$ outside a set of measure $\leq 2b_0(\psi)$.

Explicit determinations of the constants B and b_1 will be given below. Note that since η was chosen so that $\Omega \leq \eta^n/n!$, we get $F_{\gamma_j^{1/n}} \leq T_{f,\eta}/n!$. Therefore $r_1(F_{\gamma_j^{1/n}}) \geq r_1(T_{f,\eta}/n!)$. Hence we have $\Lambda < 1$. We postpone the proof of Proposition 7.2 and see immediately how the proposition implies the main theorem. First note that by assumption, for all j,

$$T_{f,\rho_j} \leq T_{f,\eta}$$

We apply Theorem 3.6 and the definitions, and we use the fact that $dd^c \log$ transforms multiplication to addition. *In the case when λ is constant, $0 < \lambda < 1$, we obtain:*

(1)
$$T_{f,\kappa}(r) + (1 - \lambda) \sum T_{f,\rho_j}(r) - (1 - \lambda) \sum N_{f,D_j}(r) + N_{f,\mathrm{Ram}}(r)$$
$$= \frac{n}{2} \int_{S(r)} (\log \gamma_\lambda^{1/n}) \sigma - \frac{1}{2} \log \gamma_\lambda(0).$$

We are interested in an inequality. Since $N_{f,D_j} \geq 0$, we can delete the factor $(1 - \lambda)$ in front to get a smaller quantity on the left hand side. We now use the special value $\lambda = \Lambda(r) = 1/qT_{f,\eta}(r)$. Then the undesirable term on the left hand side satisfies

$$-1 \leq -\Lambda(r) \sum T_{f,\rho_j}(r)$$

95

97

by the assumption on η. For the constants, we have a bound independent of λ:

$$-\frac{1}{2} \log \gamma_\lambda(0) \leq -\frac{1}{2} \log \gamma_f(0)$$

because $\gamma_f \leq \gamma_\lambda$. So, the term on the right in Theorem 7.1 is

$$-\frac{1}{2} \log \gamma_f(0) + 1.$$

Finally we take the log outside the integral over $S(r)$ to get

$$\int_{S(r)} (\log \gamma_{\Lambda(r)}^{1/n}) \sigma \leq \log \int_{S(r)} \gamma_\Lambda^{1/n} \sigma,$$

and we apply Proposition 7.2 to conclude the proof of Theorem 7.1.

There remains to give:

Proof of Proposition 7.2. We are set up to apply the calculus lemma to functions α and F_α with various choices for α, where as before

$$F_\alpha(r) = \int_0^r \frac{dt}{t^{2n-1}} \int_{B(r)} \alpha\Phi.$$

At first, we deal with an arbitrary Λ, not necessarily that of Proposition 7.2. Note that

$$F_{\gamma_\Lambda^{1/n}} \geq F_{\gamma_f^{1/n}} \qquad \text{and} \qquad F'_{\gamma_\Lambda^{1/n}} \geq F'_{\gamma_f^{1/n}}$$

So we take:

The constants b_1 and r_1: $b_1 = b_1(F_{\gamma_f^{1/n}})$ and $r_1 = r_1(F_{\gamma_f^{1/n}})$ are such that

$$F_{\gamma_\Lambda^{1/n}}(r) \geq e \qquad \text{and} \qquad b_1 r^{2n-1} F'_{\gamma_\Lambda^{1/n}}(r) \geq e \text{ for } r \geq r_1.$$

Thus we obtain the bound

$$(2) \qquad \frac{n}{2} \log \int_{S(r)} \gamma_\Lambda^{1/n} \sigma \leq \frac{n}{2} S(F_{\gamma_\Lambda^{1/n}}, b_1, \psi, r) + \frac{n}{2} \log \frac{(n-1)!}{2}.$$

96

We shall prove below the following estimate (replacing a **curvature estimate**).

Lemma 7.3. *There is a constant $b = b(\Omega, D, \eta)$ such that for any decreasing Λ with $0 < \Lambda < 1$ we have*

$$F_{\gamma_\Lambda^{1/n}} \leq (q+1) \frac{b^{1/n}}{n!} \frac{1}{\Lambda^{k/n}} T_{f,\eta} + \frac{1}{2} \frac{b^{1/n}}{n!} q \log 2 \frac{1}{\Lambda^{1+k/n}}.$$

In particular, if $\Lambda = 1/qT_{f,\eta}$ as in Proposition 7.2, then for a suitable constant B we have

$$F_{\gamma_\Lambda^{1/n}} \leq \frac{B}{(n-1)!} T_{f,\eta}^{1+k/n}.$$

Assume Lemma 7.3 for the moment, and take $\Lambda = 1/qT_{f,\eta}$ as in Proposition 7.2. Let B be whatever comes out of the lemma, namely

The constant B: $\quad B = \frac{b^{1/n}}{n}\left((q+1)q^{k/n} + \frac{1}{2}q^{2+k/n}\log 2\right).$

Then indeed

$$F_{\gamma_\Lambda^{1/n}} \leq \frac{B}{(n-1)!} T_{f,\eta}^{1+k/n}.$$

Substituting this estimate in (2) concludes the proof of Theorem 7.2, with the specific constant B as above when $c_1(\rho_j) \leq \eta$ for all j.

There remains to be proved Lemma 7.3. We base the proof on still another lemma, which also gives us a value for the constant b.

97

Lemma 7.4. *There is a constant b, depending only on Ω, η and D, via the sections s_j and metrics ρ_j, such that if $0 < \lambda < 1$ and if we put*

$$\eta_{D,\lambda} = (q+1)\lambda\eta + \sum dd^c \log(1 + |s_j|_j^{2\lambda}),$$

then $\eta_{D,\lambda}$ is closed and > 0 outside D, and

$$\lambda^{n+k}\Omega(D)_\lambda \leq b\frac{1}{n!}\eta_{D,\lambda}^n.$$

We shall first see how Lemma 7.4 implies Lemma 7.3. In the first place, it suffices to prove Lemma 7.3 with a fixed λ rather than Λ. Indeed, since Λ is a decreasing function, for a given value of r we have

$$\gamma_\Lambda(z) \leq \gamma_{\Lambda(r)}(z) \quad \text{for} \quad |z| \leq r,$$

so letting $\lambda = \Lambda(r)$ we get

$$F_{\gamma_\Lambda^{1/n}}(r) \leq F_{\gamma_\lambda^{1/n}}(r),$$

which implies Lemma 7.3 with the possibly variable Λ.

Now we have by Lemma 7.4 taking the pull back f^*, the inequality

$$\lambda^{n+k}\gamma_\lambda\Phi \leq b(\det f^*\eta_{D,\lambda})\Phi,$$

so that

$$\begin{aligned}
\lambda^{1+k/n}\gamma_\lambda^{1/n}\Phi &\leq b^{1/n}(\det f^*\eta_{D,\lambda})^{1/n}\Phi \\
&\leq \frac{b^{1/n}}{n}(\operatorname{tr} f^*\eta_{D,\lambda})\Phi \\
&\leq \frac{b^{1/n}}{n!}f^*\eta_{D,\lambda} \wedge \varphi^{n-1}.
\end{aligned}$$

98

100

Therefore

$$\frac{n!}{b^{1/n}} \lambda^{1+k/n} F_{\gamma_\lambda^{1/n}}(r)$$

$$\leq \int_0^r \frac{dt}{t^{2n-1}} \int_{\mathbf{B}(t)} f^* \eta_{D,\lambda} \wedge \varphi^{n-1}$$

$$= \int_0^r \frac{dt}{t} \int_{\mathbf{B}(t)} f^* \eta_{D,\lambda} \wedge \omega^{n-1}$$

$$= (q+1)\lambda \int_0^r \frac{dt}{t} \int_{\mathbf{B}(t)} f^* \eta \wedge \omega^{n-1} + \sum \int_0^r \frac{dt}{t} \int_{\mathbf{B}(t)} dd^c \log(1 + |s_j \circ f|_j^{2\lambda}) \wedge \omega^{n-}$$

$$= (q+1)\lambda T_{f,\eta}(r) + \sum \frac{1}{2} \int_{\mathbf{S}(r)} \log(1 + |s_j \circ f|_j^{2\lambda}) \sigma - \sum \frac{1}{2} \log(1 + |s_j \circ f(0)|$$

(by Stokes' theorem and Lemma 3.5 in Case III)

$$\leq (q+1)\lambda T_{f,\eta}(r) + \frac{1}{2} q \log 2$$

(because the expression inside the log is ≤ 2). This proves Lemma 7.3 with λ instead of Λ, but as we have already remarked, it also proves Lemma 7.3 in full.

Proof of Lemma 7.4. This is the curvature computation. We recall the general formula already given in Chapter I, §5 in a similar context:

$$(3) \qquad dd^c \log(1+u) = \frac{u \, dd^c \log u}{1+u} + \frac{1}{u(1+u)^2} du \wedge d^c u.$$

Here we let $u = u_j$ be the function

$$0 \leq u_j = |s_j|_j^{2\lambda} < 1 \qquad \text{so that} \qquad \frac{u_j \, dd^c \log u_j}{(1+u_j)} = -\frac{\lambda u_j}{1+u_j} c_1(\rho_j).$$

99

101

Since $\eta \geq c_1(\rho_j)$ we also have $\lambda\eta \geq \lambda u_j(1+u_j)^{-1}c_1(\rho_j)$, and so

$$\eta_{D,\lambda} \geq \lambda\eta + \sum \left\{ \frac{\lambda u_j}{1+u_j}c_1(\rho_j) + dd^c\log(1+u_j) \right\}$$

$$(4) \qquad = \lambda\eta + \sum \frac{1}{u_j(1+u_j)^2}du_j \wedge d^c u_j \qquad \text{by (3)}$$

In particular, the right hand side is ≥ 0, so we have proved the positivity of $\eta_{D,\lambda}$. We shall now use systematically that if $\eta_1, \ldots, \eta_n$ are positive $(1,1)$ forms, then $\eta_1 \wedge \cdots \wedge \eta_n$ is also positive. Then given a point $x \in X$ which does not lie on the divisor D, we can find a small enough neighborhood U of x and a constant b_U such that

$$\eta_{D,\lambda}^n|U \geq b_U\lambda^n\Omega(D)_\lambda|U$$

because η is positive and $du_j \wedge d^c u_j \geq 0$, so we may use merely the term $(\lambda\eta)^n$ in the expansion of $\eta_{D,\lambda}^n$.

Suppose next that x lies on the divisor. We pick a small neighborhood U with complex coordinates $z_1, \ldots, z_n$ such that D is defined by

$$z_1 \cdots z_k = 0 \quad \text{with} \quad k \leq n.$$

Let

$$\eta_j = \frac{1}{u_j}du_j \wedge d^c u_j \geq 0.$$

Then by (4) and using the fact $(1+u_j)^2 \leq 4$, shrinking U further if necessary, there exists a constant $b_4 = b_4(\eta, U)$ such that

$$\eta_{D,\lambda}|U \geq \lambda b_4 \sum_{i=1}^{n} \frac{\sqrt{-1}}{2\pi}dz_i \wedge d\bar{z}_i + \frac{1}{4}\sum_{j=1}^{k}\eta_j$$

$$= \lambda b_4\varphi + \frac{1}{4}\sum_{j=1}^{k}\eta_j.$$

Hence putting $b_5 = b_4^{n-k}/4^k$, we get

$$(5) \qquad \eta_{D,\lambda}^n|U \geq \lambda^{n-k}b_5\varphi^{n-k} \wedge \eta_1 \wedge \cdots \wedge \eta_k.$$

100

The section s_j is represented by z_j and there is a positive C^∞ function α_j such that

$$|s_j|_j^2 = z_j \bar{z}_j \alpha_j \quad \text{for} \quad j = 1, \ldots, k.$$

So it is easy to compute

$$\partial |s_j|_j^2 = \partial(z_j \bar{z}_j \alpha_j) = \bar{z}_j \alpha_j dz_j + z_j \bar{z}_j \partial \alpha_j$$
$$\bar{\partial} |s_j|_j^2 = \bar{\partial}(z_j \bar{z}_j \alpha_j) = z_j \alpha_j d\bar{z}_j + z_j \bar{z}_j \bar{\partial} \alpha_j.$$

We have $du \wedge d^c u = (\sqrt{-1}/2\pi)\partial u \wedge \bar{\partial} u$, and so by direct computation

$$(6) \qquad du_j \wedge d^c u_j = \lambda^2 |s_j|_j^{4\lambda - 2} \frac{\sqrt{-1}}{2\pi} \alpha_j (dz_j \wedge d\bar{z}_j + \zeta_j)$$

where ζ_j is a C^∞ $(1,1)$ form which vanishes on $U \cap D_j$. Combining (5) and (6), and using the fact that α_j is bounded away from zero, we get

$$\eta_{D,\lambda}^n | U \geq \lambda^{n-k} b_6 \varphi^{n-k} \wedge \prod_{j=1}^{k} \left(\lambda^2 |s_j|_j^{2\lambda - 2} \frac{\sqrt{-1}}{2\pi} (dz_j \wedge d\bar{z}_j + \zeta_j) \right)$$

$$= \lambda^{n+k} b_6 \prod_{j=1}^{k} |s_j|_j^{2\lambda - 2} \varphi^{n-k} \wedge \prod_{j=1}^{k} \frac{\sqrt{-1}}{2\pi} (dz_j \wedge d\bar{z}_j + \zeta_j)$$

$$\geq b_U \lambda^{n+k} \prod_{j=1}^{k} |s_j|_j^{2\lambda - 2} (1 + \beta) \Phi$$

where β is a function vanishing at x. Shrinking U further we can omit mentioning this function β.

So for each point of X we have found a neighborhood where $\eta_{D,\lambda}^n$ restricted to this neighborhood satisfies the inequality desired in Lemma 7.4. The full lemma now follows by the compactness of X, thus concluding the proof.

Remark: The computation as above is similar to the curvature computation in Carlson-Griffiths [**C-G**], but somewhat simpler, for a

101

103

couple of reasons. First it is a variation with a simpler formal structure, and second it has fewer terms since the log log terms from Carlson-Griffiths are not present here. The simpler power $|s_j|_j^{2\lambda}$ is easier to differentiate. These are some of the advantages of the singular form of Ahlfors, Stoll and Wong. Originally, Wong proved only the inequality

$$\lambda^{2n}\Omega(D)_\lambda \ll \eta_{D,\lambda}^n.$$

I improved Wong's argument in order to get the key exponent λ^{k+n} by being more careful at the appropriate technical step in the proof when taking the n-th power of $\eta_{D,\lambda}$. The structure of the proof and the number which comes out are so natural (i.e. the $1+k/n$ in the second main theorem) that there is a good possibility that the exponent $1+k/n$ is best possible. Thus conjecturally the error term is determined by local considerations on the divisor, and the complexities of its singularities.

II, §8. VARIATIONS AND APPLICATIONS

In this section I give two further applications of the crucial Proposition 7.2 and of the calculus lemma.

As in dimension 1, we have the lemma on the logarithmic derivative. A higher dimensional version was given by Griffiths [**Gr**], following Nevanlinna's differential geometric method [**Ne**] p. 259, without paying attention to constants. I gave a more precise version in [**La 8**], but using the Ahlfors-Wong method, I can now give the conjectured estimate.

Theorem 8.1. *Let* $f: \mathbb{C}^n \to X$ *be holomorphic non-degenerate. Let* Ψ *be a meromorphic n-form with no zeros, and such that its polar divisor D has simple normal crossings. Assume, for simplicity, that* $f(0) \notin D$ *and* $0 \notin \mathrm{Ram}_f$. *Let*

$$f^*\Psi = L_f(z)dz_1 \wedge \cdots \wedge dz_n.$$

102

Define

$$\nu_f(r) = \int_{S(r)} \log^+ |L_f| \sigma.$$

Let K be a canonical divisor and assume $-K$ is ample. Then for some constants B, B' we have

$$\nu_f(r) \le \frac{n}{2} S(BT_f^{1+k/n}, \psi, r) + B'$$

for all $r \ge r_1$ outside a set of measure $\le 2b_0(\psi)$.

Proof: By assumption, $-K$ can be taken as D, which is ample. Thus D plays the role of $(0) + (\infty)$ in the one-dimensional case. The proof is then entirely similar, following exactly the same steps as in dimension one, using the crucial estimate of Proposition 7.2. Otherwise there is no change. We carry out the details for the convenience of the reader. We let

$$T_f = T_{f,D} \quad \text{and} \quad \Lambda = 1/qT_f \quad \text{for} \quad r \ge r_1,$$

where q is the number of irreducible components of D, so that we can apply Proposition 7.2 later. We let

$$\Omega(D) = \frac{\Omega}{\Pi |s_j|_j^2} \quad \text{so that} \quad \Omega(D) = h_0 \Psi \wedge \bar{\Psi},$$

where h_0 is C^∞ on X and > 0, so bounded away from zero and infinity. The holomorphic sections s_j are always selected as in the conditions preceding Proposition 7.2, so in particular $|s_j|_j \le 1/e$. Then using the notation of those conditions, we get

$$\gamma_\Lambda = \gamma_{f,\Lambda,D} = |L_f|^2 h_D^\Lambda (h_0 \circ f) \quad \text{where} \quad h_D = \prod |s_j \circ f|_j^2.$$

We let

$$u = |L_f|^{2/n} \quad \text{and} \quad v = h_D^{\Lambda/n}.$$

103

Then uv and $\gamma_\Lambda^{1/n}$ differ by a positive function bounded away from zero and infinity. We then obtain the same sequence of inequalities as when $N = 1$, namely:

$$\nu_f(r) = \frac{n}{2} \int_{S(r)} (\log^+ u)\sigma$$

$$= \frac{n}{2} \int_{S(r)} (\log^+ u + \log v)\sigma - \frac{n}{2} \int_{S(r)} (\log v)\sigma$$

$$= \frac{n}{2} \int_{S(r)} (\log e^{\log^+ u + \log v})\sigma - \frac{n}{2}\Lambda(r) \int_{S(r)} (\log h_D)\sigma$$

$$\leq \frac{n}{2} \log \int_{S(r)} \gamma_\Lambda^{1/n}\sigma + \frac{n}{2}\Lambda(r)m_{f,D}(r) + O(1)$$

As in Chapter I, Theorem 6.1 we use

$$e^{\log^+ u + \log v} \leq uv + 1 \leq \gamma_\Lambda + 1.$$

Since $m_{f,D} \leq T_{f,D}$ our choice of $\Lambda(r)$ shows that the term involving $\Lambda(r)m_{f,D}(r)$ is bounded. We can then use Proposition 7.2 to conclude the proof.

I thank Alexander Eremenko for drawing my attention to the paper [G-G], where Goldberg and Grinshtein obtain a very good error term for the logarithmic derivative.

Next we come to a variation in the non-equidimensional case. In these notes, I do not want to go fully into the theory of holomorphic curves

$$f \colon \mathbf{C} \to X.$$

However, it may be worth while to give one version of such a result, stemming from Griffiths-King [G-K] and Vojta [Vo I] Theorem 5.7.2. I state the result eliminating unnecessary hypotheses.

104

Let Y be a complex manifold, and let $f \colon \mathbf{C} \to Y$ be a non-constant holomorphic map. If η is a $(1,1)$ form on Y then we define the **height**

$$T_{f,\eta}(r) = \int_0^r \frac{dt}{t} \int_{\mathbf{D}(t)} f^*\eta.$$

We write as usual

$$f^*\eta = \gamma_f \Phi, \qquad \text{where} \quad \Phi = \frac{\sqrt{-1}}{2\pi} dz \wedge d\bar{z}.$$

Then

$$\gamma_f = |\Delta|^2 h$$

where h is C^∞. If η is positive, then $h > 0$. Indeed, write η in local coordinates $z = (z_1, \ldots, z_n)$, and let $f = (f_1, \ldots, f_n)$. Suppose w is a complex coordinate in a neighborhood of a point in the disc so that $z = f(w)$, $f(0) = 0$, and

$$f(w) = w^e(g_1, \ldots, g_n),$$

where $g_1, \ldots, g_n$ do not all vanish at the origin. Then

$$f'(w) = w^{e-1} u(w),$$

where $u(0) \neq (0, \ldots, 0)$. Hence in these coordinates,

$$\gamma_f(w) = |w^{e-1}|^2 h(w),$$

where h is C^∞ and positive. Thus we have written γ_f in the desired form locally. The zeros of γ_f define a divisor on $\mathbf{D}(R)$, and there exists a holomorphic function Δ on $\mathbf{D}(R)$ having this divisor. We then get the desired global expression.

The function Δ is holomorphic, and defines the **ramification divisor** Ram_f, which in this case is a discrete set of zeros, with multiplicities. So we have the analogous concepts that we have met previously.

105

Theorem 8.2. *Let Y be a complex manifold (not necessarily compact). Let η be a closed, positive $(1,1)$ form on Y and let $f: \mathbf{D}(R) \to Y$ be a holomorphic map. Suppose there is a constant $B > 0$ such that*

$$Bf^*\eta \leq \mathrm{Ric}\ f^*\eta.$$

Assume $0 \notin \mathrm{Ram}_f$. Then

$$BT_{f,\eta}(r) + N_{f,\mathrm{Ram}}(r) \leq \frac{1}{2}S(T_{f,\eta}, \psi, r) - \frac{1}{2}\log\ \gamma_f(0)$$

for $r \geq r_1$ outside a set of measure $\leq 2b_0(\psi)$.

Proof: The standard arguments work just as in the one-dimensional case. First observe that by definition,

$$F_{\gamma_f} = T_{f,\eta}.$$

Then by Stokes' theorem,

$$T_{\mathrm{Ric}f^*\eta}(r) + N_{f,\mathrm{Ram}}(r) = \frac{1}{2}\int\limits_0^{2\pi} \log\gamma_f(re^{i\theta})\frac{d\theta}{2\pi} - \frac{1}{2}\log\ \gamma_f(0).$$

By assumption, $Bf^*\eta \leq \mathrm{Ric}\ f^*\eta$, so

$$BT_{f,\eta} \leq T_{\mathrm{Ric}\ f^*\eta}.$$

Applying the calculus lemma to the right hand side proves the theorem.

Remarks: Several features deserve emphasis in the above theorem.

First, there is no need to assume any special property about the divisor at infinity in some compactification, e.g. normal crossings. This hypothesis is completely irrelevant here.

It is important to take Y not compact in some applications. Vojta applies the theorem to the non-compactified moduli space, for instance.

The existence of the positive $(1,1)$-form is the strongest possible assumption in the direction of hyperbolicity. For a discussion of the relation to hyperbolicity, see [La 7], end of Chapter III, §4 and Chapter IV, Theorem 3.3 among others.

Since Y is hyperbolic in the theorem, every holomorphic map of C into Y is constant. Thus in some sense Vojta's formulation is empty. But by stating the theorem for a map of a disc into Y, one gets a non-empty statement, in line with previous remarks concerning the second main theorem, which applied to maps of discs, not just C or C^n, into X as in the Landau-Schottky theorem. For the statement to make sense, it is then important that the error term be given with fairly explicit constants depending on various parameters as we have done here. Then among other things, the theorem gives a bound on the radius of a disc which can be mapped into X in a non-constant way.

107

PART TWO

NEVANLINNA THEORY
OF COVERINGS

by William Cherry

INTRODUCTION

In this second part, the results of Part One are generalized to the case of covering spaces. Chapter III is concerned with maps from branched covers over $\mathbf{C}$ into the complex projective line, and Chapter IV works with maps from analytic coverings of $\mathbf{C}^n$ into compact n-complex dimensional manifolds. Part Two is constructed to look as much like Part One as possible, and again the methods used are those of Ahlfors, Wong, Stoll, and Griffiths and King. Stoll, in particular, treats the case of maps from covers into projective linear spaces [St]. The goal in these notes is to keep careful track of constants in order to see exactly where the degree enters into the error term in the second main theorem. As in the first part, the equidimensional case is treated from the differential geometric, rather than the projective linear, point of view.

The motivation for these chapters comes from Vojta's dictionary relating theorems in Nevanlinna theory to number theory [Vo 1]. Schematically, this analogy is as follows:

Algebraic Case	Analytic Case
$Y \xrightarrow{f} X$	$Y \xrightarrow{f} X$
$\downarrow$	$\downarrow$
Spec $\mathbf{Z}$	$\mathbf{C}$

The terms involving the degree of the covering maps which appear in the error term for the second main theorem of Nevanlinna theory are analagous to the degree of an algebraic field extension in number theory. The goal is to find an expression for the error term which is uniform for

111

all coverings. In the simpler second main theorem without a divisor, the error term will be uniform in terms of both the covering map, and the map into the compact manifold, except for an additive constant depending on the values of the maps at the points lying above zero. Thus, the error term will in fact be uniform, if all maps are taken to satisfy a normalization condition on their values and derivatives above zero. This is in line with Vojta's conjecture in the number theoretic case, bounding the height in terms of the logarithmic disriminant, uniformly for all algebraic points, not just for points of bounded degree. In fact, the error term which appears here shows precisely that the degree comes in just as a factor multiplied by a universal expression, independent of the degree. However, this is not quite what is obtained in the more general second main theorem with a divisor. Here the second main theorem can be expressed either with an error term which can be written as the degree multiplied by an expression independent of the degree, but then added to terms which are essentially like the degree multiplied by log log of the degree, or it can be expressed with the desired form for the error term, but with the inequality holding outside an exceptional set, the size of which depends on the degree of the covering.

As in Part One, in the case of $\mathbf{C}$ or $\mathbf{C}^n$, the error term here is worked out with a general type function, ψ, so that it parallels the most refined general conjectures of approximation in the number theoretic case.

Finally, I would like to thank Serge Lang for introducing me to these questions, and for suggesting that I look at the error term for coverings to investigate its uniform behavior.

William Cherry

CHAPTER III

NEVANLINNA THEORY FOR MEROMORPHIC FUNCTIONS ON COVERINGS OF C

In this chapter, the Nevanlinna theorems for meromorphic functions on branched coverings of the complex plane are developed.

III, §1. NOTATION AND PRELIMINARIES

Let $p: Y \to C$ be a covering of C. That is, let Y be a connected Riemann surface and let p be a proper surjective holomorphic map. Let:

$[Y:C] =$ the degree of the covering;

$Y(r) = \{y \in Y : |p(y)| < r\}$;

$Y[r] = \{y \in Y : |p(y)| \leq r\}$;

$Y<r> = \{y \in Y : |p(y)| = r\}$;

$\sigma_Y = d^c \log |p|^2 = p^*(\dfrac{d\theta}{2\pi})$ where $d\theta$ is the usual form on C.

In a local coordinate w,

$$\sigma_Y(w) = \frac{\sqrt{-1}}{4\pi}\left(\overline{\frac{p'(w)}{p(w)}}d\bar{w} - \frac{p'(w)}{p(w)}dw\right) \quad \text{outside of } Y<0> .$$

Proposition 1.1. *Let* $f: Y \to \mathbf{P}^1$ *be holomorphic. Then for* $r \in \mathbf{R}_{>0}$,

$$\int\limits_{Y<r>} d^c \log |f|^2 = \sum_{y \in Y(r)} (\mathrm{ord}_y f)$$

Proof: The statement follows by surrounding the singularities with small circles and then applying Stoke's Theorem.

QED

Corollary 1.2. *For* $r \in \mathbf{R}_{>0}$,

$$\int\limits_{Y<r>} \sigma_Y = [Y:\mathbf{C}].$$

Proof: Applying Proposition 1.1 to p gives

$$\int\limits_{Y<r>} \sigma_Y = \int\limits_{Y<r>} d^c \log |p|^2 = \sum_{y \in Y<0>} (\mathrm{ord}_y p) = [Y:\mathbf{C}].$$

QED

Corollary 1.2 is the reason that the error term in the Second Main Theorem is multipled by the degree of the covering map. The concavity of the logarithm will be used to move a log out of an integral against σ_Y and this will introduce a factor involving the degree.

In Chapter I, the Poisson-Jensen (I.1.2) and the Green-Jensen (I.2.3) integral formulas are stated and proved on $\mathbf{C}$. The conditions under which the formulas hold are stated in terms of global polar coordinates. No such global coordinates exist on Y, so first a more general version of the Green-Jensen formula will be proved and the analog of the Poisson-Jensen formula will be derived as a consequence.

114

116

Theorem 1.3 (Green-Jensen Formula). *Let α be a C^2 function from $Y \to \mathbb{C}$ except at a discrete set of singularities Z such that $Z \cap Y{<}0{>} = \emptyset$. Assume, in addition, that the following three conditions are satisfied:*

i) *$\alpha \sigma_Y$ is absolutely integrable on $Y{<}r{>}$ for all $r > 0$.*

ii) *$d\alpha \wedge \sigma_Y$ is absolutely integrable on $Y[r]$ for all r.*

iii) $\displaystyle \lim_{\varepsilon \to 0} \int_{S(Z,\varepsilon)(r)} \alpha \sigma_Y = 0$ *for all r,*

where for sufficiently small ε, $S(Z, \varepsilon)(r)$ denotes the disjoint union of "circles" of "radius" ε around the singularities $Z \cap Y[r]$. Then

(A) $$\int_0^r \frac{dt}{t} \int_{Y{<}t{>}} d^c \alpha = \frac{1}{2} \int_{Y{<}r{>}} \alpha \sigma_Y - \frac{1}{2} \sum_{y \in Y{<}0{>}} (\mathrm{ord}_y p)\alpha(y),$$

and

(B) $$\int_0^r \frac{dt}{t} \int_{Y(t)} dd^c \alpha + \int_0^r \frac{dt}{t} \lim_{\varepsilon \to 0} \int_{S(Z,\varepsilon)(t)} d^c \alpha$$
$$= \frac{1}{2} \int_{Y{<}r{>}} \alpha \sigma_Y - \frac{1}{2} \sum_{y \in Y{<}0{>}} (\mathrm{ord}_y p)\alpha(y).$$

Proof: Note that (B) follows from (A) because

$$\int_{Y(t)} dd^c \alpha + \lim_{\varepsilon \to 0} \int_{S(Z,\varepsilon)(t)} d^c \alpha = \int_{Y{<}t{>}} d^c \alpha$$

by Stoke's Theorem and then integrating against dt/t.

1) If α and β are C^2 functions, then for degree reasons

$$d\alpha \wedge d^c \beta = d\beta \wedge d^c \alpha.$$

115

117

2) Because Z is disjoint from $Y<0>$,

$$\lim_{\varepsilon \to 0} \int_{Y<\varepsilon>} \alpha \sigma_Y = \sum_{y \in Y<0>} (\text{ord}_y p)\alpha(y).$$

Part (A) will follow by evaluating the integral

$$\frac{1}{2} \int_{Y[r]} d(\alpha \sigma_Y)$$

in two different ways. Evaluating the integral using Stoke's Theorem gives the right hand side, and using Fubini's Theorem gives the left hand side.

Applying Stoke's Theorem, one has

$$\frac{1}{2} \int_{Y[r]} d(\alpha \sigma_Y) = \frac{1}{2} \lim_{\varepsilon \to 0} \int_{Y[r]-\left(Y(\varepsilon) \cup D(Z,\varepsilon)(r)\right)} d(\alpha \sigma_Y)$$

$$\text{[Stoke's Theorem]} \quad = \frac{1}{2} \lim_{\varepsilon \to 0} \left[\int_{Y<r>} \alpha \sigma_Y - \int_{Y<\varepsilon>} \alpha \sigma_Y - \int_{S(Z,\varepsilon)(r)} \alpha \sigma_Y \right]$$

$$\text{[by 2 and iii]} \quad = \frac{1}{2} \int_{Y<r>} \alpha \sigma_Y - \frac{1}{2} \sum_{y \in Y<0>} (\text{ord}_y p)\alpha(y),$$

where $D(Z,\varepsilon)(r)$ is the disjoint union of open "discs" of "radius" ε around the singularities in $Z \cap Y<r>$. On the other hand, applying

116

118

Fubini's Theorem gives:

$$\frac{1}{2} \int_{Y[r]} d(\alpha \sigma_Y) = \frac{1}{2} \int_{Y[r]} d\alpha \wedge \sigma_Y$$

[by the definition of σ_Y] $\qquad = \dfrac{1}{2} \displaystyle\int_{Y[r]} d\alpha \wedge d^c \log |p|^2$

[by 1] $\qquad = \dfrac{1}{2} \displaystyle\int_{Y[r]} d(\log |p|^2) \wedge d^c \alpha$

$$= \int_{Y[r]} d(\log |p|) \wedge d^c \alpha$$

[from Fubini's Theorem and ii] $\qquad = \displaystyle\int_0^r \frac{dt}{t} \int_{Y<t>} d^c \alpha.$

QED

Proposition 1.4. *Let $\overline{D}$ denote the closed unit disk in $\mathbf{C}$. Let f be a non-constant meromorphic function on $\overline{D}$ such that 0 is the only zero or pole of f in $\overline{D}$. Then,*

$$\int_{\overline{D}} \left| \frac{f'}{f} \right| |dz \wedge d\bar{z}|$$

is finite.

Proof: By assumption $f(z) = z^m h(z)$ where h is never zero or infinity. Hence h'/h is continuous on $\overline{D}$ and therefore bounded, say by M.

117

119

So

$$\int_{\overline{D}} \left| \frac{f'}{f} \right| |dz \wedge d\bar{z}| = \int_{\overline{D}} \left| \frac{m}{z} + \frac{h'}{h} \right| |dz \wedge d\bar{z}|$$

$$\leq \int_{\overline{D}} \frac{|m|}{|z|} |dz \wedge d\bar{z}| + M \int_{\overline{D}} |dz \wedge d\bar{z}|$$

$$= |m| \int_0^1 \int_0^{2\pi} \frac{1}{r} r \, dr d\theta + M \int_0^1 \int_0^{2\pi} r \, dr d\theta$$

$$= 2\pi r |m| + \pi M < \infty.$$

QED

Proposition 1.5. *Let $f: Y \to \mathbf{P}^1$ be a non-constant holomorphic map such that for $y \in Y<0>$, $f(y) \neq 0, \infty$. Let $\alpha = \log|f|$. Then, α satisfies the conditons in Theorem 1.3.*

Proof:

i) Note that $\alpha \sigma_Y$ is absolutely integrable on $Y<r>$ for the same reason $\log x$ is absolutely integrable on $[-1, 1]$.

ii) Next $d\alpha \wedge \sigma_Y$ is absolutely integrable on $Y[r]$. Indeed, since $f(y) \neq 0, \infty$ for $y \in Y<0>$, there exists an $\varepsilon > 0$ such that $\overline{Y(\varepsilon)} \cap \overline{D(Z, \varepsilon)(r)} = \emptyset$. On $Y[r] - (D(Z, \varepsilon)(r) \cup Y(\varepsilon))$, $d\alpha \wedge \sigma_Y$ is bounded, f'/f is bounded on $Y(\varepsilon)$ and p'/p is bounded on $D(Z, \varepsilon)(r)$. In a local coordinate w, $d\alpha \wedge \sigma_Y$ is given by:

$$\frac{\sqrt{-1}}{2\pi} \left(\frac{f'(w)}{f(w)} \frac{\overline{p'(w)}}{\overline{p(w)}} + \frac{\overline{f'(w)}}{\overline{f(w)}} \frac{p'(w)}{p(w)} \right) dw \wedge d\bar{w}.$$

Hence

$$|d\alpha \wedge \sigma_Y| \leq \frac{1}{\pi} \left| \frac{f'(w)}{f(w)} \right| \left| \frac{p'(w)}{p(w)} \right| |dw \wedge d\bar{w}|.$$

Therefore, the absolute integrability of $d\alpha \wedge \sigma_Y$ follows from Proposition 1.4.

118

120

iii) One has

$$\lim_{\varepsilon \to 0} \int_{S(Z,\varepsilon)(r)} \alpha \sigma_Y = 0.$$

Let y_0 be a zero of f. Let w be a complex coordinate in a neighborhood of y_0 such that $w = 0$ corresponds to y_0. Then $f(w) = w^m h(w)$ in a neighborhood of y_0, where h is holomorphic and non-vanishing. Furthermore

$$\int_{|w|=\varepsilon} \log |f|^2 \sigma_Y = \int_{|w|=\varepsilon} \log |w|^{2m} \sigma_Y + \int_{|w|=\varepsilon} \log |h|^2 \sigma_Y$$

$$= \log \varepsilon^{2m} \int_{|w|=\varepsilon} \sigma_Y + \int_{|w|=\varepsilon} \log |h|^2 \sigma_Y.$$

By Stoke's Theorem,

$$\int_{|w|=\varepsilon} \sigma_Y = \int_{|w|\le\varepsilon} d\sigma_Y = 0$$

since $y_0 \notin Y<0>$ and σ_Y is C^∞ away from $Y<0>$. But, $\log |h|^2 \sigma_Y$ is also C^∞ in a neighborhood of y_0 since h does not vanish, so

$$\lim_{\varepsilon \to 0} \int_{|w|=\varepsilon} \log |h|^2 \sigma_Y = 0.$$

Therefore

$$\lim_{\varepsilon \to 0} \int_{S(y_0,\varepsilon)} \log |f|^2 \sigma_Y = 0.$$

A similar argument shows that

$$\lim_{\varepsilon \to 0} \int_{S(y_0,\varepsilon)} \log |f|^2 \sigma_Y = 0$$

where y_0 is a pole of f.
QED

119

121

Proposition 1.6. *Let* $f : Y \to \mathbf{P}^1$ *be a non-constant holomorphic map such that for* $y \in Y{<}0{>}$, $f(y) \neq 0, \infty$. *Let* $\alpha = \log(1 + |f|^2)$. *Then,* α *satisfies the conditions of Theorem 1.3.*

Proof: The singularities of α are the poles of f. Let y_0 be a pole of f, and let U be a neighborhood of y_0 such that there are holomorphic functions $f_0, f_1 : U \to \mathbf{C}$ such that $f = f_1/f_0$ on U. Then

$$\log(1 + |f|^2) = \log(|f_0|^2 + |f_1|^2) - \log|f_0|^2$$

But $\log(|f_0|^2 + |f_1|^2)$ is C^∞ and $\log|f_0|^2$ satisfies the conditions of Theorem 1.3 by Proposition 1.5.
QED

Theorem 1.7 (Poisson-Jensen Formula). *Let* $f : Y \to \mathbf{P}^1$ *be a non-constant holomorphic map such that* $f(y) \neq 0, \infty$ *for* $y \in Y{<}0{>}$. *Then*

$$\sum_{y \in Y{<}0{>}} (\mathrm{ord}_y p) \log|f(y)| = \int_{Y{<}r{>}} \log|f| \, \sigma_Y - \sum_{y \in Y(r)} (\mathrm{ord}_y f) \log\left|\frac{r}{p(y)}\right|.$$

Proof: By Proposition 1.5, Theorem 1.3 applies to $\log|f|^2$. Hence

$$\frac{1}{2} \int_{Y{<}r{>}} \log|f|^2 \sigma_Y$$

$$= \frac{1}{2} \sum_{y \in Y{<}0{>}} (\mathrm{ord}_y p) \log|f(y)|^2 + \int_0^r \frac{dt}{t} \int_{Y{<}t{>}} d^c(\log|f|^2)$$

from Theorem 1.3 (A). But

$$\int_{Y{<}t{>}} d^c(\log|f|^2) = \sum_{y \in Y(r)} (\mathrm{ord}_y f).$$

120

122

Hence

$$\int\limits_0^r \frac{dt}{t} \int\limits_{Y<t>} d^c \log |f|^2 = \sum_{y \in Y(r)} (\mathrm{ord}_y f) \log \left| \frac{r}{p(y)} \right|.$$

QED

III, §2. FIRST MAIN THEOREM

In this section, Nevanlinna's First Main Theorem on the independence of the height is shown to hold in the covering case as well.

Henceforth, let $f: Y \to \mathbf{P}^1$ be a non-constant holomorphic map such that $f(y) \neq 0, \infty$ and $f'(y) \neq 0$, for all $y \in Y<0>$.

Let $n_f(0, r)$ denote the number of zeros of f in $Y(r)$ counted with multiplicities.

For $a \in \mathbf{C}$, let $n_f(a, r) = n_{f-a}(0, r)$.

Let $n_f(\infty, r) = n_{1/f}(0, r)$.

For simplicity's sake, the Nevanlinna height and counting functions will only be defined for values $a \in \mathbf{P}^1$ such that $f(y) \neq a$ for $y \in Y<0>$. For $a \in \mathbf{P}^1$ such that $f(y) \neq a$, for all $y \in Y<0>$, let

$$N_f(a, r) = \int\limits_0^r \frac{dt}{t} n_f(a, t) = \sum_{y \in Y(r)} (\mathrm{ord}_y f) \log \left| \frac{r}{p(y)} \right|.$$

For $a \in \mathbf{P}^1$, define

$$m_f(a, r) = \int\limits_{Y<r>} -\log \|f, a\| \sigma_Y,$$

where $\| \, , \, \|$ is the "chordal distance" on $\mathbf{P}^1$ defined in Chapter I, §1. Finally, for $a \in \mathbf{P}^1$ such that $f(y) \neq a$ for $y \in Y<0>$, define

$$T_{f,a}(r) = m_f(a, r) + N_f(a, r) + \sum_{y \in Y<0>} (\mathrm{ord}_y p) \log \|f(y), a\|.$$

121

Theorem 2.1 (First Main Theorem). $T_{f,a}(r)$ *is independenet of* $a \in \mathbf{P}^1$ *provided* $f(y) \neq a$ *for* $y \in Y\langle 0\rangle$.

Proof: Let $a \in \mathbf{C}$ such that $f(y) \neq a$, for all $y \in Y\langle 0\rangle$. From the definitions of the symbols involved:

$$
\begin{aligned}
T_{f,a}(r) = \ & N_f(a,r) + m_f(a,r) + \sum_{y \in Y\langle 0\rangle} (\mathrm{ord}_y p) \log \|f(y), a\| \\
= \ & N_{f-a}(0,r) - \int_{Y\langle r\rangle} \log \|f, a\| \sigma_Y \\
& + \sum_{y \in Y\langle 0\rangle} (\mathrm{ord}_y p) \log \|f(y), a\| \\
= \ & N_{f-a}(0,r) - \int_{Y\langle r\rangle} \log |f - a| \sigma_Y \\
& + \frac{1}{2} \int_{Y\langle r\rangle} \log(1 + |f|^2) \sigma_Y \\
& + \frac{1}{2} \int_{Y\langle r\rangle} \log(1 + |a|^2) \sigma_Y \\
& + \sum_{y \in Y\langle 0\rangle} (\mathrm{ord}_y p) \log |f(y) - a| \\
& - \frac{1}{2} \sum_{y \in Y\langle 0\rangle} (\mathrm{ord}_y p) \log(1 + |f(y)|^2) \\
& - \frac{1}{2} \sum_{y \in Y\langle 0\rangle} (\mathrm{ord}_y p) \log(1 + |a|^2).
\end{aligned}
$$

Now

$$
\int_{Y\langle r\rangle} \log(1 + |a|^2) \sigma_Y = [Y:\mathbf{C}] \log(1 + |a|^2) = \sum_{y \in Y\langle 0\rangle} (\mathrm{ord}_y p) \log(1 + |a|^2),
$$

122

and from Theorem 1.7

$$\int\limits_{Y<r>} \log|f - a|\sigma_Y =$$

$$\sum_{y \in Y<0>} (\text{ord}_y p) \log |f(y) - a| + N_{f-a}(0, r) - N_{f-a}(\infty, r).$$

Furthermore

$$N_{f-a}(\infty, r) = N_f(\infty, r) \quad \text{and} \quad \log \|f(y), \infty\| = -\frac{1}{2}\log(1 + |f(y)|^2).$$

Hence

$$T_{f,a}(r) = \int\limits_{Y<r>} -\log \|f, \infty\| + N_f(\infty, r)$$

$$+ \sum_{y \in Y<0>} (\text{ord}_y p) \log \|f(y), \infty\|$$

$$= m_f(\infty, r) + N_f(\infty, r) + \sum_{y \in Y<0>} (\text{ord}_y p) \log \|f(y), \infty\|$$

$$= T_{f,\infty}(r).$$

QED

In light of Theorem 2.1, denote $T_{f,\infty}(r)$ by $T_f(r)$.

The **Ahlfors-Shimizu** expression for the height in the covering case is identical to the complex plane case. Let

$$\Phi_Y = p^* \cdot \left(\frac{\sqrt{-1}}{2\pi} dz \wedge d\bar{z} \right) = dd^c |p|^2$$

$$= |p'(w)|^2 \frac{\sqrt{-1}}{2\pi} dw \wedge d\bar{w} = d|p|^2 \wedge \sigma_Y$$

be the pseudo-volume form obtained by pulling back the Euclidean volume form on **C**.
Let

$$\gamma_f = \frac{|f'|^2}{(1 + |f|^2)^2 |p'|^2}.$$

123

125

Note the appearance of $|p'|^2$ in the denominator of γ_f will exactly cancel the same term which appears in Φ_Y when expressed in a local coordinate.

Theorem 2.2 (Ahlfors-Shimizu). *Away from singularites, one has*

$$\gamma_f \Phi_Y = dd^c \log(1 + |f|^2) = -\frac{1}{2} dd^c \log \gamma_f,$$

and hence

$$T_f(r) = \int_0^r \frac{dt}{t} \int_{Y(t)} \gamma_f \Phi_Y$$

$$= \int_0^r \frac{dt}{t} \int_{Y(t)} dd^c \log(1 + |f|^2)$$

$$= -\frac{1}{2} \int_0^r \frac{dt}{t} \int_{Y(t)} dd^c \log \gamma_f$$

Proof: The first statement follows from the fact that

$$dd^c \log |f'|^2 = dd^c \log |p'|^2 = 0$$

away from singularities, and from the fact that $dd^c \log$ transforms products into sums.

By Proposition 1.6, Theorem 1.3 (B) can be applied to $\log(1 + |f|^2)$

124

to conclude:

$$\int_0^r \frac{dt}{t} \int_{Y(t)} dd^c \log(1 + |f|^2) = \frac{1}{2} \int_{Y<r>} \log(1 + |f|^2)\sigma_Y$$

$$- \frac{1}{2} \sum_{y \in Y<0>} (\mathrm{ord}_y p) \log(1 + |f(y)|^2)$$

$$- \int_0^r \frac{dt}{t} \lim_{\varepsilon \to 0} \int_{S(Z,\varepsilon)(t)} d^c \log(1 + |f|^2).$$

Now

$$\frac{1}{2} \int_{Y<r>} \log(1 + |f|^2)\sigma_Y = \int_{Y<r>} -\log \|f, \infty\| \sigma_Y = m_f(\infty, r),$$

and

$$-\frac{1}{2} \sum_{y \in Y<0>} (\mathrm{ord}_y p) \log(1 + |f(y)|^2) = \sum_{y \in Y<0>} (\mathrm{ord}_y p) \log \|f(y), \infty\|.$$

Let y_0 be a pole of f, and let $f = f_1/f_0$ in a neighborhood of y_0. Because $\log(|f_0|^2 + |f_1|^2)$ is C^∞, one gets

$$\lim_{\varepsilon \to 0} \int_{S(y_0,\varepsilon)} d^c \log(1 + |f|^2) = \lim_{\varepsilon \to 0} \int_{S(y_0,\varepsilon)} d^c \log(|f_0|^2 + |f_1|^2)$$

$$- \lim_{\varepsilon \to 0} \int_{S(y_0,\varepsilon)} d^c \log |f_0|^2 = (\mathrm{ord}_{y_0} f).$$

Therefore, since the singularities of $\log(1 + |f|^2)$ are precisely the poles of f, one has

$$\lim_{\varepsilon \to 0} \int_{S(Z,\varepsilon)(t)} d^c \log(1 + |f|^2) = -n_f(\infty, t).$$

125

127

This implies

$$-\int_0^r \frac{dt}{t} \lim_{\varepsilon \to 0} \int_{S(Z,\varepsilon)(t)} d^c \log(1 + |f|^2) = N_f(\infty, r).$$

Therefore

$$\int_0^r \frac{dt}{t} \int_{Y(t)} dd^c \log(1 + |f|^2) = m_f(\infty, r) + N_f(\infty, r)$$

$$+ \sum_{y \in Y<0>} (\operatorname{ord}_y p) \log \|f(y), \infty\| = T_f(r).$$

QED

§3. CALCULUS LEMMAS

The calculus lemmas of Chapter I, §3 apply to the covering case without change, with one exception, and here the change is merely a change in vocabulary necessitated by the absence of polar coordinates on Y.

Given a function α on Y, define the **height transform**:

$$F_\alpha(r) = \int_0^r \frac{dt}{t} \int_{Y(t)} \alpha \Phi_Y$$

for $r > 0$. For example, $T_f = F_{\gamma_f}$, so T_f is a height transform.

Let α be a function on Y such that the following conditions are satisfied:

(a) α is continuous and > 0 except at a discrete set of points.
(b) For each r, the integral $\int_{Y<r>} \alpha \sigma_Y$ is absolutely convergent and $r \mapsto \int_{Y<r>} \alpha \sigma_Y$ is a continuous function of r.
(c) There is an $r_1 \geq 1$ such that $F_\alpha(r_1) \geq e$.

126

Note: F_α has positive derivative, so is strictly increasing.

Lemma 3.1. *If α satisfies (a), (b) and (c) above, then F_α is C^2 and*

$$\frac{1}{r}\frac{d}{dr}\left(r\frac{dF_\alpha}{dr}\right) = 2\int_{Y<r>}\alpha\sigma_Y.$$

Proof:

Since $F_\alpha = \int_0^r \frac{dt}{t}\int_{Y(t)}\alpha\Phi_Y$ one has $\dfrac{dF_\alpha}{dr} = \dfrac{1}{r}\int_{Y(r)}\alpha\Phi_Y.$

Therefore

$$r\frac{dF_\alpha}{dr} = \int_{Y(r)}\alpha\Phi_Y = \int_{Y(r)}d|p|^2 \wedge \alpha\sigma_Y$$

$$[\text{Fubini's Theorem}] \quad = 2\int_0^r tdt\int_{Y<t>}\alpha\sigma_Y.$$

Hence

$$\frac{d}{dr}\left(r\frac{dF_\alpha}{dr}\right) = 2r\int_{Y<r>}\alpha\sigma_Y \quad \text{and} \quad \frac{1}{r}\frac{d}{dr}\left(r\frac{dF_\alpha}{dr}\right) = 2\int_{Y<r>}\alpha\sigma_Y.$$

QED

Lemma 3.2. *If α satisfies (a),(b) and (c), then*

$$\log\int_{Y<r>}\alpha\sigma_Y \le S(F_\alpha, b_1(F_\alpha), \psi, r)$$

for all $r \ge r_1(F_\alpha)$ outside a set of measure $\le 2b_0(\psi)$.

Proof: Apply Lemma 3.1 and Lemma I.3.2.
QED

127

129

III, §4. RAMIFICATION AND THE SECOND MAIN THEOREM

In this section the ramification terms are defined, and an error term in Nevanlinna's Second Main Theorem is established, which is both uniform (after a normalization of the values above zero) for all non-constant holomorphic maps $f: Y \to \mathbf{P}^1$ and for all coverings $p: Y \to \mathbf{C}$.

Let $f: Y \to \mathbf{P}^1$ be a non-constant holomorphic map such that $f(y) \neq 0, \infty$ and $f'(y) \neq 0$ for all $y \in Y<0>$. Let $y_0 \in Y$, and let $f = f_1/f_0$ in a neighborhood of y_0. The **ramification index** of f at y_0 is defined to be

$$\mathbf{n}_{f,\mathrm{Ram}}(y_0) = (\mathrm{ord}_{y_0}(f_0 f_1' - f_0' f_1)).$$

Define

$$n_{f,\mathrm{Ram}}(t) = \sum_{y \in Y(t)} \mathbf{n}_{f,\mathrm{Ram}}(y) \quad \text{and} \quad N_{f,\mathrm{Ram}}(r) = \int_0^r \frac{dt}{t} n_{f,\mathrm{Ram}}(t).$$

Note that since p is holomorphic, $N_{p,\mathrm{Ram}}(r) = N_{p'}(0, r)$.

Theorem 4.1. *Let $p: Y \to \mathbf{C}$ be a proper surjective holomorphic map such that $p'(y) \neq 0$ for all $y \in Y<0>$. Let $f: Y \to \mathbf{P}^1$ be a non-constant holomorphic map such that $f(y) \neq 0, \infty$ and $f'(y) \neq 0$ for all $y \in Y<0>$. Then, one has*

$$-2T_f(r) + N_{f,\mathrm{Ram}}(r) - N_{p,\mathrm{Ram}}(r) + \frac{1}{2}[Y: \mathbf{C}] \log[Y: \mathbf{C}]$$

$$\leq \frac{1}{2}[Y: \mathbf{C}] S(T_f, \psi, r) - \frac{1}{2} \sum_{y \in Y<0>} \log \gamma_f(y)$$

for all $r \geq r_1(T_f)$ outside a set of measure $\leq 2b_0(\psi)$, where

$$\gamma_f = \frac{|f'|^2}{\left(1 + |f|^2\right)^2 |p'|^2}.$$

128

130

Remark: The term on the right in the above inequality involving $\log \gamma_f$ depends only on the values of f, f' and p' above zero, so the right hand side is uniform in f and in p if they are normalized above zero. Furthermore, the term with $[Y:C]\log[Y:C]$ is positive and therefore actually improves the inequality.

Proof:

1) From Theorem 2.2

$$\int_0^r \frac{dt}{t} \int_{Y(t)} dd^c \log \gamma_f = -2T_f(r).$$

2) Let y_0 be a ramification point of f, and let $f = f_1/f_0$ in a neighborhood U of y_0, where $f_0, f_1 : U \to C$ are holomorphic without common zeros. Then, on U

$$\frac{|f'|^2}{(1+|f|^2)^2} = \frac{|f_0 f_1' - f_0' f_1|^2}{(|f_0|^2 + |f_1|^2)^2}.$$

Therefore

$$\lim_{\varepsilon \to 0} \int_{S(y_0,\varepsilon)} d^c \log \frac{|f'|^2}{(1+|f|^2)^2}$$

$$= \lim_{\varepsilon \to 0} \int_{S(y_0,\varepsilon)} d^c \log |f_0 f_1' - f_0' f_1|^2$$

$$- 2 \lim_{\varepsilon \to 0} \int_{S(y_0,\varepsilon)} d^c \log(|f_0|^2 + |f_1|^2)$$

$$= (\mathrm{ord}_{y_0}(f_0 f_1' - f_0' f_1)).$$

129

131

Hence

$$\int_0^r \frac{dt}{t} \lim_{\varepsilon \to 0} \int_{S(Z,\varepsilon)(t)} d^c \log \frac{|f'|^2}{(1+|f|^2)^2} = N_{f,\mathrm{Ram}}(r),$$

where Z is the set of singularities.

3) Adding 1 and 2 to the identity:

$$N_{p,\mathrm{Ram}}(r) = \int_0^r \frac{dt}{t} \lim_{\varepsilon \to 0} \int_{S(Z,\varepsilon)(t)} d^c \log |p'|^2$$

gives

$$N_{f,\mathrm{Ram}}(r) - N_{p,\mathrm{Ram}}(r) - 2T_f(r)$$

$$= \int_0^r \frac{dt}{t} \int_{Y(t)} dd^c \log \gamma_f + \int_0^r \frac{dt}{t} \lim_{\varepsilon \to 0} \int_{S(Z,\varepsilon)(t)} d^c \log \gamma_f.$$

4) However, $\log \gamma_f$ satisfies the conditions of Theorem 1.3 by applying Proposition 1.5 to $\log |f'|^2$ and $\log |p'|^2$ and Proposition 1.6 to $\log(1 + |f|^2)$. Hence, by Theorem 1.3 (B), the right hand side in 3 is equal to

$$\frac{1}{2} \int_{Y<r>} \log \gamma_f \sigma_Y - \frac{1}{2} \sum_{y \in Y<0>} \log \gamma_f(y)$$

$$= \frac{1}{2}[Y:C] \int_{Y<r>} \log \gamma_f \frac{\sigma_Y}{[Y:C]} - \frac{1}{2} \sum_{y \in Y<0>} \log \gamma_f(y)$$

$$[\text{Lemma I.3.5}] \quad \leq \frac{[Y:C]}{2} \log \left(\int_{Y<r>} \gamma_f \frac{\sigma_Y}{[Y:C]} \right) - \frac{1}{2} \sum_{y \in Y<0>} \log \gamma_f(y)$$

$$= \frac{1}{2}[Y:C] \log \left(\int_{Y<r>} \gamma_f \sigma_Y \right) - \frac{1}{2}[Y:C] \log[Y:C]$$

$$- \frac{1}{2} \sum_{y \in Y<0>} \log \gamma_f(y).$$

130

Notice that the degree enters into the above calculation because, in order to use Lemma I.3.5, σ_Y must be divided by the degree. This is the cause of the multiplicative factor of the degree appearing in front of the error term.

5) From Theorem 2.2, $F_{\gamma_f} = T_f$, so

$$\log \left(\int_{Y<r>} \gamma_f \sigma_Y \right) \le S(T_f, \psi, r),$$

by Lemma 3.2 for $r \ge r_1(T_f)$ and outside a set of measure $\le 2b_0(\psi)$.
QED

III, §5. A GENERAL SECOND MAIN THEOREM

In this section, some of the curvature calculations in Chapter I, §5, are modified slightly for the covering case, and then a general second main theorem is stated and proved in two forms. One desires a theorem, as in the case without a divisor, where the error term can be expressed as the degree multiplied by an expression which is independent of the degree. This is not what is obtained here. The second main theorem, in the first form stated here, has an error term which is expressed as the degree multiplied by an expression independent of the degree, but then added to a term essentially of the form

$$[Y:\mathbf{C}] \log \log [Y:\mathbf{C}]$$

when the type-function ψ is specialized to $(\log u)^{1+\varepsilon}$. In its second form, the second main theorem is stated with an error term of the desired form, but the inequality is only valid outside an exceptional set for $r \ge r_2$, where r_2 is a number which depends on the degree.

Henceforth, the covering map $p: Y \to \mathbf{C}$ will be assumed to be such that $p'(y) \ne 0$, for all $y \in Y<0>$ (i.e. Y is unramified above zero).

131

Let $f: Y \to \mathbf{P}^1$ be a non-constant holomorphic map such that $f(y) \neq 0, \infty$ and $f'(y) \neq 0$ for all $y \in Y{<}0{>}$.

Recall

$$\gamma_f = \frac{|f'|^2}{\left(1 + |f|^2\right)^2 |p'|^2}.$$

Lemma 5.1. *Let $\lambda \in \mathbf{R}_{>0}$ and let $a \in \mathbf{P}^1$. Then*

$$dd^c \|f, a\|^{2\lambda} = \left[\lambda^2 \|f, a\|^{-2(1-\lambda)} - \lambda(\lambda+1)\|f, a\|^{2\lambda} \right] \gamma_f \Phi_Y.$$

Proof: The definition of γ_f was chosen so that in local coordinates, $\gamma_f \Phi_Y$ looks just like the case of the complex plane. Hence with the new symbols, the proof of Lemma I.5.4 proves the current lemma without modification.
QED

Lemmma 5.2. *Let $\lambda \in (0,1)$ and $a \in \mathbf{P}^1$. Then,*

$$\lambda^2 \|f, a\|^{-2(1-\lambda)} \gamma_f \Phi_Y \leq 4 dd^c \log(1 + \|f, a\|^{2\lambda}) + 12\lambda \gamma_f \Phi_Y.$$

Proof: Again, the symbols have been defined so that the proof of Lemma I.5.3 works as is.
QED

Proposition 5.3. *Let $a \in \mathbf{P}^1$ such that $f(y) \neq a$ for all $y \in Y{<}0{>}$. The function $\alpha = \log(1 + \|f, a\|^{2\lambda})$ satisfies the conditions of Theorem 1.3 and*

$$\int_0^r \frac{dt}{t} \lim_{\varepsilon \to 0} \int_{S(Z,\varepsilon)(t)} d^c \alpha = 0.$$

132

134

Proof: Since α is continuous, conditions i) and iii) of Theorem 1.3 are automatically satisfied, and since α is C^∞ away from points y such that $f(y) = a$, the remaining questions can be resolved locally. Let y_0 be a point such that $f(y_0) = a$, and let w be a complex coordinate in a neighborhood of y_0 such that $w = 0$ corresponds to the point y_0. Locally, $\|f, a\|^2$ can be written

$$\|f, a\|^2 = |w|^{2m} h(w).$$

where h is positive and C^∞. Hence, locally

$$\alpha = \log(1 + u^\lambda) \quad \text{where } u = |w|^{2m} h(w).$$

Now, let $w = \rho\, e^{i\xi}$ be local polar coordinates. Then

$$d\alpha = \frac{\lambda u^{\lambda - 1} du}{1 + u^\lambda}$$

$$du = m|w|^{2(m-1)} h(w)(\bar{w}\, dw + w\, d\bar{w}) + |w|^{2m} dh$$

$$\Rightarrow \quad |u^{\lambda-1} du| \leq M |w|^{2m\lambda - 1} \qquad \text{for some constant } M$$

$$\Rightarrow \quad |d\alpha \wedge \sigma_Y| \leq M' \frac{|w|^{2m\lambda - 1}}{1 + |w|^{2m\lambda}} |dw \wedge d\bar{w}|$$

$$\Rightarrow \quad |d\alpha \wedge \sigma_Y| \leq M'' \frac{\rho^{2m\lambda - 1}}{1 + \rho^{2m\lambda}} \rho\, d\rho d\xi.$$

Therefore, $|d\alpha \wedge \sigma_Y|$ is absolutely integrable in a neighborhood of y_0. Hence α satisfies the conditions of Theorem 1.3.

The statement that

$$\lim_{\varepsilon \to 0} \int_{S(y_0, \varepsilon)} d^c \alpha = 0$$

follows from the proof of Proposition I.2.6.
QED

133

Now the degree enters into a calculation. The fact that the following estimate depends on the degree is precisely what prevents us from obtaining the error term in its desired form. When r is large compared with the degree, this problem can be overcome as will be seen in the proof of the second version of the second main theorem.

Proposition 5.4. *Let $a \in \mathbf{P}^1$. Let Λ be a decreasing function of r with $0 < \Lambda < 1$. Let*

$$\alpha_\Lambda = \|f, a\|^{-2(1-\Lambda)} \gamma_f.$$

Then

$$F_{\alpha_\Lambda} \leq \frac{[Y:\mathbf{C}]\log 4}{\Lambda^2} + \frac{12}{\Lambda} T_f(r).$$

Proof: Let $\lambda \in (0,1)$, and let $\alpha_\lambda = \|f, a\|^{-2(1-\lambda)} \gamma_f$. Then

$$F_{\alpha_\lambda}(r) = \int_0^r \frac{dt}{t} \int_{Y(t)} \|f, a\|^{-2(1-\lambda)} \gamma_f \Phi_Y$$

$$\leq \frac{4}{\lambda^2} \int_0^r \frac{dt}{t} \int_{Y(t)} dd^c \log(1 + \|f, a\|^{2\lambda}) + \frac{12}{\lambda} \int_0^r \frac{dt}{t} \int_{Y(t)} \gamma_f \Phi_Y$$

[From Lemma 5.2]

$$= \frac{4}{\lambda^2} \int_0^r \frac{dt}{t} \int_{Y(t)} dd^c \log(1 + \|f, a\|^{2\lambda}) + \frac{12}{\lambda} T_f(r).$$

[By Theorem 2.2]

But, by Proposition 5.3 and Theorem 1.3 (B),

134

$$\int\limits_0^r \frac{dt}{t} \int\limits_{Y(t)} dd^c \log(1 + \|f, a\|^{2\lambda})$$

$$= \frac{1}{2} \int\limits_{Y<r>} \log(1 + \|f, a\|^{2\lambda})\sigma_Y - \frac{1}{2} \sum_{y \in Y<0>} \log(1 + \|f(y), a\|^{2\lambda})$$

$$\leq \frac{1}{2}[Y:C] \log 2.$$

Hence

$$F_{\alpha\lambda}(r) \leq \frac{[Y:C] \log 4}{\lambda^2} + \frac{12}{\lambda} T_f(r).$$

Since Λ is a decreasing function of r, one has for all $y \in Y[r]$

$$\|f(y), a\|^{-2(1-\Lambda(|p(y)|))} \leq \|f(y), a\|^{-2(1-\Lambda(r))}.$$

Therefore, $F_{\alpha_\Lambda}(r) \leq F_{\alpha\lambda}(r)$ where $\lambda = \Lambda(r)$.

QED

Let $q \geq 1$ and let $a_1, ..., a_q$ be a finite set of distinct points in $\mathbf{P}^1$
Assume that $f(y) \neq a_j$ for all j and all $y \in Y<0>$.
Let:

$r_1 \geq 1$ such that $T_f(r_1) \geq e$;

Λ be a decreasing function of r with $0 < \Lambda < 1$;

$$\gamma_\Lambda = \prod_{j=1}^q \|f, a_j\|^{-2(1-\Lambda)} \gamma_f;$$

$$\alpha_\Lambda = \sum_{j=1}^q \|f, a_j\|^{-2(1-\Lambda)} \gamma_f.$$

135

For the convenience of the reader, Lemma I.4.2 is restated here.

Lemma 5.5. *Let:*

$$b_3 = s^{-2(q-1)} \quad where \quad s = \frac{1}{3} \min_{i \neq j} \|a_i, a_j\|$$

Then $\gamma_\Lambda \leq b_3 \alpha_\Lambda$, and b_3 depends only on $a_1, ..., a_q$.

Lemma 5.6. *Let*

$$\Lambda_1(r) = \begin{cases} \dfrac{1}{qT_f(r)} & for\ r \geq r_1 \\ constant & for\ r \leq r_1. \end{cases}$$

Let r_2 be such that $qT_f(r) > [Y:C]^{1/2}$ for all $r \geq r_2$, and let

$$\Lambda_2(r) = \begin{cases} \dfrac{[Y:C]^{1/2}}{qT_f(r)} & for\ r \geq r_2 \\ constant & for\ r \leq r_2. \end{cases}$$

Let $B = q^3 \log 4 + 12q^2$. (Note: B depends only on q.) Then

$$\log \int_{Y<r>} \gamma_{\Lambda_1} \sigma_Y \leq S([Y:C]BT_f^2, b_1(T_f), \psi, r) + \log b_3$$

for $r \geq r_1$, outside a set of measure $\leq 2b_0(\psi)$ and

$$\log \int_{Y<r>} \gamma_{\Lambda_2} \sigma_Y \leq S(BT_f^2, b_1(T_f), \psi, r) + \log b_3$$

for $r \geq r_2$, outside a set of measure $\leq 2b_0(\psi)$, where b_3 is the constant of Lemma 5.5, which depends only on $a_1, ..., a_q$.

Proof: Let $\Lambda = \Lambda_1$ or Λ_2. Then

$$\log \int_{Y<r>} \gamma_\Lambda \sigma_Y \leq \log \int_{Y<r>} \alpha_\Lambda \sigma_Y + \log b_3 \qquad [\text{By Lemma 5.5}]$$

$$\leq S(F_{\alpha_\Lambda}, b_1(F_{\alpha_\Lambda}), \psi, r) + \log b_3$$

136

138

for $r \geq r_1(F_{\alpha_\Lambda})$ outside a set of measure $\leq 2b_0(\psi)$. But, $\alpha_\Lambda \geq \gamma_f$ so $F_{\alpha_\Lambda} \geq T_f$, which implies that $r_1(F_{\alpha_\Lambda}) \leq r_1(T_f)$, and similarly, $\alpha_\Lambda \geq \gamma_f$ implies $b_1(F_{\alpha_\Lambda}) \leq b_1(T_f)$. Hence

$$S(F_{\alpha_\Lambda}, b_1(F_{\alpha_\Lambda}), \psi, r) \leq S(F_{\alpha_\Lambda}, b_1(T_f), \psi, r) \quad \text{for} \quad r \geq r_1(T_f).$$

By Proposition 5.4,

$$F_{\alpha_\Lambda} \leq \frac{q[Y:C]\log 4}{\Lambda^2} + \frac{12q}{\Lambda}T_f.$$

For $r \geq r_1$, substituting Λ_1 for Λ, one has

$$([Y:C]q^3 \log 4 + 12q^2)\, T_f^2 \leq [Y:C]BT_f^2,$$

and for $r \geq r_2$, substituting Λ_2, one has

$$\left(q^3 \log 4 + \frac{12q^2}{[Y:C]^{1/2}} \right) T_f^2 \leq BT_f^2,$$

where $B = q^3 \log 4 + 12q^2$.
QED

Theorem 5.7 (Second Main Theorem). *Let* $p:Y \to C$ *be a proper, surjective holomorphic map such that* $p'(y) \neq 0$ *for all* $y \in Y<0>$. *Let* $f:Y \to P^1$ *be a non-constant holomorphic map such that* $f(y) \neq 0, \infty$ *and* $f'(y) \neq 0$ *for all* $y \in Y<0>$. *Let:*

$$\delta(Y/C) = \frac{1}{2}[Y:C]\log[Y:C] - [Y:C]^{1/2};$$

$$S_1(F, c, \psi, r) = \log \psi(F(r)) + \log \psi(crF(r)\psi(F(r)));$$

$$B = q^3 \log 4 + 12q^2;$$

$$b_3 = \text{ the constant of Lemma 5.5};$$

$$\gamma_f = \frac{|f'|^2}{\left(1 + |f|^2\right)^2 |p'|^2}.$$

137

Then, for all $r \geq r_1$ outside a set of measure $\leq 2b_0(\psi)$ and for all $b_1 \geq b_1(T_f)$, one has (first version)

$$(q-2)T_f(r) - \sum_{j=1}^{q} N_f(a_j, r) + N_{f,\mathrm{Ram}}(r) - N_{p,\mathrm{Ram}}(r)$$

$$\leq \frac{1}{2}[Y:\mathbf{C}]\{\log(BT_f^2) + S_1([Y:\mathbf{C}]BT_f^2, b_1, \psi, r)\}$$

$$+ \frac{1}{2}[Y:\mathbf{C}]\log b_3 - \frac{1}{2}\sum_{y \in Y<0>} \log \gamma_f(y) + 1.$$

Furthermore, for all $r \geq r_2$ outside of a set of measure $\leq 2b_0(\psi)$ and for all $b_1 \geq b_1(T_f)$, one has (second version)

$$(q-2)T_f(r) - \sum_{j=1}^{q} N_f(a_j, r) + N_{f,\mathrm{Ram}}(r) - N_{p,\mathrm{Ram}}(r)$$

$$\leq \frac{1}{2}[Y:\mathbf{C}]S(BT_f^2, b_1, \psi, r) + \frac{1}{2}[Y:\mathbf{C}]\log b_3$$

$$- \frac{1}{2}\sum_{y \in Y<0>} \log \gamma_f(y) - \delta(Y/\mathbf{C}).$$

Remark: The constant B depends only on q, and the term

$$\frac{1}{2}[Y:\mathbf{C}]\log b_3 - \frac{1}{2}\sum_{y \in Y<0>} \log \gamma_f(y)$$

depends on the degree, the points $a_1, ..., a_q$, and the values of f, f', and p' at the points in Y which lie above zero. Furthermore, the term $\delta(Y/\mathbf{C})$, which appears in the second version, is bounded from below by -1 and is positive when $[Y:\mathbf{C}] \geq 4$, in which case the inequality is improved. A third version of the second main theorem could be stated in which the size of the exceptional set depends on the degree. The S_1 terms in the first version, since they all involve the function ψ, can be viewed as correcting for the fact that the size of the exceptional set

138

140

does not grow with the degree, as in the second version. It is not known whether the dependence on the degree in the S_1 terms is necessary in the sharpest form of the inequality possible if the exceptional set is to remain uniform in size for all coverings.

Proof: From Theorem 2.2,

$$T_f(r) = \int_0^r \frac{dt}{t} \int_{Y(t)} dd^c \log(1 + |f|^2).$$

Since $f - a_j$ is holomorphic,

$$\int_0^r \frac{dt}{t} \int_{Y(t)} dd^c \log|f - a_j|^2 = 0.$$

Also

$$N_f(a_j, r) - N_f(\infty, r) = \int_0^r \lim_{\varepsilon \to 0} \int_{S(Z,\varepsilon)(t)} d^c \log|f - a_j|^2,$$

and by the proof of Theorem 2.2,

$$N_f(\infty, r) = -\int_0^r \frac{dt}{t} \lim_{\varepsilon \to 0} \int_{S(Z,\varepsilon)(t)} d^c \log(1 + |f|^2).$$

Adding the above and using the definition of $\| \ , \ \|$, one gets

$$T_f(r) - N_f(a_j, r) =$$

$$-\int_0^r \frac{dt}{t} \int_{Y(t)} dd^c \log\|f, a_j\|^2 - \int_0^r \frac{dt}{t} \lim_{\varepsilon \to 0} \int_{S(Z,\varepsilon)(t)} d^c \log\|f, a_j\|^2.$$

If λ is a constant, one has the following:

$$(1 - \lambda)T_f(r) - (1 - \lambda)N_f(a_j, r)$$

$$= -\int_0^r \frac{dt}{t} \int_{Y(t)} dd^c \log\|f, a_j\|^{2(1-\lambda)} - \int_0^r \frac{dt}{t} \lim_{\varepsilon \to 0} \int_{S(Z,\varepsilon)(t)} d^c \log\|f, a_j\|^{2(1-\lambda)}.$$

139

where $Z = \{y \in Y : f(y) = a_j\}$. Applying Theorem 4.1 and its proof to the above yields

$$q(1-\lambda)T_f(r) - (1-\lambda)\sum_{j=1}^{q} N_f(a_j, r) + N_{f,\mathrm{Ram}}(r) - N_{p,\mathrm{Ram}}(r) - 2T_f(r)$$

$$= \int_0^r \frac{dt}{t} \int_{Y(t)} dd^c \log \gamma_\lambda + \int_0^r \frac{dt}{t} \lim_{\varepsilon \to 0} \int_{S(Z,\varepsilon)(t)} d^c \log \gamma_\lambda$$

$$= \frac{1}{2} \int_{Y<r>} \log \gamma_\lambda \sigma_Y - \frac{1}{2} \sum_{y \in Y<0>} \log \gamma_\lambda(y)$$

$$\leq \frac{1}{2} \int_{Y<r>} \log \gamma_\lambda \sigma_Y - \frac{1}{2} \sum_{y \in Y<0>} \log \gamma_f(y)$$

because $-\log \|\,,\,\|^{-2(1-\Lambda)} \leq 0$ since $\|\,,\,\| \leq 1$. But r remains constant in the last integral in the above inequality, so one can replace λ by one of the functions Λ_1 or Λ_2. Now the degree enters into the calculation again; using Lemma I.3.5 and putting $\Lambda = \Lambda_1$ or Λ_2, one has

$$\frac{1}{2} \int_{Y<r>} \log \gamma_\Lambda \sigma_Y - \sum_{y \in Y<0>} \log \gamma_f(y)$$

$$\leq \frac{[Y:\mathbf{C}]}{2} \log \left(\int_{Y<r>} \gamma_\Lambda \sigma_Y \right) - \frac{[Y:\mathbf{C}]}{2} \log[Y:\mathbf{C}] - \frac{1}{2} \sum_{y \in Y<0>} \log \gamma_f(y).$$

Then, by Lemma 5.6

$$\frac{1}{2}[Y:\mathbf{C}] \log \left(\int_{Y<r>} \gamma_{\Lambda_1} \sigma_Y \right) \leq \frac{1}{2}[Y:\mathbf{C}]S([Y:\mathbf{C}]BT_f^2, b_1, \psi, r) + \frac{1}{2}[Y:\mathbf{C}] \log b_3$$

for $r \geq r_1$ outside a set of measure $\leq 2b_0(\psi)$, and

$$\frac{1}{2}[Y:\mathbf{C}] \log \left(\int_{Y<r>} \gamma_{\Lambda_2} \sigma_Y \right) \leq \frac{1}{2}[Y:\mathbf{C}]S(BT_f^2, b_1, \psi, r) + \frac{1}{2}[Y:\mathbf{C}] \log b_3$$

140

for $r \geq r_2$ outside a set of measure $\leq 2b_0(\psi)$. Now, since $-1 \leq -(1-\Lambda)$ the term in front of $\sum N_f(a_j, r)$ can be replaced by -1.

Substituting for Λ_1 when $r \geq r_1$, one has

$$q\left(1 - \frac{1}{qT_f(r)}\right)T_f(r) - \sum_j N_f(a_j, r) + N_{f,\mathrm{Ram}}(r) - N_{p,\mathrm{Ram}}(r)$$

$$-2T_f(r) \leq \frac{1}{2}[Y:C]S([Y:C]BT_f^2, b_1, \psi, r) + \frac{1}{2}[Y:C]\log b_3$$

$$-\frac{1}{2}[Y:C]\log[Y:C] - \frac{1}{2}\sum_{y \in Y<0>}\log \gamma_f(y).$$

Cancelling the $[Y:C]\log[Y:C]$ terms leaves:

$$(q-2)T_f(r) + N_{f,\mathrm{Ram}}(r) - N_{p,\mathrm{Ram}}(r) - 1 - \sum_j N_f(a_j, r)$$

$$\leq \frac{1}{2}[Y:C]\left(\log(BT_f^2) + S_1([Y:C]BT_f^2, b_1, \psi, r)\right)$$

$$+\frac{1}{2}[Y:C]\log b_3 - \frac{1}{2}\sum_{y \in Y<0>}\log \gamma_f(y)$$

which concludes the proof of the first version.

On the other hand, substituting for Λ_2 when $r \geq r_2$, one has

$$q\left(1 - \frac{[Y:C]^{1/2}}{qT_f(r)}\right)T_f(r) - \sum_j N_f(a_j, r) + N_{f,\mathrm{Ram}}(r) - N_{p,\mathrm{Ram}}(r)$$

$$-2T_f(r) \leq \frac{1}{2}[Y:C]S(BT_f^2, b_1, \psi, r) + \frac{1}{2}[Y:C]\log b_3$$

$$-\frac{1}{2}[Y:C]\log[Y:C] - \frac{1}{2}\sum_{y \in Y<0>}\log \gamma_f(y).$$

141

which leaves:

$$(q-2)T_f(r) + N_{f,\text{Ram}}(r) - N_{p,\text{Ram}}(r) - [Y:\mathbf{C}]^{1/2} - \sum_j N_f(a_j, r)$$

$$\leq \frac{1}{2}[Y:\mathbf{C}]S(BT_f^2, b_1, \psi, r) + \frac{1}{2}[Y:\mathbf{C}]\log b_3$$

$$-\frac{1}{2}[Y:\mathbf{C}]\log[Y:\mathbf{C}] - \frac{1}{2}\sum_{y \in Y<0>} \log \gamma_f(y),$$

proving the second version of the theorem.
QED

142

CHAPTER IV

EQUIDIMENSIONAL NEVANLINNA THEORY
ON COVERINGS OF C^n

As is the case with maps from C^n, the equidimensional covering case is essentially the same as the one dimensional case, once the additional necessary language for working in higher dimensions has been added. One new feature which appears here is that the covering spaces are not assumed to be manifolds, but only analytic spaces. However, as integration is defined over the regular part of the cover, this introduces only one minor technical difficulty involving the absolute integrability of functions on the cover. Second, the determinant and trace formulas show up here, much like they did in the C^n case, only this time the formulas are a bit more complicated.

IV, §1. NOTATION AND PRELIMINARIES

Let $p: Y \to C^n$ be a normal analytic covering of C^n. Therefore (Y, p) satisfies the following conditions:

a) Y is a connected locally compact Hausdorff space;

b) p is a proper, surjective, continuous map such that for all $z \in C^n$, $p^{-1}(z)$ is a finite set of points;

c) The set of singularities, denoted Y_{sing}, in Y is of complex codimension ≥ 2.

143

The subset, $Y_{\text{reg}} = Y - Y_{\text{sing}}$, is called the regular part of Y, and integration over Y is defined to be integration over the regular part. The assumption that Y is normal is precisely what is needed to ensure that Stoke's Theorem can be applied. For details see Lang [**La 5**]. For simplicity, Y is assumed to be nonsingular over the origin which means that there exists an open neighborhood U of $\mathbf{C}^n$ such that $p(Y_{\text{sing}}) \cap U = \emptyset$. The Jacobian of the map p is also assumed to be non-vanishing in a neighborhood above the origin.

Let:

$[Y : \mathbf{C}^n] = $ the degree of the covering;

$\quad z = (z_1, ..., z_n)$ be the complex coordinates of $\mathbf{C}^n$;

$$\|z\|^2 = \sum_{j=1}^{n} z_j \bar{z}_j;$$

$Y(r) = \{y \in Y : \|p(y)\| < r\};$

$Y[r] = \{y \in Y : \|p(y)\| \le r\};$

$Y\!<\!r\!> = \{y \in Y : \|p(y)\| = r\};$

Consider the following differential forms:

$\omega(z) = dd^c \log \|z\|^2;$

$\varphi(z) = dd^c \|z\|^2;$

$\sigma(z) = d^c \log \|z\|^2 \wedge \omega^{n-1}(z);$

$$\Phi(z) = \prod_{j=1}^{n} \left(\frac{\sqrt{-1}}{2\pi} dz_j \wedge d\bar{z}_j \right).$$

The pullback of these forms to Y via p will be denoted by a subscript Y:

$$\omega_Y = p^*\omega, \qquad \varphi_Y = p^*\varphi, \qquad \sigma_Y = p^*\sigma, \qquad \Phi_Y = p^*\Phi.$$

Let $\omega_{\mathbf{P}^n}$ be the Fubini-Study form on $\mathbf{P}^n$ defined in Chapter II, §2,

144

and recall that:

$$\omega = \pi^* \omega_{\mathbf{P}^{n-1}},$$

where $\pi \colon \mathbf{C}^n - \{0\} \to \mathbf{P}^{n-1}$ is the natural map representing a point in projective space by its homogeneous coordinates.

Proposition 1.1. *For $r \in \mathbf{R}_{>0}$, one has*

$$\int_{Y<r>} \sigma_Y = [Y \colon \mathbf{C}^n].$$

Proof: Outside of $Y<0>$, one has

$$d\sigma_Y = d(p^* \sigma) = p^*(d\sigma) = p^*(\omega^n) = p^*(\pi^*(\omega_{\mathbf{P}^{n-1}}^n)) = 0,$$

with the last equality holding for degree reasons. Then, by Stoke's Theorem, $\int_{Y<r>} \sigma_Y$ is independent of r. Because Y is assumed regular above zero, let r be small enough so that $p^{-1}(0)$ can be covered by $[Y \colon \mathbf{C}^n]$ disjoint open sets U_j such that $Y[r] \subseteq \bigcup U_j$ and such that $\bigcup U_j$ does not contain any of the ramification points of p. Then, $p \colon Y(r) \cap U_j \to \mathbf{C}^n$ is biholomorphic with a neighborhood of the origin, and hence

$$\int_{Y<r>\cap U_j} p^* \sigma = \int_{S(r)} \sigma = 1.$$

Summing over U_j gives the result.
QED

Again, Proposition 1.1 is the reason that the error term in the Second Main Theorem is multiplied by the degree of the covering map. As in Chapter III, the next step is to prove the Green-Jensen integral formula.

Theorem 1.2 (Green-Jensen Formula). *Let α be a C^2 function from $Y \to \mathbf{C}$ except on a negligible set of singularities Z such*

145

that $Z \cap Y{<}0{>} = \emptyset$. Assume, in addition, that the following three conditions are satisfied:

  i) $\alpha \sigma_Y$ is absolutely integrable on $Y{<}r{>}$ for all $r > 0$.

  ii) $d\alpha \wedge \sigma_Y$ is absolutely integrable on $Y[r]$ for all r.

  iii) $\displaystyle\lim_{\varepsilon \to 0} \int_{S(Z,\varepsilon)(r)} \alpha \sigma_Y = 0$ for all r,

where for sufficiently small ε, $S(Z,\varepsilon)(r)$ denotes the boundry of the tubular neighborhood of radius ε around the singularities $Z \cap Y[r]$, which is regular for all but a discrete set of values ε. Then

(A) $\displaystyle \int_0^r \frac{dt}{t} \int_{Y{<}t{>}} d^c\alpha \wedge \omega_Y^{n-1} = \frac{1}{2} \int_{Y{<}r{>}} \alpha \sigma_Y - \frac{1}{2} \sum_{y \in Y{<}0{>}} \alpha(y),$

and

(B) $\displaystyle \int_0^r \frac{dt}{t} \int_{Y(t)} dd^c\alpha \wedge \omega_Y^{n-1} + \int_0^r \frac{dt}{t} \lim_{\varepsilon \to 0} \int_{S(Z,\varepsilon)(t)} d^c\alpha \wedge \omega_Y^{n-1}$

$\displaystyle \qquad\qquad = \frac{1}{2} \int_{Y{<}r{>}} \alpha \sigma_Y - \frac{1}{2} \sum_{y \in Y{<}0{>}} \alpha(y).$

Proof: Note that (B) follows from (A) because

$$\int_{Y(t)} dd^c\alpha \wedge \omega_Y^{n-1} + \lim_{\varepsilon \to 0} \int_{S(Z,\varepsilon)(t)} d^c\alpha \wedge \omega_Y^{n-1} = \int_{Y{<}t{>}} d^c\alpha \wedge \omega_Y^{n-1}$$

by Stoke's Theorem since $d\omega_Y = 0$ and then integrating against dt/t.

1) If α and β are C^2 functions then, as in the proof of Theorem II.3.1,

$$d\alpha \wedge d^c\beta \wedge \omega_Y^{n-1} = d\beta \wedge d^c\alpha \wedge \omega_Y^{n-1}$$

146

for degree reasons.

2) Because Z is disjoint from $Y<0>$, one has

$$\lim_{\varepsilon \to 0} \int_{Y<\varepsilon>} \alpha \sigma_Y = \sum_{y \in Y<0>} \alpha(y)$$

by the proof of Proposition 1.1.

Part (A) will follow by evaluating the integral

$$\frac{1}{2} \int_{Y[r]} d(\alpha \sigma_Y)$$

in two different ways. Evaluating the integral using Stoke's Theorem gives the right hand side, and using Fubini's Theorem gives the left hand side.

Applying Stoke's Theorem, one has

$$\frac{1}{2} \int_{Y[r]} d(\alpha \sigma_Y) = \frac{1}{2} \lim_{\varepsilon \to 0} \int_{Y[r] - \left(Y(\varepsilon) \cup V(Z,\varepsilon)(r) \right)} d(\alpha \sigma_Y)$$

[Stoke's Theorem] $\quad = \dfrac{1}{2} \lim_{\varepsilon \to 0} \left[\displaystyle\int_{Y<r>} \alpha \sigma_Y - \int_{Y<\varepsilon>} \alpha \sigma_Y - \int_{S(Z,\varepsilon)(r)} \alpha \sigma_Y \right]$

[by 2 and iii] $\quad = \dfrac{1}{2} \displaystyle\int_{Y<r>} \alpha \sigma_Y - \dfrac{1}{2} \sum_{y \in Y<0>} \alpha(y),$

where $V(Z,\varepsilon)(r)$ is the tubular neighborhood of radius ε around the singularities in $Z \cap Y<r>$. On the other hand, applying Fubini's

147

Theorem gives:

$$\frac{1}{2}\int_{Y[r]} d(\alpha\sigma_Y) = \frac{1}{2}\int_{Y[r]} d\alpha \wedge \sigma_Y$$

[by the definition of σ_Y]
$$= \frac{1}{2}\int_{Y[r]} d\alpha \wedge d^c \log\|p\|^2 \wedge \omega_Y^{n-1}$$

[by 1]
$$= \frac{1}{2}\int_{Y[r]} d(\log\|p\|^2) \wedge d^c\alpha \wedge \omega_Y^{n-1}$$

$$= \int_{Y[r]} d(\log\|p\|) \wedge d^c\alpha \wedge \omega_Y^{n-1}$$

[from Fubini's Theorem and ii]
$$= \int_0^r \frac{dt}{t} \int_{Y<t>} d^c\alpha \wedge \omega_Y^{n-1}.$$

QED

Recall, that a function α on Y is **admissible** if it is a linear combination of the following three types of functions:

I. Functions which are C^∞.

II. Functions locally of the form $\log(h|g|^2)$, where g is holomorphic, and h is C^∞ and positive.

III. Functions locally of the form $\log(1+(|g|^2 h)^\lambda)$, where g is holomorphic, h is C^∞ and positive, and $0 < \lambda \le 1$.

Proposition 1.3 *Admissible functions on Y which are C^∞ in a neighborhood of $Y<0>$ satisfy the conditions of Theorem 1.2.*

Proof: Functions of type I trivially satisfy all the conditions.

Since σ_Y is C^∞ away from $Y<0>$ and since the functions in question are C^∞ near $Y<0>$, all the conditions of Theorem 1.2 can be checked locally around their singularities.

148

Let α be a function of type II. Let y_0 be a singularity for α. Locally, α can be assumed to be of the form $\log |g|^2$, where g is holomorphic and $g(y_0) = 0$.

First, assume that y_0 is in the regular part of Y. If g can be taken as a local coordinate around y_0, say w_1, then one can assume $\alpha = \log |w_1|^2$. When α is restricted to the line $w_2 = \ldots = w_n = 0$, it satisfies the conditions of Theorem 1.2 because this is simply the one-dimensional case. Therefore, α in fact satisfies the conditions of Theorem 1.2 because it does not depend on $w_2, \ldots, w_n$. If g can not be taken as a coordinate function in a neighborhood of y_0, but its divisor has normal crossings, then there exist coordinate functions $w_1, \ldots, w_n$ in a neighborhood of y_0 such that $g(w) = w_1^{m_1} \cdots w_n^{m_n}$. In this case, α also satisfies the conditions of the theorem because the log converts products into sums and by what was said above. Finally, if the divisor of g does not have normal crossings, then by resolving the singularity y_0, see [**Hi**], a neighborhood U of y_0 can be covered by an analytic space $q \colon \tilde{U} \to U$ such that q is proper and such that locally g lifts to $\tilde{g}$, a holomorphic function whose divisor has normal crossings at each point above y_0. Then, the necessary integrals can be estimated on $\tilde{U}$ by what was said above, and since q is proper, everything is still finite down below.

The same technique is used when the point y_0 is not a regular point of Y. By first resolving the singularity of p, and then, if necessary, resolving any singularities of α, a general function of type II satisfies all the conditions of Theorem 1.2.

The statements for functions of type III are proved in an analogous fashion, because the singularities of these functions are also a divisor on Y.
QED

149

Proposition 1.4 *If α is an admissible function of type I or III, which is C^∞ in a neighborhood of $Y<0>$, then*

$$\lim_{\varepsilon \to 0} \int_{S(Z,\varepsilon)(t)} d^c\alpha \wedge \omega_Y^{n-1} = 0.$$

Proof: The statement is local around the singularities of α, and since ω_Y is C^∞ away from $Y<0>$, Lemma II.3.4 applies directly.
QED

Proposition 1.5 *If α is an admissible function of type II, which is C^∞ in a neighborhood of $Y<0>$, then*

$$\lim_{\varepsilon \to 0} \int_{S(Z,\varepsilon)(t)} d^c\alpha \wedge \omega_Y^{n-1} = \int_{Z(t)} \omega_Y^{n-1}.$$

Proof: Again, the question is local around the singularities of α and ω_Y is C^∞ away from $Y<0>$, so Lemma II.3.3 gives the result.
QED

Putting Theorem 1.2 together with Proposition 1.5 yields the following.

Theorem 1.6 *If α is admissible of type II with singular set Z, and C^∞ near $Y<0>$, then*

$$\int_0^r \frac{dt}{t} \int_{Y(t)} dd^c\alpha \wedge \omega_Y^{n-1} + \int_0^r \frac{dt}{t} \int_{Z(t)} \omega_Y^{n-1}$$

$$= \frac{1}{2} \int_{Y<r>} \alpha\sigma_Y - \frac{1}{2} \sum_{y \in Y<0>} \alpha(y).$$

150

152

IV, §2. FIRST MAIN THEOREM

In this section, the Nevanlinna functions are defined, and the First Main Theorem is proved in the covering case. Henceforth, let $f: Y \to X$ be a non-degenerate (i.e. not contained in any divisor on X) holomorphic map where X is a compact n-complex dimensional manifold. Let D be a divisor on X, and L_D a line bundle with a meromorhpic section s with $(s) = D$. As in Chapter II, all line bundles are holomorphic. Let ρ be a metric on L_D. Assume that $f(y) \notin D$ for all $y \in Y<0>$. (See Chapter II, §1.)

Pre height and height

If η is a $(1,1)$ form on X, then define

$$t_{f,\eta}(t) = \int_{Y(t)} f^*\eta \wedge \omega_Y^{n-1} \quad \text{and} \quad T_{f,\eta}(r) = \int_0^r t_{f,\eta}(t)\frac{dt}{t}$$

and similarly

$$t_{f,\rho}(t) = \int_{Y(t)} f^*c_1(\rho) \wedge \omega_Y^{n-1} \quad \text{and} \quad T_{f,\rho}(r) = \int_0^r t_{f,\rho}(t)\frac{dt}{t}$$

Proximity function

Let

$$m_{f,D,\rho}^0(r) = -\frac{1}{2}\int_{Y<r>} (\log|s \circ f|_\rho^2)\sigma_Y + \frac{1}{2}\sum_{y \in Y<0>} \log|s \circ f(y)|_\rho^2$$

Counting functions

Let

$$n_{f,D}(t) = \int_{(f^*D)(t)} \omega_Y^{n-1} \quad \text{and} \quad N_{f,D}(r) = \int_0^r n_{f,D}(t)\frac{dt}{t}$$

151

Proposition 2.1. *Let ρ and ρ' be two metrics on L, then*

$$T_{f,\rho'} = T_{f,\rho} + O(1).$$

Proof: The result follows from Theorem 1.2 as in the proof of Proposition II.4.2.
QED

Proposition 2.2. *Let η and η' be two $(1,1)$ forms on X, and assume that η is positive. Then*

$$T_{f,\eta'} = O(T_{f,\eta}).$$

Proof: The result follows from the compactness of X as in Proposition II.4.3.
QED

Theorem 2.3 (First Main Theorem). *For any metric ρ on L_D,*

$$T_{f,\rho} = N_{f,D} + m^0_{f,D,\rho}.$$

Proof: Since $\log |s \circ f|^2_\rho$ is an admissible function of type II on Y, the result is a restatement of Theorem 1.6 with new notation.
QED

Let Φ be the Euclidean volume form on $\mathbf{C}^n$ and let $\Phi_Y = p^*(\Phi)$ be the pullback to a pseudo-volume form on Y. Let Ω be a volume form on X, and let γ_f be the non-negative C^∞ function such that

$$f^*\Omega = \gamma_f \Phi_Y.$$

152

154

Note that γ_f vanishes on the ramification divisor of f, defined locally by the zeros of the Jacobian determinant of f, and is singular along the ramification divisor of p.

The volume form Ω defines a metric κ on the canonical line bundle K on X. Since,

$$f^* \mathrm{Ric}\, \Omega = f^* c_1(\kappa) = dd^c \log \gamma_f$$

the **height associated to the volume form** Ω is defined as:

$$T_{f,\kappa} = \int_0^r \frac{dt}{t} \int_{Y(t)} dd^c \log \gamma_f \wedge \omega_Y^{n-1}$$

Henceforth, assume that the ramification divisor for f does not intersect $Y<0>$.

Applying Theorem 1.2 and Proposition 1.5 to the new notation gives the following theorem.

Theorem 2.4. *Assume that $p: Y \to \mathbf{C}^n$ is unramified above zero, and let $f: Y \to X$ be a non-degenerate holomorphic map such that the ramification divisor of f does not intersect $Y<0>$. Then*

$$T_{f,\kappa}(r) + N_{f,\mathrm{Ram}}(r) - N_{p,\mathrm{Ram}}(r)$$
$$= \frac{1}{2} \int_{Y<r>} (\log \gamma_f) \sigma_Y - \frac{1}{2} \sum_{y \in Y<0>} \log \gamma_f(y).$$

153

Again, the calculus lemmas of II, §5 apply to the covering case almost without change.

Lemma 3.1.

$$\Phi_Y = \frac{\|p\|^{2(n-1)}}{(n-1)!} \, d\|p\|^2 \wedge \sigma_Y.$$

Proof: This is simply the pullback of the statement:

$$\Phi = \frac{\|z\|^{2(n-1)}}{(n-1)!} \, d\|z\|^2 \wedge \sigma$$

on $\mathbf{C}^n$, which is verified by direct computation using the fact that

$$d\|z\|^2 \wedge d\|z\|^2 = d^c\|z\|^2 \wedge d^c\|z\|^2 = 0.$$

Indeed,

$$d\|z\|^2 \wedge \sigma = d\|z\|^2 \wedge d^c \log\|z\|^2 \wedge \left(dd^c \log\|z\|^2\right)^{n-1}$$

$$= \frac{d\|z\|^2 \wedge d^c\|z\|^2}{\|z\|^2} \wedge \left(\frac{dd^c\|z\|^2}{\|z\|^2} - \frac{d\|z\|^2 \wedge d^c\|z\|^2}{\|z\|^4}\right)^{n-1}$$

$$= \frac{1}{\|z\|^{2n}} \, d\|z\|^2 \wedge d^c\|z\|^2 \wedge \left(dd^c\|z\|^2\right)^{n-1}$$

$$= \frac{1}{\|z\|^{2n}} \, d\|z\|^2 \wedge d^c\|z\|^2 \wedge \varphi^{n-1}$$

$$= \frac{(n-1)!}{\|z\|^{2n}} \operatorname{tr}\left(d\|z\|^2 \wedge d^c\|z\|^2\right) \Phi$$

$$= \frac{(n-1)!}{\|z\|^{2(n-1)}} \Phi.$$

QED

154

Given a function α on Y, define the **height transform**:

$$F_\alpha(r) = \int_0^r \frac{dt}{t^{2n-1}} \int_{Y(t)} \alpha \Phi_Y$$

for $r > 0$.

Let α be a function on Y such that the following conditions are satisfied:

(a) α is continuous and > 0 except on a divisor of Y.

(b) For each r, the integral $\int_{Y<r>} \alpha \sigma_Y$ is absolutely convergent and $r \mapsto \int_{Y<r>} \alpha \sigma_Y$ is a piecewise continuous function of r.

(c) There is an $r_1 \geq 1$ such that $F_\alpha(r_1) \geq e$.

Note: F_α has positive derivative, so is strictly increasing.

Lemma 3.2 *If α satisfies* (a),(b) *and* (c) *above, then F_α is C^2 and*

$$\frac{1}{r^{2n-1}} \frac{d}{dr}(r^{2n-1} F_\alpha'(r)) = \frac{2}{(n-,1)!} \int_{Y<r>} \alpha \sigma_Y.$$

Proof:

Since $F_\alpha = \int_0^r \frac{dt}{t^{2n-1}} \int_{Y(t)} \alpha \Phi_Y$ one has $\frac{dF_\alpha}{dr} = \frac{1}{r^{2n-1}} \int_{Y(r)} \alpha \Phi_Y.$

Hence

$$r^{2n-1} \frac{dF_\alpha}{dr} = \int_{Y(r)} \alpha \Phi_Y$$

$$[\text{Lemma 3.1}] \quad = \frac{1}{(n-1)!} \int_{Y(r)} \|p\|^{2(n-1)} d\|p\|^2 \wedge \alpha \sigma_Y$$

$$[\text{Fubini's Theorem}] \quad = \frac{2}{(n-1)!} \int_0^r t^{2n-1} dt \int_{Y<t>} \alpha \sigma_Y.$$

155

Therefore

$$\frac{d}{dr}\left(r^{2n-1}\frac{dF_\alpha}{dr}\right) = \frac{2}{(n-1)!}r^{2n-1}\int_{Y<r>}\alpha\sigma_Y.$$

QED

Lemma 3.3. *If α satisfies* (a),(b) *and* (c) *above, then*

$$\log\int_{Y<r>}\alpha\sigma_Y \leq S(F_\alpha, b_1^-(F_\alpha), \psi, r) + \log\frac{(n-1)!}{2}$$

for all $r \geq r_1(F_\alpha)$ outside a set of measure $\leq 2b_0(\psi)$.

Proof: Apply Lemma 3.2 and Lemma II.5.1.

QED

IV, §4. SECOND MAIN THEOREM WITHOUT A DIVISOR

Given a $(1,1)$ form η on Y, define the **trace** and **determinant** outside the ramification points of p as follows:

$$(\det(\eta))\,\Phi_Y = \frac{1}{n!}\eta^n$$
$$(n-1)!\,\mathrm{tr}(\eta)\,\Phi_Y = \eta\wedge\varphi_Y^{n-1}.$$

Lemma 4.1. *If η is a semi-positive $(1,1)$ form on Y, then*

$$(\det(\eta))^{1/n} \leq \frac{1}{n}\mathrm{tr}(\eta)$$

for the regular points in Y which are not ramification points of p.

Proof: This is a point-wise condition, and if y_0 is a regular point of Y which is not a ramification point for p, then p is biholomorphic

156

in a neighborhood of y_0. Since everything is defined via pull-back, the statement follows from the relationship on $\mathbf{C}^n$. See the remark preceeding Theorem II.6.3.

QED

Let η be a closed, positive $(1,1)$ form such that

$$\Omega = \frac{1}{n!}\eta^n.$$

Since $f^*\Omega = \gamma_f \Phi_Y$, one finds that $\gamma_f = \det(f^*\eta)$.

Proposition 4.2. *Let $\tau_f = \operatorname{tr}(f^*\eta)$. Then*

$$T_{f,\eta} = (n-1)! F_{\tau_f}.$$

Proof: All the symbols have been defined so that the proof is identical to that of Proposition II.6.2.

QED

Theorem 4.3. *Assume that $p\colon Y \to \mathbf{C}^n$ is unramified above zero, and let $f\colon Y \to X$ be a non-degenerate holomorphic map which is also unramified above zero. Let $T_{f,\kappa}$ be the height associated to the volume form Ω on X. Then*

$$T_{f,\kappa}(r) + N_{f,\mathrm{Ram}}(r) - N_{p,\mathrm{Ram}}(r) + \frac{n}{2}[Y\colon\mathbf{C}^n]\log[Y\colon\mathbf{C}^n]$$

$$\leq [Y\colon\mathbf{C}^n]\frac{n}{2}S(T_{f,\eta}, b_1(T_{f,\eta}/(n-1)!), \psi, r) - \frac{1}{2}\sum_{y\in Y<0>}\log\gamma_f(y)$$

for all $r \geq r_1(T_{f,\eta}/(n-1)!)$ outside a set of measure $\leq 2b_0(\psi)$.

Remark: The term on the right involving $\log\gamma_f$ in the above inequality depends only on the values of f, the Jacobian of f, and the Jacobian of p above zero, so the right hand side is uniform in f and p if they are normalized above zero. Furthermore, the term with $[Y\colon\mathbf{C}^n]\log[Y\colon\mathbf{C}^n]$ is positive and therefore actually improves the inequality.

157

Proof:

$$T_{f,\kappa}(r) + N_{f,\mathrm{Ram}}(r) - N_{p,\mathrm{Ram}}(r) + \frac{1}{2} \sum_{y \in Y<0>} \log \gamma_f(y)$$

$$= \frac{1}{2} \int_{Y<r>} (\log \gamma_f) \sigma_Y \qquad \text{[Theorem 2.4]}$$

$$= \frac{n}{2} \int_{Y<r>} \log \gamma_f^{1/n} \sigma_Y$$

$$\leq [Y:\mathbf{C}^n] \frac{n}{2} \log \int_{Y<r>} \gamma_f^{1/n} \sigma_Y - [Y:\mathbf{C}^n] \frac{n}{2} \log[Y:\mathbf{C}^n]$$

$$\text{[Lemma I.3.5]}$$

$$\leq [Y:\mathbf{C}^n] \frac{n}{2} \log \int_{Y<r>} \tau_f \sigma_Y - [Y:\mathbf{C}^n] \frac{n}{2} \log[Y:\mathbf{C}^n]$$

$$\text{[Proposition 4.1]}$$

$$\leq [Y:\mathbf{C}^n] \frac{n}{2} S(F_{\tau_f}, b_1(F_{\tau_f}), \psi, r) + \log(n-1)!$$

$$- [Y:\mathbf{C}^n] \frac{n}{2} \log[Y:\mathbf{C}^n]$$

$$\text{[Proposition 3.3]}$$

$$\leq [Y:\mathbf{C}^n] \frac{n}{2} S(T_{f,\eta}, b_1(T_{f,\eta}/(n-1)!), \psi, r) - [Y:\mathbf{C}^n] \frac{n}{2} \log[Y:\mathbf{C}^n]$$

$$\text{[Proposition 4.2]}$$

for all $r \geq r_1(F_{\tau_f}) = r_1(T_{f,\eta}/(n-1)!)$ outside a set of measure $\leq 2b_0(\psi)$.
QED

IV, §5. A GENERAL SECOND MAIN THEOREM

In this section, a more general second main theorem, involving a divisor, is stated and proved. Ideally, one whould have a theorem where the error term can be expressed as the degree multiplied by an expression which is independent of the degree. This is not quite what is obtained here. The second main theorem is stated in this section in

158

two forms. In the first form, the error term is expressed as the degree multiplied by an expression independent of the degree added to a term essentially of the form

$$[Y:C^n]\log\log[Y:C^n]$$

when the type function ψ is specialized to $(\log u)^{1+\epsilon}$. In its second form, the second main theorem is stated with an error term which is of the desired form, but the inequality is only required to hold outside of an exceptional set for $r \geq r_2$, where r_2 is a number which depends on the degree.

For the rest of this section, let:

X be a compact n-complex dimensional manifold;

Ω be a volume form on X with associated metric κ on K the canonical line bundle;

$T_{f,\kappa} = T_{f,\mathrm{Ric}\,\Omega}$ be the associated height;

γ_f be the function such that $f^*\Omega = \gamma_f \Phi_Y$;

$D = \sum_{j=1}^q D_j$ be a divisor on X with simple normal crossings of complexity k;

$L_j = L_{D_j}$ the holomorphic line bundle associated to D_j with hermitian metric ρ_j;

η be a closed, positive $(1,1)$ form on X such that $\eta \geq c_1(\rho_j)$ for all j, and $\eta^n/n! \geq \Omega$;

s_j be a holomorphic section of L_j such that $(s_j) = D_j$;

Since X is compact, after possibly multiplying s_j by a constant, assume without loss of generality that

$$|s_j|_{\rho_j} \leq 1/e \leq 1.$$

For convenience, also assume that $f(y) \notin D$ for all $y \in Y<0>$, and that $Y<0>$ does not intersect the ramification divisor of f. If λ is a constant with $0 < \lambda < 1$, then define the **Ahlfors-Wong** singular volume form

$$\Omega(D)_\lambda = \left(\prod |s_j|_j^{-2(1-\lambda)}\right)\Omega,$$

and define

$$\gamma_\lambda = \prod |s_j \circ f|_j^{-2(1-\lambda)}\gamma_f.$$

Given Λ a positive decreasing function of r with $0 < \Lambda < 1$, define

$$\gamma_\Lambda = \prod |s_j \circ f|_j^{-2(1-\Lambda)}\gamma_f.$$

Note that because of the assumption $|s_j|_j \leq 1/e \leq 1$, one has $\gamma_f \leq \gamma_\Lambda$.

The next lemma uses the curvature computation of Lemma II.7.4, and is similar to Lemma II.7.3, but since a factor involving the deree enters the calculation, it is repeated here. The appearance of the degree in this lemma is the obstacle to obtaining an error term in the second main theorem that can be expressed as the degree multiplied by an expression independent of the degree.

Lemma 5.1 *Let b be the constant of Lemma II.7.4, which depends only on Ω, D and η. Then for any decreasing function Λ with $0 < \Lambda < 1$, one has*

$$F_{\gamma_\Lambda^{1/n}}(r) \leq (q+1)\frac{b^{1/n}}{n!}\frac{T_{f,\eta}(r)}{(\Lambda(r))^{k/n}} + \frac{qb^{1/n}}{2n!}\frac{[Y:\mathbf{C}^n]\log 2}{(\Lambda(r))^{1+k/n}}$$

for all r.

Proof: Let $0 < \lambda < 1$ be constant. From Lemma II.7.4, the constant b, depending only on Ω, D and η (and in particular, not λ) is such that

$$\lambda^{n+k}\Omega(D)_\lambda \leq \frac{b}{n!}\eta_{D,\lambda}^n,$$

where $\eta_{D,\lambda}$ is the $(1,1)$ form on X given by

$$\eta_{D,\lambda} = (q+1)\lambda\eta + \sum_{j=1}^{q} dd^c \log(1 + |s_j|_j^{2\lambda}),$$

160

162

and $\eta_{D,\lambda}$ is closed and positive outside of D. Pulling this back via f gives

$$\lambda^{n+k}\gamma_\lambda \Phi_Y \le b(\det(f^*\eta_{D,\lambda}))\Phi_Y,$$

and hence

$$\lambda^{1+k/n}\gamma_\lambda^{1/n}\Phi_Y \le b^{1/n}(\det(f^*\eta_{D,\lambda}))^{1/n}\Phi_Y$$

$$[\text{Proposition 4.1}] \qquad \le \frac{b^{1/n}}{n}(\text{tr}(f^*\eta_{D,\lambda}))\Phi_Y$$

$$= \frac{b^{1/n}}{n!}f^*\eta_{D,\lambda} \wedge \varphi_Y^{n-1}.$$

Therefore

$$\frac{n!}{b^{1/n}}\lambda^{1+k/n}F_{\gamma_\lambda^{1/n}}(r)$$

$$\le \int_0^r \frac{dt}{t^{2n-1}} \int_{Y(t)} f^*\eta_{D,\lambda} \wedge \varphi_Y^{n-1}$$

$$= \int_0^r \frac{dt}{t} \int_{Y(t)} f^*\eta_{D,\lambda} \wedge \omega_Y^{n-1}$$

$$= (q+1)\lambda \int_0^r \frac{dt}{t} \int_{Y(t)} f^*\eta \wedge \omega_Y^{n-1}$$

$$+ \sum_{j=1}^q \int_0^r \frac{dt}{t} \int_{Y(t)} dd^c \log(1 + |s_j \circ f|_j^{2\lambda}) \wedge \omega_Y^{n-1}$$

$$= (q+1)\lambda T_{f,\eta}(r)$$

$$+ \frac{1}{2}\sum_{j=1}^q \left[\int_{Y<r>} \log(1 + |s_j \circ f|_j^{2\lambda})\sigma_Y - \sum_{y \in Y<0>} \log(1 + |s_j \circ f(y)|_j^{2\lambda}) \right]$$

[Proposition 1.4 and Theorem 1.2 (B)]

$$\le (q+1)\lambda T_{f,\eta}(r) + \frac{q\log 2}{2}[Y:\mathbf{C}^n].$$

[the expression inside the log is ≤ 2]

161

163

This proves Lemma 5.1 for a constant λ, but since Λ is a decreasing function of r, one has

$$\gamma_\Lambda(y) \leq \gamma_\lambda(y) \text{ for } \|p(y)\| \leq r,$$

where $\lambda = \Lambda(r)$.

QED

In light of Lemma 5.1, two positive decreasing functions Λ_1 and Λ_2 will be defined. Let $r_1 = r_1(F_{\gamma_f^{1/n}})$ and let

$$\Lambda_1(r) = \begin{cases} \dfrac{1}{qT_{f,\eta}(r)} & \text{for } r \geq r_1 \\ \text{constant} & \text{for } r \leq r_1. \end{cases}$$

Note that since $\eta^n/n! \geq \Omega$, one has $F_{\gamma_f^{1/n}} \leq T_{f,\eta}/n!$. Therefore $r_1(F_{\gamma_f^{1/n}}) \geq r_1(T_{f,\eta}/n!)$, and hence one has $\Lambda_1 \leq 1$. Let r_2 be such that $qT_{f,\eta}(r) > [Y:C^n]^{n/(n+k)}$ for all $r \geq r_2$. Let

$$\Lambda_2(r) = \begin{cases} \dfrac{[Y:C^n]^{n/(n+k)}}{qT_{f,\eta}(r)} & \text{for } r \geq r_2 \\ \text{constant} & \text{for } r \leq r_2 \end{cases}$$

Note that r_2 was chosen so that $\Lambda_2 < 1$.

Lemma 5.2 *Let b be the constant of Lemma 5.1 and let*

$$B = \frac{b^{1/n}}{n}\left((q+1)q^{k/n} + \frac{1}{2}q^{2+k/n}\log 2\right).$$

Then

$$F_{\gamma_{\Lambda_1}^{1/n}}(r) \leq \frac{B}{(n-1)!}[Y:C^n]T_{f,\eta}^{1+k/n}$$

for $r \geq r_1$, and

$$F_{\gamma_{\Lambda_2}^{1/n}}(r) \leq \frac{B}{(n-1)!}T_{f,\eta}^{1+k/n}$$

for $r \geq r_2$.

162

164

Proof: The first statement follows from applying Lemma 5.1 to Λ_1, and the second statement follows from applying Lemma 5.1 to Λ_2 since the numerator in Λ_2 was chosen to exactly cancel the appearance of the degree in Lemma 5.1.
QED

Lemma 5.3 *One has*

$$\log \int_{Y<r>} \gamma_{\Lambda_1}^{1/n} \sigma_Y \leq S([Y:C^n]BT_{f,\eta}^{1+k/n}, b_1, \psi, r)$$

for all $r \geq r_1$ *outside a set of measure* $\leq 2b_0(\psi)$, *and*

$$\log \int_{Y<r>} \gamma_{\Lambda_2}^{1/n} \sigma_Y \leq S(BT_{f,\eta}^{1+k/n}, b_1, \psi, r)$$

for all $r \geq r_2$, *outside a set of measure* $\leq 2b_0(\psi)$, *where*

$$B = \frac{b^{1/n}}{n}((q+1)q^{k/n} + \frac{1}{2}q^{2+k/n}\log 2)$$
$$b_1 = b_1(F_{\gamma_f^{1/n}}) \quad and \quad r_1 = r_1(F_{\gamma_f^{1/n}})$$

Proof: Let $\Lambda = \Lambda_1$ or Λ_2. Because $\gamma_\Lambda \geq \gamma_f$, one has

$$F_{\gamma_\Lambda^{1/n}} \geq F_{\gamma_f^{1/n}} \qquad and \qquad F'_{\gamma_\Lambda^{1/n}} \geq F'_{\gamma_f^{1/n}}$$

Hence $b_1 = b_1(F_{\gamma_f^{1/n}})$ and $r_1 = r_1(F_{\gamma_f^{1/n}})$ are such that for $r \geq r_1$,

$$F_{\gamma_\Lambda^{1/n}}(r) \geq e \qquad and \qquad b_1 r^{2n-1} F'_{\gamma_\Lambda^{1/n}} \geq e.$$

From Lemma 3.3, one has

$$\log \int_{Y<r>} \gamma_\Lambda^{1/n} \sigma_Y \leq S(F_{\gamma_\Lambda^{1/n}}, b_1, \psi, r) + \log \frac{(n-1)!}{2}$$

163

for $r \geq r_1$ outside an exceptional set of measure $\leq 2b_0(\psi)$. Now from Lemma 5.2, one has

$$S(F_{\gamma_{\Lambda_1}^{1/n}}, b_1, \psi, r) + \log \frac{(n-1)!}{2} \leq S([Y : C^n] BT_{f,\eta}^{1+k/n}, b_1, \psi, r)$$

for $r \geq r_1$, and

$$S(F_{\gamma_{\Lambda_2}^{1/n}}, b_1, \psi, r) + \log \frac{(n-1)!}{2} \leq S(BT_{f,\eta}^{1+k/n}, b_1, \psi, r)$$

for $r \geq r_2$.
QED

Remark: The $(n-1)!$ in the definition of the trace cancels the $\log(n-1)!$ in Lemma 3.3.

Theorem 5.4. *Assume $p : Y \to C^n$ is unramified above zero. Let $f : Y \to X$ be a non-degenerate holomorphic map such that f is unramified above zero. Let D, η, κ, and ρ_j be as in the beginning of this section, and assume that $f(y) \notin D$ for all $y \in Y{<}0{>}$. Let:*

$$\delta(Y/C^n, k) = \frac{n}{2}[Y : C^n] \log[Y : C^n] - [Y : C^n]^{k/(n+k)};$$

$$S_1(F, c, \psi, r) = \log \psi(F(r)) + \log \psi(cr^{2n-1} F(r) \psi(F(r)));$$

$$B = \frac{b^{1/n}}{n}((q+1)q^{k/n} + \frac{1}{2}q^{2+k/n} \log 2);$$

$$b_1 = b_1(F_{\gamma_f^{1/n}}) \quad and \quad r_1 = r_1(F_{\gamma_f^{1/n}}),$$

where b is the constant of Lemma 5.1. Then, one has (first version)

$$T_{f,\kappa}(r) + \sum_{j=1}^{q} T_{f,\rho_j}(r) - N_{f,D}(r) + N_{f,\mathrm{Ram}}(r) - N_{p,\mathrm{Ram}}(r)$$

$$\leq \frac{n}{2}[Y : C^n]\{\log(BT_{f,\eta}^{1+k/n}) + S_1([Y : C^n]BT_{f,\eta}^{1+k/n}, \psi, b_1, r)\}$$

$$- \frac{1}{2} \sum_{y \in Y{<}0{>}} \log \gamma_f(y) + 1$$

164

for $r \geq r_1$ outside of a set of measure $\leq 2b_0(\psi)$. Furthermore, one has (second version)

$$T_{f,\kappa}(r) + \sum_{j=1}^{q} T_{f,\rho_j}(r) - N_{f,D}(r) + N_{f,\mathrm{Ram}}(r) - N_{p,\mathrm{Ram}}(r)$$

$$\leq \frac{n}{2}[Y:\mathbf{C}^n]S(BT_{f,\eta}^{1+k/n}, \psi, b_1, r) - \frac{1}{2}\sum_{y \in Y<0>} \log \gamma_f(y) - \delta(Y/\mathbf{C}^n, k)$$

for $r \geq r_2$ outside of a set of measure $\leq 2b_0(\psi)$.

Remark: Note that the constant B depends only on D and η, the term

$$\sum_{y \in Y<0>} \log \gamma_f(y)$$

depends only on the values of f, the Jacobian of f, and the Jacobian of p above zero, and the term $\delta(Y/\mathbf{C}^n, k)$, which appears in the second version of the inequality, is bounded from below by -1, and is positive when $[Y:\mathbf{C}^n] \geq 4$, in which case it actually improves the inequality. A third version of the inequality could also be stated with the size of the exceptional set growing with the degree. The relationship between these versions is best seen by noting that the term involving S_1 in the first version contains only the error terms involving the type function ψ, and, thus, in some sense this term corrects for the fact that the exceptional set is not enlarged in the first version as it is in the second (via r_2). It is not known whether the dependence on the degree can be removed from the S_1 terms in the sharpest form of the inequality if the size of the exceptional set is to remain independent of the degree.

Proof: Let λ be a constant with $0 < \lambda < 1$. Using Theorem 1.2 (B), Proposition 1.5, and the fact that $dd^c \log$ transforms products into

165

sums, one obtains:

$$T_{f,\kappa}(r) + (1-\lambda)\sum_{j=1}^{q} T_{f,\rho_j}(r) - (1-\lambda)\sum_{j=1}^{q} N_{f,D_j}(r)$$

$$+ N_{f,\mathrm{Ram}}(r) - N_{p,\mathrm{Ram}}(r)$$

$$= \int_0^r \frac{dt}{t} \int_{Y(t)} dd^c \log \gamma_\lambda + \int_0^r \frac{dt}{t} \lim_{\varepsilon \to 0} \int_{S(Z,\varepsilon)(t)} d^c \log \gamma_\lambda$$

$$= \frac{n}{2} \int_{Y<r>} \log \gamma_\lambda^{1/n} \sigma_Y - \frac{1}{2} \sum_{y \in Y<0>} \log \gamma_\lambda(y)$$

Because of the assumption that $|s_j|_j \leq 1$, one also has

$$-\frac{1}{2} \sum_{y \in Y<0>} \log \gamma_\lambda(y) \leq -\frac{1}{2} \sum_{y \in Y<0>} \log \gamma_f(y).$$

Also, since Λ_1 and Λ_2 are constant on $Y<r>$, the function $\Lambda = \Lambda_1$ or Λ_2 can replace λ in the above equality. Furthermore, $N_{f,D_j} \geq 0$ and $-1 \leq -(1-\lambda)$, so the factor $(1-\lambda)$ in front can be deleted. When $r \geq r_1$, one has

$$-1 \leq -\Lambda_1(r) \sum_{j=1}^{q} T_{f,\rho_j}(r),$$

and when $r \geq r_2$, one has

$$-[Y:\mathbf{C}^n]^{n/(n+k)} \leq -\Lambda_2(r) \sum_{j=1}^{q} T_{f,\rho_j}(r),$$

from the definitions of Λ_1 and Λ_2, and from the fact that η was chosen so that

$$T_{f,\eta} \geq T_{f,\rho_j} \quad \text{for all } j.$$

Finally, by moving the log out of the integral, one has

$$\int_{Y<r>} \log \gamma_{\Lambda(r)}^{1/n} \sigma_Y \leq [Y:\mathbf{C}^n] \log \left(\int_{Y<r>} \gamma_{\Lambda(r)}^{1/n} \sigma_Y \right) - [Y:\mathbf{C}^n] \log[Y:\mathbf{C}^n].$$

166

Applying the estimate in Lemma 5.3 for Λ_2 to the term with the integral on the right and collecting terms yields the second version of the theorem. To arrive at the first version, apply the estimate in Lemma 5.3 for Λ_1 to get

$$\log\left(\int\limits_{Y<r>} \gamma_{\Lambda_1(r)}^{1/n} \sigma_Y\right) \le S([Y:\mathbf{C}^n]BT_{f,\eta}^{1+k/n}, b_1, \psi, r)$$

for $r \ge r_1$ outside a set of measure $\le 2b_0(\psi)$. But

$$S([Y:\mathbf{C}^n]BT_{f,\eta}^{1+k/n}, b_1, \psi, r)$$
$$= \log[Y:\mathbf{C}^n] + \log(BT_{f,\eta}^{1+k/n}) + S_1([Y:\mathbf{C}^n]BT_{f,\eta}^{1+k/n}, b_1, \psi, r)$$

After multiplying by $[Y:\mathbf{C}^n](n/2)$, the $\log[Y:\mathbf{C}^n]$ term in the above equality is exactly canceled by the $\log[Y:\mathbf{C}^n]$ term which came from moving the log out of the integral above. Collecting terms then yields the first version.
QED

IV, §6. A VARIATION

In this section the analog of Theorem II.8.2, a non-equidimensional statement will be proved.

In this section, let $p: Y \to \mathbf{C}$ be a branched covering of $\mathbf{C}$ which is unramified over the origin, and let Φ_Y be the pull-back to Y of the standard Euclidean volume form on $\mathbf{C}$.

Theorem 6.1. *Let M be a complex manifold which is not necessarily compact. Let η be a closed, positive $(1,1)$ form on M, and let $f: Y(R) \to M$ be a holomorphic map which is unramified above zero. Suppose there exists a constant $B > 0$ such that*

$$Bf^*\eta \le \operatorname{Ric} f^*\eta.$$

167

Let

$$T_{f,\eta}(r) = \int\limits_0^r \frac{dt}{t} \int\limits_{Y(t)} f^*\eta \ \ and \ \ f^*\eta = \gamma_f \Phi_Y.$$

Then

$$BT_{f,\eta}(r) + N_{f,\mathrm{Ram}}(r) - N_{p,\mathrm{Ram}}(r) + \frac{1}{2}[Y:\mathbf{C}]\log[Y:\mathbf{C}]$$

$$\le \frac{1}{2}[Y:\mathbf{C}]S(T_{f,\eta},\psi,r) - \frac{1}{2}\sum_{y\in Y<0>}\log\gamma_f(y)$$

for $r \ge r_1$ *outside a set of measure* $\le 2b_0(\psi)$.

Proof: By definition, one has $F_{\gamma_f} = T_{f,\eta}$. By Stoke's Theorem,

$$T_{\mathrm{Ric}(f^*\eta)}(r) + N_{f,\mathrm{Ram}}(r) - N_{p,\mathrm{Ram}}(r)$$

$$= \frac{1}{2}\int\limits_{Y<r>}\log\gamma_f\sigma_Y - \frac{1}{2}\sum_{y\in Y<0>}\log\gamma_f(y).$$

But, by assumption,

$$Bf^*\eta \le \mathrm{Ric}\ f^*\eta, \quad so \quad BT_{f,\eta} \le T_{\mathrm{Ric}(f^*\eta)}.$$

The theorem is concluded by applying the calculus lemma to the right hand side.

QED

168

170

References

[Ad 1] W. ADAMS, *Asymptotic diophantine approximations to e*, Proc. Nat. Acad. Sci. USA **55**, (1966), 28-31.

[Ad 2] W. ADAMS, *Asymptotic diophantine approximations and Hurwitz numbers*, Amer. J. Math. **89** (1967), 1083-1108.

[A-L] W. ADAMS and S. LANG, *Some computations in diophantine approximations*, J. reine angew. Math. **220** (1965), 163-173.

[Ah] L. AHLFORS, *The theory of meromorphic curves*, Acta. Soc. Sci. Fenn. Nova Ser. A **3** (1941), 1-31.

[Br] A.D. BRYUNO, *Continued fraction expansion of algebraic numbers*, Zh. Vychisl. Mat. i Mat. Fiz. 4 nr. **2**, 211-221; translated USSR Comput. Math. and Math. Phys 4 (1964), 1-15.

[C-G] J. CARLSON and P. GRIFFITHS, *A defect relation for equidimensional holomorphic mappings between algebraic varieties*, Ann. Math. **95** (1972), 557-584.

[Ch] S.S. CHERN, *Complex analytic mappings of Riemann surfaces I*, Amer. J. Math. **LXXXII No. 2** (1960), 323-337.

[G-G] A. GOLDBERG and V. GRINSHTEIN, *The logarithmic derivative of a meromorphic function* , Mathematical Notes **19** (1976), AMS Translation, 320-323.

[Gr] P. GRIFFITHS, *Entire Holomorphic Mappings in one and Several Complex Variables*, Ann. of Math. Studies Vol 85, Princeton University Press, Princeton, NJ, 1976.

[G-K] P. GRIFFITHS and J. KING, *Nevanlinna theory and holomorphic mappings between algebraic varieties*, Acta Mathematica **130** (1973), 145-220.

[GuR] R. GUNNING and H. ROSSI, *Introduction to Holomorphic Functions of Several Variables* (Replacing *Analytic Functions of Several Complex Variables*, Prentice Hall, 1965) (coming out in 1990)

[Hi] H. HIRONAKA, *Resolution of singularities of an algebraic variety over a field of characteristic zero:* I, II, Annals of Mathematics **79** (1964), 109-326.

169

[Kh]   A. KHINTCHINE, *Continued fractions*, Chicago University Press, 1964

[La 1] S. LANG, *Report on diophantine approximations*, Bull. Soc. Math. France **93** (1965), 177-192.

[La 2] S. LANG, *Asymptotic diophantine approximations*, Proc. NAS **55 No. 1** (1966), 31-34.

[La 3] S. LANG, *Introduction to Diophantine Approximations*, Addison Wesley, 1966.

[La 4] S. LANG, *Transcendental numbers and diophantine approximations*, Bull. AMS **77 No. 5** (1971), 635-677.

[La 5] S. LANG, *Real Analysis*, Addison Wesley, 1969; Second Edition, 1983.

[La 6] S. LANG, *Differential Manifolds*, 1972; reprint by Springer Verlag, 1985.

[La 7] S. LANG, *Introduction to Complex Hyperbolic Spaces*, Springer Verlag, 1987.

[La 8] S. LANG, *The error term in Nevanlinna theory*, Duke Math. J. (1988), 193-218.

[La 9] S. LANG, *The error term in Nevanlinna theory* II, Bull. AMS (1990) pp. 115-125.

[L-T 1] S. LANG and H. TROTTER, *Continued fractions of some algebraic numbers*, J. reine angew. Math. **255** (1972), 112-123.

[L-T 2] S. LANG and H. TROTTER, *Addendum to the above*, J. reine angew. Math. **267** (1974), 219-220.

[Ne]   R. NEVANLINNA, *Analytic Functions*, Springer Verlag, 1970; (revised translation of the German edition, 1953).

[vN-T] von NEUMANN and B. TUCKERMAN, *Continued fraction expansion of* $2^{1/3}$, Math. Tables Aids Comput **9** (1955), 23-24.

[Os 1] C.F. OSGOOD, *A number theoretic-differential equations approach to generalizing Nevanlinna theory*, Indian J. of Math. **23** (1981), 1-15.

170

[Os 2] C.F. OSGOOD, *Sometimes effective Thue-Siegel Roth-Schmidt-Nevanlinna bounds, or better*, J. Number Theory **21** (1985), 347-389.

[RDM] R. RICHTMYER, M. DEVANEY and N. METROPOLIS, *Continued fraction expansions of algebraic numbers*, Numer. Math. **4** (1962), 68-84.

[Sc] W. SCHMIDT, *Diophantine Approximations*, Lecture Notes in Mathematics, Springer Verlag, 1980.

[Sh] S.V. SHABAT, *Distribution of Values of HolomorphicMappings*, translated from the Russian, AMS, 1985 (Russian edition 1982).

[St] W. STOLL, *The Ahlfors-Weyl theory of meromorphic maps on parabolic manifolds*, Lecture Notes in Mathematics 981, Springer Verlag, 1981.

[Vo 1] P. VOJTA, *Diophantine Approximations and Value Distribution Theory*, Lecture Notes in Mathematics 1239, Springer Verlag, 1987.

[Vo 2] P. VOJTA, *A refinement of Schmidt's subspace theorem*, Am. J. Math 111(1989), pp. 489-518.

[Wo] P.M. WONG, *On the second main theorem of Nevanlinna theory*, Am. J. Math. 111(1989), pp. 549-583.

Added in proof: I thank Alexander Eremenko for drawing my attention to the paper "The Logarithmic Derivative of a Meromorphic Function" by A.A. Goldberg and V.A. Grinshtein. Mathematical Notes Vol. 19, 1976, AMS translation pp. 320–323. In that paper they obtain a very good error term for the logarithmic derivative.

INDEX

Adams, 10

Admissible functions, 70, 148

Ahlfors-Shimizu, 26, 124

Ahlfors-Wong, 37, 47, 91, 94, 160

Calculus Lemma, 29, 85

Carlson-Griffiths, 11, 57, 78, 92, 101

Chern, 11, 40

Chern form, 62

Complexity, 92

Counting function, 15, 78, 121, 151

Cover, 111, 113, 143

Curvature, 45, 65, 97, 99

Degree, 113, 144

Determinant, 87, 156

Error term, 30, 40, 47, 85, 91, 102 112, 131, 158

Euclidean form, 20, 32, 63, 123, 154

First Main Theorem, 19, 83, 122, 152

Fubini-Study form, 20

Fubini-Study metric, 63

Fubini theorem, 67

Gauss curvature, 65

Green-Jensen, 24, 115, 145

Griffiths, 102

Griffiths function, 65

Griffiths-King, 101, 111

Height, 9, 19, 79, 81, 84, 105, 121, 151, 153

Height transform, 32, 86, 126, 155

Khintchine, 9, 30

Log derivative, 48, 102

Mean proximity, 18, 78, 81, 151

Metric, 60

Negligible, 71

173

175

Nevanlinna functions, 15, 19, 78, 80, 121, 151

Normal, 143

Osgood, 9

Poincaré lemma, 74

Poisson-Jensen, 12, 13, 120

Positive metric, 62

Proximity function, 18, 78, 81, 151

Ramification divisor, 34, 83, 105

Ramification index, 34, 128

Ramification theorem, 35

Regular part, 25

Ricci form, 64, 83

Richtmyer, 10

Roth, 10

Second Main Theorem, 36, 55, 90, 93, 12 137, 157, 164

Simple normal crossings, 92

Singular part, 25, 73

Stokes' theorem, 23, 70

Stoll, 37, 47, 76, 94, 111

Totally ramified, 49

Trace, 87, 156

Type, 9, 42

Vojta, 5, 10, 104, 106, 111

Volume form, 64

Weil function, 18

Wong, 11, 37, 40, 44, 47, 53, 91, 102, 11

Wronskian, 21

Ye, 53

174

Notices AMS (1995) 339–350

Mordell's Review, Siegel's Letter to Mordell, Diophantine Geometry, and 20th Century Mathematics

Serge Lang

In 1962, I published *Diophantine Geometry*. Mordell reviewed this book [Mor 1964],[1] and the review became famous. Immediately after the review appeared, in that same year, Siegel wrote a letter to Mordell to express his agreement with Mordell's review concerning the overall nature of the book, and to express more generally his negative reaction to trends in mathematics of the 1950's and 1960's. I learned of Siegel's letter to Mordell only in the seventies by hearsay, without knowing its precise content. At that time, in a letter dated 11 December 1975, I wrote to Siegel to tell him I got the message, and I sent a copy of my letter to many people. There was considerable gossip about Siegel's letter to Mordell, but I saw the letter for the first time only in March 1991, when I received from Michel Waldschmidt a copy which he made from the original in the Cambridge Library of St. John College.

Siegel's letter is a historical document of interest from many points of view. I would like to deal here with one of these points of view having to do with the relation between number theory and algebraic geometry, or what has come to be known as the number field case and the function field case. I shall document part of the 20th century history of the way these two cases have benefited from each other, and the extent to which both Mordell and Siegel failed to understand the accomplishments of the fifties and sixties in connection with them.[2] Among other things, Siegel wrote to Mordell:

---

[1] I reproduced Mordell's review *in toto* as an appendix to the greatly expanded version *Fundamentals of Diophantine Geometry* [Lan 1983] because I wanted future generations to evaluate his position for themselves. I also reproduced my review of his book [Lan 1970], including a letter which I wrote to Mordell in 1966.

[2] For an account of results and conjectures in current diophantine geometry, much more systematic and complete than I can give here, as well as looking to the future rather than the past, see my book *Number Theory III: Diophantine Geometry*, Encyclopedia of Mathematics Vol. 60, Springer-Verlag, 1991; also the later version *Survey of Diophantine Geometry*, 1997.

— that "the whole style of the author [of *Diophantine Geometry*] contradicts the sense for simplicity and honesty which we admire in the works of the masters in number theory. . . ";

— that "just now, Lang has published another book on algebraic numbers which, in my opinion, is still worse than the former one. I see a pig broken into a beautiful garden and rooting up all flowers and trees";

— that "unfortunately there are many 'fellow travelers' who have already disgraced a large part of algebra and function theory";

— that "these people remind [Siegel] of the impudent behaviour of the national socialists who sang: 'Wir werden weiter marschieren, bis alles in Scherben zerfällt!' ";

— and that "mathematics will perish before the end of this century if the present trend for senseless abstraction—I call it: theory of the empty set—cannot be blocked up."

I shall also deal with some concrete instances of the more general problem mathematicians face in dealing with advances in mathematics which may pass them by.

§1. From Dedekind-Weber to the Riemann Hypothesis in function fields over finite fields

The analogy between number fields and function fields has been realized since the latter part of the 19th century. Kronecker was already in some sense aware of some of its aspects. Dedekind originated a terminology in his study of number fields which he and Weber applied to function fields in one variable [Ded-W 1882]. Hensel-Landsberg then provided a first systematic book treatment of basic facts concerning these function fields [Hen-L 1902], using the Dedekind-Weber approach. Artin in his thesis [Art 1921] translated the Riemann hypothesis to the function field analogue (actually for quadratic fields). Several years later F.K. Schmidt treated general analytic number theory including the functional equation of the zeta function for function fields of arbitrary genus [Schm 1931]. However, Artin thought that the Riemann hypothesis in the function field case would be as difficult as in the classical case of the ordinary Riemann or Dedekind zeta function (he told me so around 1950). It was Hasse in 1934 and 1936 who pointed out the "key for the problem" in the function field case through the theory of correspondences, as Weil writes in [Wei 1940]. (Hasse also indicated another way through reduction mod p using complex multiplication in characteristic zero.) Hasse himself proved the Riemann hypothesis (Artin's conjecture) for curves of genus 1

[Has 1934], [Has 1936].[3] Then Deuring pursued the higher dimensional generalization of Hasse's theory of endomorphisms on elliptic curves and correspondences [Deu 1937], [Deu 1940], by showing that some results of Severi [Sev 1926] could be proved so that they applied in characteristic p, especially to curves over finite fields. Weil went much further than Hasse and Deuring in this direction. "Directly inspired" [Wei 1948a, p. 28] by works of Severi [Sev 1926] and Castelnuovo [Cas 1905], [Cas 1906], [Cas 1921], Weil developed a purely algebraic theory of correspondences and abelian varieties; and he formulated the positive definiteness of his trace (which he related to Castelnuovo's equivalence defect), thus yielding the Riemann-Hypothesis in the function field case for curves of higher genus [Wei 1940], [Wei 1948a], [Wei 1948b].

Hasse also defined a zeta function for arbitrary varieties over number fields and conjectured its analytic continuation and functional equation. This point of view was promoted by Weil in the fifties. There was a serious problem of algebraic geometry even dealing with varieties of higher dimension over finite fields, let alone number fields, because as Weil conjectured, the analogue of the functional equation and Riemann hypothesis in this case would depend on finding algebraic analogues of homology groups (homology functors) satisfying the Lefschetz fixed point formula [Wei 1949].

In the forties and fifties several subjects in mathematics, including algebraic topology and algebraic geometry, systematically developed new foundations and internal results. Indeed, homological algebra developed first from algebraic topology, but soon saw its domain of applications extend to several other fields including algebraic geometry. This algebra was affectionately called "abstract nonsense" by Steenrod (with a quite different intent and meaning from Siegel's "senseless abstraction"). A large body of material, recognized to be fairly dry by some of its creators, had to be systematically worked out to provide appropriate background for more extensive applications. The dryness was unavoidable.

One also saw the simultaneous development of commutative algebra. One of its motivations was that in the context of algebraic geometry, a

---

[3] In 1932–1933 Davenport and Hasse started collaborating on a classical paper concerning Gauss sums [Dav-H 1934]. Davenport had previously been concerned with Gauss sums, and he learned from Hasse the connection with the Riemann hypothesis in function fields as formulated by Artin. I find it appropriate to quote here a historical comment made by Halberstam, who edited Vol IV of Davenport's collected works, and states p. 1553: "In fact, Davenport spent part of the academic session 1932–33 with Hasse in Marburg; he obviously learnt a great deal from Hasse (cf. [8], [18], [27])—in later years he would say that he had not learnt nearly as much as he would have done if he had been 'less pig-headed'—and it seems that he in turn sharpened Hasse's interest in the arithmetical questions discussed above. ... According to Mordell, Hasse was led to his proof of (5) [*RH in elliptic function fields*] in response to a challenge from Davenport to produce a concrete application of abstract algebra."

curve over a number field can be defined by equations whose coefficients lie in the ring of algebraic integers, and so can be viewed as a family of curves obtained by reducing mod p for all primes p, or even mod p^n for higher n so as to include infinitesimal properties. This case can be unified with the case of algebraic families of curves over an arbitrary field, including curves over the complex numbers. Furthermore, one wants to treat higher dimensional varieties in the same fashion, in an algebraic and analytic context. For the analytic context one is led to work over power series rings, and more generally over complete local rings because of the presence of singularities and the infinitesimal aspects. One was also led to globalize from modules to sheaves, in a context involving both homological algebra and commutative algebra, thus leading to further abstractions.

These developments were a prelude to the subsequent conceptual unification of topology, complex differential geometry and algebraic geometry during the sixties, the seventies, and beyond. For such a unification to take place, it was necessary to develop not only a language, but an extensive theory containing very substantial results as well, starting with commutative algebra and merging into algebraic geometry. In the fifties and sixties, these developments appeared as "senseless abstractions" to some people, including Siegel, who writes as if these developments deal only with the "theory of the empty set." But it is precisely the insights of Grothendieck which led to an extension—including an abstraction—of algebraic geometry whereby he defined the cohomology functors algebraically; whereby he proved the Lefschetz formula [Gro 1964]; and whereby finally a decade later, Deligne finally proved the analogue of the Riemann hypothesis for varieties in the higher dimensional case [Del 1974]. Deligne also proved related applications, because in a Bourbaki seminar talk [Del 1969], he had previously shown how to reduce the Ramanujan-Petersson conjecture for eigenvalues of modular forms under Hecke operators to this Riemann hypothesis, in a direction first foreseen by Sato and also using some insights of Kuga-Shimura, to whom he refers at the beginning of his Bourbaki seminar talk. A very short and clear account of the ideas, leading from a classical problem involving the partition function to the most advanced uses of Grothendieckian algebraic geometry, is given in the first two pages of [Del 1969].

§2. Some implications

We now pause a moment to consider some implications of these great developments. As I wrote in my 1961 review of Grothendieck's *Éléments de géometrie algébrique* [Lan 1961] a decade before Deligne's applications of Grothendieckian geometry occurred: "The present work... is one of the major landmarks in the development of algebriac geometry.... Before we go into a closer description of the contents of Chapter 0 and I [*which were just appearing and prompted the review*] it is necessary to say a few words explaining why the present treatise differs radically in its point of view from previous ones." I then mentioned four specific points like those already listed above: the need to deal with algebraic families of varieties, applications to number theory and reduction modulo a prime power, defining algebraically the functors from topology such as homology and homotopy, and the study of non-abelian coverings. I also emphasized throughout the importance and far reaching implications of Grothendieck's functorial point of view.

I ended my review as follows: "To conclude this review, I must make a remark intended to emphasize a point which might otherwise lead to misunderstanding. Some may ask: If Algebraic Geometry really consists of (at least) 13 Chapters, 2,000 pages [*it turned out to be more like 10,000*], all of commutative algebra, then why not just give up? The answer is obvious. On the one hand, to deal with special topics which may be of particular interest only portions of the whole work are necessary, and shortcuts can be taken to arrive faster to specific goals.... But even more important, theorems and conjectures still get discovered and tested on special examples, for instance elliptic curves or cubic forms over the rational numbers. And to handle these, the mathematician needs no great machinery, just elbow grease and imagination to uncover their secrets. Thus as in the past, there is enough stuff lying around to fit everyone's taste. Those whose taste allow them to swallow the *Elements*, however, will be richly rewarded." Thus I did not see the developments of Grothendieck's algebraic geometry as incompatible with doing beautiful or deep mathematics with only a minimum of knowledge.

Five years later, when I wrote to Mordell the letter reproduced in [Lan 1970] and [Lan 1983], I continued to take such a balanced view. Since Mordell had written in his review: "When proof of an extension makes it exceedingly difficult to understand the simpler cases, it might sometimes be better if the generalizations were left in the journals" (see below for the context of this judgment), I replied:

"I see no reason why it should be prohibited to write very advanced monographs, presupposing substantial knowledge in some fields, and thus allowing certain expositions at a level which may be appreciated only by a few, but achieves a certain coherence which would not otherwise be possible.

"This of course does not preclude the writing of elementary monographs. For instance, I could rewrite Diophantine Geometry by working entirely on elliptic curves, and thus make the book understandable to any first year graduate student (not mentioning you...) Both books would then coexist amicably, and neither would be better than the other. Each would achieve different ends. [*In fact, I eventually wrote Elliptic Curves: Diophantine Analysis, Springer Verlag, 1978.*]

"... When I write a standard text in Algebra, I attempt something very different from writing a book which for the first time gives a systematic point of view on the relations of diophantine equations and the advanced contexts of algebraic geometry. The purpose of the latter *is* to jazz things up as much as possible. The purpose of the former is to educate someone in the first steps which might eventually culminate in his knowing the jazz too, if his tastes allow him that path. And if his tastes don't, then my blessings to him also. This is known as aesthetic tolerance. But just as a composer of music (be it Bach or the Beatles), I have to take my responsibility as to what *I* consider to be beautiful and write my books accordingly, not just with the intent of pleasing one segment of the population. Let pleasure then fall where it may."

Thus I advocated "aesthetic tolerance"—which is certainly absent from Siegel's letter, to say the least.

It is of course not only a matter of "taste" or "aesthetic tolerance." It may also have to do with one's natural limitations. For instance, I had my own limitations vis a vis Grothendieck's work (and other works). Having gone through Weil's *Foundations* right after my Ph.D., I myself was unable later to absorb completely Grothendieck's work, and I was unable to read much of that work, as well as some of its applications, such as those by Deligne [Del 1969], [Del 1974]. However, I did not put down Grothendieck's work. I admired it (as quoted above), and merely regretted my own limitations. I also could not read the Italian geometers myself and I needed van der Waerden and Weil as intermediaries in algebraicizing and modernizing italian geometry.

§3.   Diophantine Results over Number Fields and Function Fields

Next we consider diophantine questions over the rationals or over number fields. At the turn of the century, Poincaré defined the "rank" of the group of rational points on an elliptic curve over the rational numbers [Poi 1901]. By "rank" he actually meant something different from what we mean today. Roughly speaking, he meant the smallest number of generators of the set of rational points using the secant and tangent method to generate points. Poincaré wrote as if this rank is always finite. The finite generation was proved by Mordell [Mor 1921], and again Weil extended this result to abelian varieties over number fields using more algebraic geometry in his thesis at the end of that decade [Wei 1928]. The analytic parametrization of abelian varieties, and especially Jacobians of curves, was a convenient tool at the time, and for this particular application a complete algebraization of curves and their Jacobians was not yet needed.

At the purely algebraic level, the fifties saw a clarification of the Mordell-Weil theorem and its relations to the algebraic-geometric situation in the function field case. The Artin-Whaples product formula of the forties [Ar-W 1945] was the number theoretic analogue of the geometric theorem that a rational function on a curve has the same number of zeros and poles (counting multiplicities), or in higher dimension that the degree in projective space of the divisor of a rational function is zero. I used this product formula as the basic axiom for the theory of heights in *Diophantine Geometry*, applicable simultaneously to the number field and function field case, in any dimension. Mordell complained that here "we have definitions which many other authors do not find necessary." However, varieties over number fields have their analogues in algebraic families of varieties over any field, especially over the complex numbers. Rational points have their analogues in sections of such families, and in fact *are* sections when the proper language and setting has been defined. The analogy has been interesting and fruitful not only because it has allowed techniques to go back and forth enriching the two cases, but because for instance in the study of algebraic surfaces, cases occur systematically when these varieties are generically fibered by curves of genus 1. One then wants to know which fibers have rational points, and how many. In case the generic fiber of an algebraic family is an abelian variety, the sections form a group, and Lang-Néron proved that this group is finitely generated modulo the subgroup of "constant" sections [La-Ne 1959], this being the function field analogue of the Mordell-Weil theorem. Furthermore, Severi long ago conjectured that the algebraic part of the first cohomology group, i.e. the group of divisors modulo algebraic equivalence, was finitely generated (Theorem of the Base), and he had the intuition that such a result was also analogous in some way to the Mordell-Weil theorem. Néron proved Severi's conjecture [Ner 1952], and Lang-Néron established an actual isomorphism between the Néron-Severi group of a variety, and a subgroup of the group of sections (modulo constant sections) of the Jacobian of the generic curve in some projective imbedding. These results of the fifties formed the backbone of my book *Diophantine Geometry*, but were viewed as "senseless abstraction... the theory of the empty set" by Siegel.

Using Weil's results, and his own results on diophantine approximations (Thue-Siegel theorem), Siegel proved that an affine curve of genus at least 1 over a number field has only a finite number of integral points [Sie 1929]. In [Lan 60] and in another part of *Diophantine Geometry*, I also showed how the Thue-Siegel-Schneider-Roth theorem and Siegel's theorem on integral points had analogues in the function field case. The interdependence between the number field case and the function field case lies not only in the analogy of results and methods applicable to both cases, but also in the fact that when, say, a curve depending on parameters defined by a family of equation $f_t(x, y) = 0$ has solutions in polynomials $x = x(t)$ and $y = y(t)$, such polynomials may have complex coefficients, or in a more arithmetic setting they may have ordinary integer coefficients. In the latter case, by specialization of the parameter t in integers, one obtains integral solutions of the specialized equation. It is a problem to classify all surfaces which admit such a generic fibration by rational curves, over the complex numbers and over the ordinary integers. More generally, one can consider the case when x and y are integral affine algebraic functions rather than polynomials. In order to treat both the number field and function field case simultaneously, there developed a language and results which are now natural throughout the world. At the time, this language and results appeared unnatural or worse to some people. As Siegel wrote to Mordell: "The whole style of the author contradicts the sense for simplicity and honesty which we admire in the works of the masters of number theory."

In part of the proof of Roth's theorem, it is necessary to solve certain linear equations with upper bounds on the size of the solution. A lower bound on the number of solutions is required in the number field case, and a lower bound on the dimension of the space of solutions is required in the function field case. Classically, the Riemann-Roch theorem on curves provides the desired estimates in the function field case, and I drew the analogy explicitly with the number field case by an appropriate axiomatization, whereby I treated both cases simultaneously. But Mordell states in his review: "The author claims to follow Roth's proof. The reader might prefer to read this which requires only a knowledge of elementary algebra and then he need not be troubled with axioms which are very weak forms of the Riemann-Roch theorem." But drawing closer together various manifestations of what goes under the trade name of Riemann-Roch has been a very fruitful viewpoint over decades. Already in [Schm 1931] we see the Riemann-Roch theorem closely related to the functional equation of the zeta function in the function field case. In the thirties, Artin recognized the functional equation of the theta function as an analogue of Riemann-Roch in the number field case. Following the ideas in a course of Artin, Weissinger gave the connection between Riemann-Roch and the functional equation of L-functions in the function field case [Weis 1938]. Weil went further by giving an analogy of Riemann-Roch not only to the problem of counting lattice points in parallelotopes, but also by formulating an analogue for Cauchy's residue formula in the number field case [Wei 1939]. In my book on algebraic number theory, I emphasized the Riemann-Roch viewpoint in these ways. First I gave a formula for the number of lattice points in adelic parallelotopes, asymptotic with respect to the normalized volume; and second, I reproduced the formulation and proof of the functional equation for the zeta function and L-functions via the adelic method in Tate's thesis, especially the adelic Poisson summation formula having a corollary what was properly called by Tate a number theoretic Riemann-Roch theorem.[4] But Siegel found my book

---

[4]Be it said in passing that the adelic method of Tate's thesis was to become standard in the treatment of analogous situations on linear algebraic groups.

on algebraic numbers "still worse than the former one." Nevertheless, I shall continue below to describe the ever expanding extent to which the Riemann-Roch umbrella covers aspects of number theory and algebraic geometry.

Naturally, to deal simultaneously with the number field and function field case in diophantine geometry, I had to assume the basic language and results of algebraic geometry and abelian varieties. Mordell in his review of the book complained: "Let us note some of the concepts required in the chapter. There are a 'K/k-trace of A', a 'Theorem of Chow', 'Chow's Regularity Theorem', 'Chow Coordinates', 'compatibility of projections and specializations', 'blowing up a point', 'Albanese Variety', 'Picard variety', 'Jacobian of a curve', 'Chow's theory of the $k(u)/k$-trace'. When proof of an extension makes it exceedingly difficult to understand the simpler cases, it might sometimes be better if the generalizations were left in the Journals." I ask: exceedingly difficult to whom? Current readers and subsequent generations can evaluate for themselves Mordell's admonition to leave what he calls "generalizations" to the journals. But Mordell went on: "The reviewer was reminded of Rip Van Winkle, who went to sleep for a hundred years and woke up to a state of affairs and a civilization (and perhaps a language) completely different from that to which he had been accustomed." Siegel accepted the comparison with Rip Van Winkle when he wrote to Mordell: "My feeling is very well expressed when you mention Rip Van Winkle." In particular both Siegel and Mordell had difficulty understanding some basic notions of algebraic geometry as recalled above. But these notions were of course accepted without further ado by younger mathematicians and by other schools of mathematics and algebraic geometry, notably by the Russian school, whose contributions to diophantine geometry were to dominate the sixties and seventies, as we shall now indicate.

Mordell himself in [Mor 1922] had conjectured that a curve of genus at least 2 over the rational numbers has only a finite number of rational points. In [Lan 1960] and in *Diophantine Geometry* I translated this conjecture into the function field analogue, to the effect that for an algebraic family of such curves, there is only a finite number of sections unless the family is constant, in a suitable sense. Independently, Manin had already started his investigations of the Picard-Fuchs differential equations and their connections with algebraic families of curves, their Jacobians and their periods, via horizontal differentiation and the Gauss-Manin connection [Man 1958]. Manin put these two mathematical threads together by proving the function field analogue of the Mordell conjecture via his differential methods [Man 1963]. We note in passing that the function field analogue of Siegel's theorem on integral points is needed to complete that proof. (See [Col 1990].) Manin's work kindled various people's interests in various directions lying between algebraic geometry and the theory of algebraic differential equations. Furthermore in 1970–1971 Deligne proved the semisimplicity of the action of the monodromy group on the cohomology of a family of projective smooth varieties [Del 1972]. After Coleman pointed out that Manin's "theorem of the kernel" had not been completely proved [Col 1990], Deligne's theorem was applied by Chai to complete the proof, independently of the application to the Mordell conjecture in the function field case [Cha 1990]. Even more recently, Buium has pursued the application of differential algebra in this direction and he has obtained a substantial extension of results showing that the intersection of a curve with certain subsets of its Jacobian defined by algebraic differential conditions is finite [Bui 1991].

I learned of Manin's proof on a trip to Moscow in 1963, and I lectured on it at the Arbeitstagung in Bonn upon returning. Grauert was in the audience, and was then led to find another proof of the function field case of Mordell's conjecture [Gra 1965] (see among others the final remarks of the introduction to his paper). Grauert's method also involved horizontal differentiation, taking the derivative of a section into the projectivized tangent bundle. For the latest development of this method in a quantitative direction, see Vojta [Vo 1991]. Grauert's proof also worked in characteristic p, as pointed out by Samuel [Sa 1966]. For further insight in the problem in characteristic p, see Voloch [Vo 1990].

To this day, no one has seen how to translate Manin's or Grauert's proofs of Mordell's conjecture from the function field case to the number field case. However, in the early sixties, Shafarevich conjectured that over a number field, given a finite set of places, there exists only a finite number of isomorphism classes of curves of given genus at least 1 and having good reduction outside this finite set [Sha 1963]. In 1968, Parshin showed how Shafarevich's conjecture implied Mordell's conjecture [Par 1968], and he proved the analogue of Shafarevich's conjecture in the function field case (under an additional technical condition, later removed by Arakelov [Ara 1971]). Parshin's proof was based entirely on the intersection theory of surfaces, without making use of horizontal differentiation. This provided hope for an eventual translation to the number field case. As we have already mentioned, a curve over the ring of integers of a number field can be viewed as a family of curves obtained by reduction mod p for all primes p. In a fundamental paper, Arakelov showed how to complete such a family of curves over a number field by introducing the components at infinity, and by defining a new type of divisor class group taking the components at infinity into account [Ara 1974]. With this point of view, a curve over the ring of integers of a number field is called an arithmetic surface. Whereas the Artin-Whaples product formula has been the starting point for unifying the case of number fields and function fields in one variable, Arakelov theory laid the foundations for unifying intersection theory on arithmetic surfaces and the classical intersection theory, thus making Parshin's method more accessible to the number field case.[5]

Arakelov defined intersection numbers at infinity as the values of Green's functions, and made extensive use of hermitian metrics on line bundles [Ara 1974]. His foundations could lead in several directions. In one direction, inspired by the basic idea of carrying out algebraic geometry with complete objects, including the components at infinity and the metrized line bundles, Faltings gave his proof of Mordell's conjecture a decade later [Fal 1983].[6] Be it noted that Faltings also depended on the full-

---

[5]Parshin himself was quite aware of the historical context in which he was writing, and gives a very different perspective from Mordell and Siegel, as we find in the introduction of [Par 1968]: "Finally when $g > 1$, numerous examples provide a basis for Mordell's conjecture that in this case $X(\mathbf{Q})$ is always finite. The one general result in line with this conjecture is the proof by Siegel that the number of integral points (i.e., points whose affine coordinates belong to the ring $\mathbf{Z}$ of integers) is finite. These results are also true for arbitrary fields of finite type over $\mathbf{Q}$. Fundamentally this is because the fields are global, i.e., there is a theory of divisors with a product formula, which makes it possible to construct a theory of the height of quasi-projective schemes of finite type over K. Lang's book [*Diophantine Geometry*] contains a description of that theory and its application to the proof of the Mordell and Siegel theorems. It appears that further progress in diophantine geometry involves a deeper use of the specific nature of the ground field. This is confirmed by Ju. I. Manin's proof of the functional analogue of Mordell's conjecture."

[6]In his thesis [Wei 1928] Weil refers explicitly to "Mordell's conjecture," and states that "it seems confirmed to some extent" by Siegel's theorem on the finiteness of integral points on curves of genus at least 1. In [Wei 1936] he makes a similar evaluation without reference to Mordell: "On the other hand, Siegel's theorem, for curves of genus > 1, is only the first step in the direction of the following statement: *On every curve of genus > 1, there are only finitely many rational points.*" However, some forty years later, he inveighed against "conjectures," when he wrote [Wei 1974]: "For instance, the so-called 'Mordell conjecture' on Diophantine equations says that a curve of genus at least two with rational coefficients has at most finitely many rational points. It would be nice if this were so, and I would rather bet for it than against it. But it is no more than wishful thinking because there is not a shred of evidence for it, and also none against it." Finally in comments in his collected works made in 1979 (Vol. III, p. 454), he goes one better. "Nous sommes moins avancés à l'égard de la 'conjecture de Mordell'. Il s'agit à d'une question qu'un arithméticien ne peut guère manquer de se poser; on n'apperçoit d'ailleurs aucun motif sérieux de parier pour ou contre." First, concerning a "question which an arithmetician can hardly fail to raise," I would ask when? It's quite a different matter to raise the question in 1921, as did Mordell, or decades later. As for the statements in 1974 and 1979 that there is no "shred of evidence" or "motif sérieux" for Mordell's conjecture, they not only went against Weil's own evaluations in earlier decades, but they were made after Manin proved the function field analogue in 1963; after Grauert gave his other proof in 1965; after Parshin gave his other proof in 1968, while indicating that Mordell's conjecture follows from Shafarevich's conjecture (which Shafarevich himself had proved

fledged abstractions of contemporary algebraic geometry, for instance by using techniques of Raynaud [Ray 1974], reducing modulo a prime power (actually mod p^2), to bound the degrees of certain isogenies of abelian varieties.

We now come back to the Riemann-Roch theme. In the direction of algebraic geometry, the Italian algebraic geometers dealt classically with the Riemann-Roch theorem on algebraic surfaces. Hirzebruch in the early fifties gave an entirely new slant to the theorem by his formula expressing the (holomorphic-algebraic) Euler characteristic as a polynomial in the Chern classes, for non-singular projective varieties of arbitrary dimension [Hir 1956]. Thus Hirzebruch drew together algebraic geometry, topology, and complex differential geometry. Siegel did not appreciate Hirzebruch's mathematics any more than some other mathematics of the period. Indeed, Siegel was the principal factor causing the collapse of negotiations between Göttingen and Hirzebruch in the fifties, when Hirzebruch was in the process of returning to Germany after his stay in America. Furthermore, in 1960, there was an eary attempt to create a Max Planck Institute to be headed by Hirzebruch. Siegel wrote negatively about Hirzebruch and his mathematics in this connection.[7]

---

for curves of genus 1); at the same time that Arakelov theory was being developed and that Zarhin was working actively on the net of conjectures in those directions; and within four years of Faltings' proof.

[7]In June 1991 I wrote to the President of the Max Planck Society to ask for a copy of Siegel's letter so that one has primary sources on which to base factual historical reporting. I received a friendly answer and the letter was sent to me. Siegel wrote four and a half pages, discussing institutes in general, and giving his evaluation of Hirzebruch in particular, as follows: "Was den zum Schluss vorgeschlagenen Leiter des zu gründenden Institute betrifft, so habe ich auch darüber eine abweichende Meinung.... Seine [Hirzebruchs] mathematischen Leistungen wurden allerdings damals auch hier ziemlich hoch bewertet, insbesondere wegen seiner Jugend. Jetzt erscheint es mir aber zweifelhaft, ob sich das von ihm bisher bearbeitete sehr abstrakte Gebiet weiter erschliessen und fruchtbar machen lässt, und ich halte es für möglich, ja sogar für wahrscheinlich, dass diese ganze Richtung sich schon in wenigen Jahren totlaufen wird. Nach den vorhergehenden Ausführungen möchte ich Ihre Fragen 1), 2), 3) und 5) mit *Nein* beantworten."

As for others, according to a letter from Behnke to Hirzebruch dated 7 September 1960: "Im übrigen liegen von Ihnen ausser von Siegel nur die glänzendsten Gutachten vor. Es gibt jetzt zwei Hauptbedenken: 1) Man darf Sie nicht aus dem Universitätsleben nehmen, weil die Lücke nicht zu ersetzen ist.... 2) Es würde nur die abstrakte Mathematik gepflegt...." Courant was among those who wrote along these lines: "Hirzebruch is sicherlich einer der allerbesten unter den Mathematikern der jüngeren Generation. Ich bin stets für ihn eingetreten und hege sehr freundschaftliche Gesinnungen für ihn. Er ist einer des besten Dozenten, die ich kenne. Nach meiner Meinung würde es ein schweres Unrecht an der Mathematik sein, ihn aus seiner produktiven Lehrtätigkeit herauszureissen. Ausserdem würde er als Hauptleiter des Max Planck Institutes die Präponderanz der abstrakten Richtung weithin sichtbar symbolisieren. Leistungen und Renommée würden dies in Moment wohl rechtfertigen. Aber,

Later in the fifties, Grothendieck vastly extended Hirzebruch's Riemann-roch theorem partly by formulating it in such a way that it applies to families and partly by making the theorem more functorial [Bor-S 1957]. Still later, he further expanded the formulation of the theorem so that in particular, it applied over arbitrary Noetherian rings, and therefore could be used in the number theoretic context over the ring of algebraic integers of a number field [Gro 1971]. These extensions required the full fledged abstractions of algebraic geometry and algebraic topology, which he had developed, including both the cohomology functors and the K-theory functors [Gro EGA], [Gro SGA]. An especially interesting application of Grothendieck Riemann-Roch was made by Mumford in his contributions to the theory of moduli spaces for curves and abelian varieties [Mum 1977].

Just before Faltings proved Mordell's conjecture, he developed Arakelov theory so far as to give an arithmetic version of the Riemann-Roch theorem on arithmetic surfaces [Fal 1984]. This version was vastly extended recently by Gillet-Soulé, for varieties of arbitrary dimension, putting together the Hirzebruch-Grothendieck Riemann-Roch theorems, the complex differential geometry inherent in the components at infinity, and also the theories of real partial differential equations most recently developed by Bismut, necessary to handle the analogues of Green's functions in the higher dimensional case [Gi-S 1990], [Gi-S 1991]. Thus comes a grand unification of several fields of mathematics, under the heading of the code-word Riemann-Roch. At the moment, a complete translation of Parshin's proof of Mordell's conjecture from the function field case has not yet taken place. It still requires a proof of an inequality conjectured by Parshin in the number field case, whose known analogue in the case of algebraic surfaces evolved from work of van de Ven, Bogomolov, Parshin, Miyaoka and Yau [Par 1989] (see also [Voj 1988]). Such an inequality is related to the so-called Noether formula in the theory of algebraic surfaces. It is known that such an inequality implies Fermat's theorem for all but a finite number of cases, which cases depending on how effectively the Parshin inequality can be proved.

---

auch in Hirzebruchs eigenem Interesse, und sicherlich in dem der Wissenschaft rate ich dringend davon ab. Es ist nicht nötig, das Institut in einer solchen persönlichen Art zu organisieren, um den höchsten Grad der Wirksamkeit zu erreichen...." But Courant also added: "Meine Bemerkungen sind nicht sorgfältig ausgearbeitet. Sie brauchen nicht vertraulich behandelt zu werden...." Thus Courant also expressed himself with caution. Courant made his letter public at the time.

Some letters to the Max Planck Society were unreservedly for the creation of the Institute, for instance van der Waerden's. After listing Hirzebruch's qualities in all directions (mathematical, personal, and administrative), he asks: "Was will man mehr?" For more on the history of the Max Planck Institute, see Schappacher [Sch 1985].

The Riemann-Roch story in its arithmetic context does not end there. Vojta, in a major development, showed how to globalize and sheafify on curves of higher genus the basic ideas of the proof of Roth's theorem, in such a way that he found an entirely new proof of Mordell's conjecture (Faltings' theorem) [Voj 1990]. Be it noted that Vojta first gave his proof in the function field case, using intersection theory on surfaces [Voj 1989]. He then translated his proof to the number field case using the Arakelov type intersection theory and the newly found (asymptotic) arithmetic Riemann-Roch theorem of Gillet-Soulé. Although Bombieri subsequently simplified Vojta's proof by eliminating the Arakelov part [Bom 1990], he still used the classical Riemann-Roch theorem on surfaces. The use of Riemann-Roch in one form or another occurs at the same point in the pattern of proof as in Roth's theorem, but of course in the more sophisticated context of curves of higher genus and their products, rather than the projective or affine line. Vojta's idea and a heavy dose of algebraic geometry were then used by Faltings to prove a conjecture of mine dating back to [Lan 1960], concerning higher dimensional diophantine analogues for subvarieties of abelian varieties [Fal 1990]. Neither Vojta, Bombieri nor Faltings has shown that he is "troubled" about using Riemann-Roch theorems, and major breakthroughs have thus been made by expanding the perspectives on old problems, rather than by narrowing the viewpoint to "simpler cases."

Thus we see that since the translation of the Riemann hypothesis in the twenties and the very first translations of the Mordell-Weil theorem from the number field case into the function field analogue in the fifties, there has been constant interaction between the number field case and the function field case. A number of subsequent results have been proved first in the function field case, using geometric intuition and methods from algebraic geometry as well as differential geometry. In some, but not yet all cases, these proofs could then be translated back to the number field case, thus giving new results in number theory.

§4. Further Implications

Mordell and Siegel were great mathematicians, a fact which is made obvious once more by their great theorems cited repeatedly in this article. But their lack of vision and understanding at certain periods of their life obstructed the development of certain areas of mathematics in their own countries. Of course they did not have absolute power. In England, Atiyah could develop the Riemann-Roch theme in the topological and analytic direction for elliptic operators on vector bundles, for instance, but the direction of number theory in England was seriously affected by Mordell's obstructions. In Germany, Hirzebruch could create an independent center in Bonn, but Siegel did have an effect in Göttingen and some other places, although his influence has waned to the point where I don't see it explicitly any more.

In the Soviet Union and France, the obstructing influence of Mordell and Siegel in algebra and algebraic geometry was nil. The development of Grothendieck's school in France needs no further comment. In the Soviet Union, one sees the absence of obstructing influence in the existence of the school of algebraic geometry created by Shafarevich. One also sees the absence of obstructing influence in concrete instances, such as the introductions to Manin's and Parshin's papers [Man 1963] and [Par 1968] (as mentioned in footnote 5). Furthermore, for the Russian translation of *Fundamentals of Diophantine Geometry*, I was asked if it was OK with me to omit the appendices consisting of Mordell's review and my review of his book, and to replace them with an appendix by Parshin and Zarhin describing previous work of theirs on a net of conjectures (Mordell-Shafarevich-Tate), as well as the latest developments concerning Faltings' proof of these conjectures. I agreed without reservations.

In the United States, the influence is more complex to evaluate. Be it noted here only that as recently as December 1989, in the context of my continued activities concerning the non-election of Samuel P. Huntington to the National Academy of Sciences, MacLane wrote me a letter commenting in part on my own 1986 election to the NAS: "I welcomed your election to the NAS. But please observe that if some social scientist had then known and used Mordell's famous comments on your Diophantine book plus the silly mistakes in the last chapter of your Differential Manifolds plus... you too would have been soundly defeated on the floor of the Academy."

Mordell used to pull out Siegel's letter from his wallet to show people, to my knowledge without receiving comments that both his and Siegel's attitudes were parochial and blind (if not worse).[8] Thus some members of the mathematical community behaved with "collegiality" and bowed to authority, in the face of claims such as those quoted at the beginning of this article.

Those members of the mathematical community who did not stand up to Mordell and Siegel are not entirely blameless for the obstructing influence of Mordell's review and Siegel's letter, such as it was.

---

[8]For instance, Mostow remembers distinctly Mordell showing Siegel's letter about me to those members of the math department at Yale in the sixties, when they were at dinner at the local restaurant Mori's. An instructor at Yale today, Jay Jorgenson, heard gossip about this letter when he was a freshman at the University of Minnesota several years ago. An official of the National Science Foundation was shown the letter by Mordell some 25 years ago in Washington. And so it goes on and on.

Bibliography

[Ara 1971]   S. J. ARAKELOV, Families of algebraic curves with fixed degeneracies, *Izv. Akad. Nauk SSSR Ser. Mat.* **35** (1971) pp. 1269–1293; translation *Math USSR Izv.* **5** (1971) pp. 1277–1302.

[Ara 1974]   S. J. ARAKELOV, Intersection theory of divisors on an arithmetic surface, *Izv. Akad. Nauk SSSR Ser. Mat.* **38** No. 6 (1974) pp. 1179–1192; translation *Math. USSR Izv.* **8** No. 6 (1974) pp. 1167–1180.

[Art 1921]   E. ARTIN, Quadratische Körper im Gebiete der höheren Kongruenzen, I and II, *Math. Zeit.* **19** (1924) pp. 153–246.

[Ar-W 1945]   E. ARTIN and G. WHAPLES, Axiomatic characterization of fields by the product formula for valuations, *Bull. AMS* **51** (1945) pp. 469–492.

[Bom 1990]   E. BOMBIERI, The Mordell conjecture revisited, preprint, 1990.

[Bor-S 1957] E. BOREL and J. P. SERRE, Le théorème de Riemann-Roch, *Bull. Soc. Math. France* **86** (1958) pp. 97–136.

[Bui 1991]   A. BUIUM, Intersections in jet spaces and a conjecture of S. Lang, to appear.

[Cas 1905]   G. CASTELNUOVO, Sugli integrali semplici appartenenti ad una superficie irregolare, *Rend. Acad. Lincei* (5) **14** (1905) pp. 545–556, 593–598, 655–663.

[Cas 1906]   G. CASTELNUOVO, Sulle serie algebriche di gruppi di punti appartenenti ad una curva algebrica, *Rend. Acad. Linc* (5) **15** (1906) pp. 337–344.

[Cas 1921]   G. CASTELNUOVO, Sulle funzione abeliane: I. Le funzioni intermediarie II. La geometria sulle varietà abeliane III. Le varietà di Jacobi IV. Applicazioni alle serie algebriche di gruppi sopra una curva, *Rend. Accad. Lincei* (5) **30** (1921) pp. 50–55, 99–103, 195–200, 355–359. See also the review by Krazer in *Jahrbuch für die Fortschritte der Mathematik* (1921) p. 445.

[Cha 1990]   C. L. CHAI, A note on Manin's theorem of the kernel, *Am. J. Math.* **113** (1991) pp. 387–389.

[Col 1990]   R. COLEMAN, Manin's proof of the Mordell conjecture over function fields, *Enseignement Mathématique* **36** (1990) pp. 393–427.

[Dav-H 1934] H. DAVENPORT and H. HASSE, die Nullstellen der Kongruenz-zetafunktionen in gewissen zyklischen Fällen, *J. reine angew. Math.* **172** (1934) pp. 151–182.

[Ded-W 1882] R. DEDEKIND and H. WEBER, Theorie der algebraischen Funktionen einer Veränderlichen, *J. reine angew. Math.* **92** (1882) pp. 181–290.

[Del 1969] P. DELIGNE, Formes modulaires et représentations *l*-adiques, *Séminaire Bourbaki* No. 355, 34, February 1969.

[Del 1970] P. DELIGNE, Équations différentielles à points singuliers réguliers, *Springer Lecture Notes* **163** (1970).

[Del 1972] P. DELIGNE, Théorie de Hodge II, *Pub. Math. IHES* **40** (1972) pp. 5–58.

[Del 1974] P. DELIGNE, La conjecture de Weil, *Pub. Math. IHES* **43** (1974) pp. 273–307.

[Del 1975] P. DELIGNE, Théorie de Hodge III, *Pub. Math. IHES* **44** (1975) pp. 5–77.

[Deu 1937] M. DEURING, Arithmetischer Theorie der Korrespondenzen algebraischer Funktionenkörper I, *J. reine angew. Math.* **177** (1937) pp. 161–191; see also the review by H. L. Schmid in *Jahrbuch über die Fortschritte der Mathematik* (1937) p. 97.

[Deu 1940] M. DEURING, Arithmetischer Theorie der Korrespondenzen algebraischer Funktionenkörper II, *J. reine angew. Math.* **183** (1940) pp. 25–36.

[Fal 1983] G. FALTINGS, Endlichkeitssätze für abelsche Varietäten über Zahlkörpern, *Invent. Math.* **73** (1983), pp. 349–366.

[Fal 1984] G. FALTINGS, Calculus on arithmetic surfaces, *Ann. Math.* **119** (1984) pp. 387–424.

[Fal 1990] G. FALTINGS, Diophantine approximations on abelian varieties, to appear.

[Gi-S 1990a] H. GILLET and C. SOULÉ, Arithmetic intersection theory, *Pub. IHES* **72** (1990) pp. 93–174.

[Gi-S 1990b] H. GILLET and C. SOULÉ, Characteristic classes for algebraic vector bundles with hermitian metrics I and II, *Ann. Math.* **131** (1990) pp. 163–203 and pp. 205–238.

[Gi-S 1991]   H. GILLET and C. SOULÉ, Analytic torsion and the arithmetic Todd genus, *Topology* **30** No. **1** (1991) pp. 21–54.

[Gra 1965]   H. GRAUERT, Mordell's Vermutung über rational Punkte auf algebraischen Kurven und Funktionenkörper, *Pub. IHES* **25** (1965) pp. 131–149.

[Gro 1964]   A. GROTHENDIECK, Formule de Lefschetz et rationalité des fonctions *L*, *Séminaire Bourbaki* **279** (1964).

[Gro EGA]   A. GROTHENDIECK, Éléments de géometrie algébrique, *Pub. Math. IHES* **4, 8, 11, 17, 20, 24, 28, 32** (1961–1967).

[Gro SGA]   A. GROTHENDIECK *et al.* Séminaire de géometrie algébrique, *Springer Lecture Notes* **151, 152, 153, 224, 225, 269, 270, 288, 305, 340, 569** (1960–1974).

[Gro 1971]   A. GROTHENDIECK, with P. BERTHELOT, L. ILLUSIE *et al.*: Théorie des intersections et théorème de Riemann-Roch, **SGA 6**, *Springer Lecture Notes* **225** (1971). [*Note:* The seminar SGA 6 was actually held in 1966–1967, but it took a while to get published.]

[Has 1934]   H. HASSE, Abstrakte Begrundung der Komplexen Multiplikation und Riemannsche Vermutung in Funktionenkörpern, *Abh. Math. Sem.* Hamburg **10** (1934) pp. 325–348.

[Has 1936]   H. HASSE, Zur Theorie der abstrakten elliptischen Funktionenkörper I, II, III, *J. reine angew. Math.* **175** (1936).

[Hen-L 1902]   K. HENSEL and G. LANDSBERG, *Theorie der algebraischen Funktionen einer Variablen und ihre Anwendung auf algebraische Kurven und Abelsche Integrale*, Leipzig 1902.

[Hir 1956]   F. HIRZEBRUCH, Neue topologische Methoden in der algebraischen Geometrie, *Ergebnisse der Mathematik*, Springer-Verlag, 1956; translated and expanded to the English edition, Topological methods in algebraic geometry, *Grundlehren der Mathematik*, Springer-Verlag, 1966.

[Lan 1960]   S. LANG, Integral points on curves, *Pub. Math. IHES* **6** (1960) pp. 27–43.

[Lan 1961]   S. LANG, Review of Grothendieck's *Éléments de géometrie algébrique*, *Bull. AMS* **67** (1961) pp. 239–246.

[Lan 1970]   S. LANG, Review of Mordell's *Diophantine Equations*. *Bull. AMS* **76** (1970) pp. 1230–1234.

[Lan 1983]   S. LANG, *Fundamentals of Diophantine Geometry*, Springer-Verlag, 1983.

[La-N 1959]   S. LANG and A. NÉRON, Rational points on abelian varieties over function fields, *Amer. J. Math.* **81** (1959) pp. 95–118.

[Man 1958]   J. MANIN, Algebraic curves over fields with differentiation, *Izv. Akad. Nauk SSSR Ser. Mat.* **22** (1958) pp. 373–756.

[Man 1963]   J. MANIN, Rational points of algebraic curves over function fields, *Izv. Akad. Nauk SSSR Ser. Mat.* **27** (1963) pp. 1395–1440; *AMS Translation* **37** (1966) pp. 189–234.

[Mor 1922]   L. J. MORDELL, On the rational solutions of the indeterminate equation of the third and fourth degrees, *Proc. Cambridge Philos. Soc.* **21** (1922) pp. 179–192.

[Mor 1964]   L. J. MORDELL, Review of Lang's *Diophantine Geometry*, *Bull. AMS* **70** (1964) pp. 491–498.

[Mum 1977]   D. MUMFORD, Stability of projective varieties, *Enseignement Mathématique* **XXIII** fasc. 1–2 (1977) pp. 39–110. See Theorem 5.10.

[Ner 1952]   A. NÉRON, Probèmes arithmétiques et géometriques rattachés à la notion de rang d'une courbe algébrique dans un corps, *Bull. Soc. Math. France* **80** (1952) pp. 101–106.

[Par 1968]   A. N. PARSHIN, Algebraic curves over function fields, *Izv. Akad. Nauk SSSR Ser. Mat.* **32** (1968); translation *Math. USSR Izv.* **2** (1968) pp. 1145–1170.

[Par 1989]   A. N. PARSHIN, On the application of ramified coverings to the theory of Diophantine Geometry, *Math. Sbornik* **180** No. 2 (1989) pp. 244–259.

[Poi 1901]   H. POINCARÉ, Sur les propriétés arithmétiques des courbes algébriques, *J. de Liouville* (V) **7** (1901) pp. 161–233.

[Ray 1974]   M. RAYNAUD, Schémas en groupes de type $(p, p, \ldots, p)$, *Bull. Soc. Math. France* **102** (1974) pp. 241–280.

[Sa 1966]   P. SAMUEL, Compléments à un article de Hans Grauert sur la conjecture de Mordell, *Pub. Math. IHES* **29** (1966) pp. 55–62.

[Scha 1985]   N. SCHAPPACHER, Max Planck Institut für Mathematik—Historical Notes on the New Research Institute at Bonn, *Math. Intelligencer* **Vol. 7 No. 2** (1985) pp. 41–52.

[Schm 1931]   F. K. SCHMIDT, Analytische Zahlentheorie in Körpern der Charakteristik *p*, *Math. Zeit.* **33** (1931) pp. 1–32.

[Sev 1926]   F. SEVERI, *Trattato di geometria algebrica*, Vol. I, Parte 1 (1926).

[Sha 1963]   I. SHAFAREVICH, Algebraic number fields, *Proc. Inter. Congress Math. Stockholm* (1963); *AMS translation* (2) **31** (1963) pp. 25–39.

[Sie 1929]   C. L. SIEGEL, Über einige Anwendungen Diophantischer Approximationen, *Abh. Preuss. Akad. Wiss. Phys. Math. Kl.* (1929) pp. 41–69.

[Voj 1988]   P. VOJTA, Diophantine inequalities and Arakelov theory, appendix in Lang: *Introduction to Arakelov Theory*, Springer Verlag (1988) pp. 155–178.

[Voj 1989]   P. VOJTA, Mordell's conjecture over function fields, *Invent. Math* **98** (1989) pp. 115–138.

[Voj 1990]   P. VOJTA, Siegel's theorem in the compact case, *Ann. of Math.* **133** (1991) pp. 509–548.

[Voj 1991]   P. VOJTA, On algebraic points on curves, *Compositio Math.* **789** (1991) pp. 29–36.

[Vol 1990]   J. VOLOCH, On the conjectures of Mordell and Lang in positive characteristic, *Invent. Math.* **104** (1991) pp. 643–646.

[Wei 1928]   A. WEIL, L'arithmétique sur les courbes algébriques, *Acta Math.* **52** (1928) pp. 281–315.

[Wei 1936]   A. WEIL, Arithmetic on algebraic varieties, *Uspekhi Mat. Nauk* **3** (1936) pp. 101–112; see also *Collected Papers* Vol. I [1936b], p. 126.

[Wei 1939]   A. WEIL, Sur l'analogie entre les corps de nombres algébriques et les corps de fonctions algébriques, *Revue Scient.* **77** (1939) pp. 104–106.

[Wei 1940]   A. WEIL, Sur les fonctions algébriques à corps de constantes fini, *C. R. Acad. Sci. Paris* **210** (1940) pp. 592–594.

[Wei 1948a]   A. WEIL, *Sur les courbes algébriques et les variétés qui s'en déduisent*, Hermann, Paris, 1948.

[Wei 1948b]   A. WEIL, *Variétés abéliennes et courbes algébriques*, Hermann, Paris 1948.

[Wei 1949]   A. WEIL, Number of solutions of equations in finite fields, *Bull. AMS* **55** (1949) pp. 497–508.

[Wei 1974]   A. WEIL, Two lectures on number theory, past and present, *Collected Papers* [1974a] p. 294.

[Weis 1938]   J. WEISSINGER, Theorie der Divisorenkongruenzen, *Abh. Math. Sem. Hamburg* **12** (1938) pp. 115–126.

Notices of the AMS
November 1995
pp. 1301–1307

Some History of the Shimura-Taniyama Conjecture

Serge Lang

I shall deal specifically with the history of the conjecture which asserts that every elliptic curve over **Q** (the field of rational numbers) is modular. In other words, it is a rational image of a modular curve $X_0(N)$, or equivalently of its Jacobian variety $J_0(N)$. This conjecture is one of the most important of the century. The connection of this conjecture with the Fermat problem is explained in the introduction to Wiles's paper (*Ann. of Math.* May 1995), and I shall not return here to this connection. However, over the last thirty years, there have been false attributions and misrepresentations of the history of this conjecture, which has received incomplete or incorrect accounts on several important occasions. For ten years, I have systematically gathered documentation which I have distributed as the "Taniyama-Shimura File." Ribet refers to this file and its availability in [Ri 95]. It is therefore appropriate to publish a summary of some relevant items from this file, as well as some more recent items, to document a more accurate history. I call the conjecture the Shimura-Taniyama conjecture for specific reasons which will be made explicit.

Serre's Bourbaki Seminar. To start, I quote from Serre's Bourbaki Seminar of June 1995, when Serre wrote:

> Une courbe elliptique sur Q pour laquelle la conjecture 1′ est vraie a été longtemps appelée une courbe "de Weil." On dit maintenant que c'est une courbe elliptique "modulaire."

> Le terme de "conjecture de Weil" a été d'abord utilisé pour désigner l'ensemble des conjectures du n° 1.1; c'était un peu facheux, vu le risque de confusion avec d'autres conjectures de Weil. On est passé de là à "conjecture de Taniyama-Weil"; c'est la terminologie utilisée ici. Plus récemment, on trouve "conjecture de Shimura-Taniyama-Weil," ou même "conjecture de Shimura-Taniyama," le nom de Shimura étant ajouté en hommage à son étude des quotients de $J_0(N)$. Le lecteur choisira. L'essential est qu'il sache qu'il s'agit du même énoncé.

Serre's statement that "Shimura's name was added in homage to his study of the quotients of $J_0(N)$" is false. Serre misrepresents other people's reasons for associating Shimura's name to the conjecture, namely, that the conjecture is due principally to Shimura. An "Erratum" in the *Notices* (January 1994) corrected the previous use of the expression "Taniyama conjecture" in two previous articles (July/August 1993 and October 1993) and concluded that these articles "should have used the standard name, 'Taniyama-Shimura Conjecture'." Wiles in his e-net message of 4 December 1993 called it the Taniyama-Shimura conjecture. In the article [DDT 95] by Darmon, Diamond, and Taylor, it is called the Shimura-Taniyama conjecture. Faltings in his account of Wiles's proof in the *Notices* July 1995 refers to "the conjecture of Taniyama-Weil (which essentially is due to Shimura)." Thus Faltings points to a contradiction in the way some people have called the conjecture.

So what happened which led to such contradictions?

§0. Preliminaries: Hasse's Conjecture

In the 1920s and 1930s until about 1940–1941, zeta functions and L-functions had been extensively studied by Artin, Hasse, and Hecke from various points of view. There is no need to go extensively into this preliminary history here; but to understand the context of what follows, it is worth recalling that in the thirties, Hasse defined the zeta function of a variety over a number field by taking the product over all prime ideals of the zeta functions of this variety reduced modulo the primes. He conjectured that this product has a meromorphic continuation over the whole plane and a functional equation. In an influential address and paper, Weil brought the conjecture to the attention of the mathematical community at the International Congress in 1950. He attributed this "very interesting conjecture" to Hasse [We 1950b], cf. *Collected Papers* Vol. I, p. 451. Weil commented: "In a few simple cases, this function [previously defined by Hasse] can actually be computed; e.g., for the curve $Y^2 = X^3 - 1$ it can be expressed in terms of Hecke's L-functions for the field $k(\sqrt[3]{1})$; this example also shows that such functions have infinitely many poles, which is a clear indication of the very considerable difficulties that one may expect in their study." In 1950, as far as I know, Hasse had not published his conjecture, but he did publish it in 1954; see his comments on the first page of [Ha 54].

§1. The Situation in 1955

The Taniyama problems. Renewed interest in modular curves in the post–WW II period of mathematics occurred in the fifties as a result of work of Taniyama and Shimura. Taniyama at the conference on number theory in Tokyo-Nikko in 1955 was interested in obtaining various zeta functions and L-series as Mellin transforms of some type of automorphic forms. He formulated four problems along these lines—problems 10, 11, 12, 13—in a collection of 36 problems passed out in English at this conference, which was attended by both Serre and Weil. Although these problems were published in Japanese in Taniyama's collected works, they were not, unfortunately, published in English. However, many people, including Serre, had copies. Serre drew attention to these problems in the early 1970s. Taniyama's problem 10 was concerned with Dedekind zeta functions and Hecke L-series, as follows (incorrect English being reproduced as in the original here and subsequently):

10. Let k be a totally real number field, and $F(\tau)$ be a Hilbert modular form to the field k. Then, choosing $F(\tau)$ in a suitable manner, we can obtain a system of Hecke's L-series with "Grössencharaktere" λ, which corresponds one-to-one to this $F(\tau)$ by the process of Mellin-transformation. This can be proved by a generalization of the theory of operator T of Hecke to Hilbert modular functions (cf. Herrmann).

The problem is to generalize this theory in the case where k is a general (not necessarily totally real) number field. Namely, to find an automorphic form of several variables from which L-series with "Grössencharaktere" λ of k may be obtained, and then to generalize Hecke's theory of operator T to this automorphic form.

One of the aim of this problem is to characterize L-series with "Grössen oder Klassencharaktere" of k; especially to characterize the Dedekind zeta function of k in this method, which is not yet done even if k is totally real.

Problem 11 shifts to elliptic curves with complex multiplication, but is less relevant to the questions considered here. Then Taniyama formulates two problems which begin the process of identifying the zeta function of an elliptic curve with the Mellin transform of some automorphic form, namely, problems 12 and 13, which we quote in full.

12. Let C be an elliptic curve defined over an algebraic number field k, and $L_C(s)$ denote the L-function of C over k. Namely,

$$\zeta_C(s) = \frac{\zeta_k(s)\zeta_k(s-1)}{L_C(s)}$$

is the zeta function of C over k. If a conjecture of Hasse is true for $\zeta_C(s)$, then the Fourier series obtained from $L_C(s)$ by the inverse Mellin transformation must be an automorphic form of dimension -2, of some special type (cf. Hecke). If so, it is very plausible that this form is an elliptic differential of the field of that automorphic functions. The problem is to ask if it is possible to prove

· Hasse's conjecture for C, by going back to this considerations, and by finding a suitable automorphic form from which $L_C(s)$ may be obtained.

13. Concerning the above problem, our new problem is to characterize the field of elliptic modular functions of "Stufe" N, and especially, to decompose the Jacobian variety J of this function field into simple factors, in the sense of isogeneity.

It is well known, that, in case $N = q$ is a prime number, satisfying $q = 3 \pmod 4$, J contains elliptic curves with complex multiplication. Is this true for general N?

As Shimura has pointed out, there were some questionable aspects to the Taniyama formulation in problem 12. First, the simple Mellin transform procedure would make sense only for elliptic curves defined over the rationals; the situation over number fields is much more complicated and is not properly understood today, even conjecturally. Second, Taniyama had in mind automorphic forms much more general than what are now called "modular forms" which belong to the modular curves $X_0(N)$.[1]

The "mysterious" elliptic curves over Q. In any case, at the time the matter was an enigma. In a letter to me dated 13 August 1986, Shimura brought to my attention notes taken from Taniyama of an informal discussion session held 12 September 1955, 7:30–9:30 p.m. These notes were

---

[1] At my request to get clarification, Shimura wrote me on 22 September 1986:

I think Taniyama wasn't very careful when he stated his problem No. 12. He referred to Hecke and as I wrote, *he was thinking about Hecke's paper No. 33 (1936) which concerns automorphic function fields of dimension one*. The functional equation treated there involves only one $\Gamma(s)$, so that the *problem doesn't make sense unless the curve is defined over Q*.

Also, he specifically speaks of a form of weight 2, and of an elliptic differential of the function field. These make sense only in the one-dimensional case for the following reasons.

First of all, you have to remember, in 1955, the results of Hecke, Maass, and Hermann were the only relevant things. Obviously the Maass theory can be eliminated, because it doesn't produce function fields nor elliptic differentials. In the Hilbert modular case, if an elliptic differential means a holomorphic 1-form, then its weight must be $(2, 0, \ldots, 0)$, $(0, 2, 0, \ldots, 0)$, or $(0, \ldots, 0, 2)$, but such a form cannot be called a form of weight 2. In fact, such a nonvanishing form doesn't exist. If it is a form of weight $(2, \ldots, 2)$, then it defines a differential form of highest degree, so that you cannot call it an elliptic differential.

For these reasons, I think he was not completely careful, and if someone had pointed out this, he would have agreed that the problem would have to be revised accordingly.

published in Japanese in *Sugaku*, May 1956, pp. 227–231, giving the following exchange between Taniyama and Weil (Shimura's translation):

Weil asks Taniyama: Do you think all elliptic functions are uniformized by modular functions?

Taniyama: Modular functions alone will not be enough. I think other special types of automorphic functions are necessary.

Weil: Of course some of them can probably be handled that way. But in the general case, they look completely different and mysterious. But, for the moment, it seems effective to use Hecke operators. Eichler employed the Hecke theory, and certain elliptic curves with no complex multiplication are contained (in his results). Infinitely many such elliptic curves...

Deuring: No, only a finite number of such curves are known.

In this letter of 13 August 1986, Shimura also write me:

The same issue [of *Sugaku*] contains also Taniyama's problem No. 12 (p. 269), in which he says that the Mellin transform of the zeta function of an elliptic curve must be an automorphic form of weight 2 of a special type (cf. Hecke). He doesn't say modular form. That explains why he said other automorphic functions were necessary. I am sure he was thinking of Hecke's paper No. 33 (1936) which involves some Fuchsian groups not necessarily commensurable with $SL_2(Z)$. Of course, in 1955, our understanding of the subject was incomplete, and he wasn't bold enough to speculate that modular functions were enough.

As for Weil, he was far from the conjecture. (It seems that strictly speaking, Weil has never made the conjecture; see item 4 below.) Indeed, in his lecture titled "On the breeding of bigger and better zeta functions" at the University of Tokyo sometime in August or September 1955, he mentions Eichler's result and adds: "But already in the next simplest case, that is, the case of an elliptic curve which cannot be connected with modular functions in Eichler's fashion, the properties of its zeta function are completely mysterious...." (loc. cit. p. 199)

§2. The Sixties

Shimura's conjecture. Shimura himself in the late fifties and sixties extended Eichler's results and proved that elliptic curves which are modular have zeta functions which have an analytic continuation. (Cf. the three papers [Sh 58], [Sh 61], and [Sh 67].) But except for Shimura, it was universally accepted in the early sixties that most elliptic curves over the rationals are not modular. In a letter to Freydoon Shahidi (16 September 1986), Shimura gave evidence to this effect when he wrote:

> At a party given by a member of the Institute in 1962–64, Serre came to me and said that my results on modular curves (see below) were not so good since they didn't apply to an arbitrary elliptic curve over Q. I responded by saying that I believed such a curve should always be a quotient of the Jacobian of a modular curve. Serre mentioned this to Weil who was not there. After a few days, Weil asked me whether I really made that statement. I said: "Yes, don't you think it plausible?"

At this point, Weil replied: "I don't see any reason against it, since one and the other of these sets are denumerable, but I don't see any reason either for this hypothesis." (For a confirmation by Weil of this conversation, see below.)[2]

In the middle sixties, Shimura was giving lectures on the arithmetic theory of modular forms. Especially, he gave a version of the functional equation satisfied by a modular elliptic curve, which he communicated to Weil in 1964–65. This version was extended to higher dimensional factors in his book *Introduction to the Arithmetic Theory of Automorphic Functions*, Theorems 7.14 and 7.15.

Weil's 1967 paper. No attribution of the conjecture to Shimura. After thinking about the conjecture told to him by Shimura, Weil published his 1967 paper "Über die Bestimmung Dirichletscher Reihen durch Functionalgleichungen" [We 1967a] in which he proved that if the zeta function of an elliptic curve and sufficiently many "twists" have a functional equation, then it is the Mellin transform of a modular form. However, nowhere in this paper does Weil mention Taniyama's or Shimura's role in the conjecture. In the letter to Shahidi, Shimura also stated that he explained to Weil "perhaps in 1965" how the zeta function of a modular elliptic curve has an analytic continuation. At the end of his 1967 paper, Weil acknowledges this ("nach eine mitteilung von G. Shimura..."). But Shimura added:

> I even told him [Weil] at that time that the zeta function of the curve C' mentioned there is the Mellin transform of the cusp form in question, but he spared that statement. Eventually I published a more general result in my paper in *J. Math. Soc. Japan* 25 (1973), as well as in my book (Theorem 7.14 and Theorem 7.15).

---

[2]The rationale for Shimura's conjecture was precisely the conjectured functional equation (Hasse), along the lines indicated in Taniyama's problem 12, suitably corrected. Shimura's bolder insight was that the ordinary modular functions for a congruence subgroup of $SL_2(Z)$ suffice to uniformize elliptic curves defined over the rationals.

Of course Weil made a contribution to this subject on his own, but he is not responsible for the result on the zeta functions of modular elliptic curves, nor for the basic idea that such curves will exhaust all elliptic curves over Q.

Weil calls the modularity "still problematic."   Actually, at the very end of his 1967 paper, written in German, Weil concludes: "Ob die Dinge immer, d.h. für jede über Q definierte Kurve C, sich so verhalten, scheint im Moment noch problematisch zu sein und mag dem interessierten Leser als Übungsaufgabe empfohlen werden." By "sich so verhalten," Weil meant whether every elliptic curve over Q is modular, and so even then, he did not outright make the conjecture; he called it "at the moment still problematic" and left it as an "exercise for the interested reader"!

Weil's 1979 account of the conversation with Shimura.   A decade later, Weil gave an account of previous work on the subject, and he translated into French the answer he gave to Shimura when Shimura expressed the conjecture to him. I reproduce here these historical comments from *Weil's Collected Papers* Vol. III (1979), p. 450.

> ...D'autre part, Eichler en 1954, puis Shimura en 1958 dans des cas plus généraux, avaient déterminé les fonctions zêta de courbes définies par des sous-groupes de congruence du groupe modulaire; la célèbre courbe de Fricke définie par le group $\Gamma_0(11)$ (cf. [1971a], pp. 143–144) en était un exemple typique.
>
> Dans les cas traités par Eichler et par Shimura, on savait d'avance que la fonction zêta de la courbe est transformée de Mellin d'une forme modulaire. Déja en 1955, au colloque de Tokyo-Nikko, Taniyama avait proposé de montrer que la fonction zêta de toute courbe elliptique définie sur un corps de nombres algébriques est la transformée de Mellin d'une forme automorphe d'un type approprié; c'est le contenu du problème 12 de la collection de problèmes déja cité (v. [1959a]*). Quelques années plus tard, à Princeton, Shimura me demanda si je trouvais plausible que toute courbe elliptique sur Q fût contenue dans la jacobienne d'une courbe définie par un sous-groupe de congruence du groupe modulaire; je lui répondis, il me semble, que je n'y voyais pas d'empêchement, puisque l'un et l'autre ensemble est dénombrable, mais que je ne voyais rien non plus qui parlât en faveur de cette hypothèse.

When I first read Weil's answer about "one and the other set being denumerable," I characterized it as "stupid." I have since also called it inane. But actually, Weil's answer gives further evidence that he did not think of the conjecture himself. Indeed, as a result of his conversations with Serre and Weil, Shimura was directly responsible for changing the prevailing psychology about elliptic curves over Q. Weil's account of the conversation with Shimura in his collected papers as quoted above confirms in the published record Shimura's report of the conversation.

Thus a major source of confusion and contradictions in the way the conjecture has been reported for three decades lies in the fact that the above historical comments were not made in [We 1967a], *let alone in the introduction to that paper, but were made only in 1979 in Weil's collected papers.*

§3. The Seventies: Weil Inveighs against Conjectures

As a result of a wider distribution of Taniyama's problems in the early seventies, the terminology of "Weil curves" shifted to "modular curves," and the conjecture that all elliptic curves over the rationals are modular became the "Taniyama-Weil conjecture" rather than the "Weil conjecture." Throughout the sixties and seventies, there was still incomplete knowledge of Shimura's role, partly because he himself did not have anything in print about the conjecture. For Shimura's explanation, see § 4 below. However, when others brought out the role of Taniyama and Shimura, Weil started inveighing against conjectures in general, especially in two instances in 1974 and 1979. He did it first in his "Two lectures on number theory, past and present" [We 74], when he wrote:

> For instance, the so-called "Mordell conjecture" [*why "so-called"?
> S.L.*] on Diophantine equations says that a curve of genus at least
> two with rational coefficients has at most finitely many rational
> points. It would be nice if this were so, and I would rather bet for
> it than against it. But it is no more than wishful thinking because
> there is not a shred of evidence for it, and also none against it.

Weil picked up the same theme against conjectures in 1979 comments about [We 1967a], *Collected Papers* Vol. III, p. 453, when he wrote specifically about the way he avoided mentioning conjectures when he lectured about the results of his 1967 paper [We 1967a]:

> Néanmoins, dans l'exposé que je fis de mes résultats à Munich en
> Juin 1965, à Berkeley en Février 1966, puis dans [1967a], j'évitais
> de parler de "conjectures." Ceci me donne l'occasion de dire mon
> sentiment sur ce mot dont on a tant usé et abusé. . .

Weil went on, mentioning first Burnside's conjecture, which turned out to be false, and then inveighing against Mordell's conjecture (as documented further below). Thus Weil had a quite different attitude from the one he had when he brought Hasse's conjecture to the attention of the mathematical community in 1950. In writing this way, Weil conveniently (consciously or not) set a self-justifying stage for avoiding to mention the Shimura-Taniyama conjecture as such. About Weil being against the idea of making conjectures as on page 454 of his collected works Vol. III, Shimura wrote in his letter to Shahidi: "For this reason, I think, he avoided to say in a straightforward way that I stated the conjecture."

The extent to which Weil went out on a limb in 1979, going one better than in 1974, is shown by the following passage:

Nous sommes moins avancés à l'égard de la "conjecture de Mordell." Il s'agit la d'une question qu'un arithméticien ne peut guère manquer de se poser; on n'aperçoit d'ailleurs aucun motif sérieux de parler pour ou contre.

First, concerning Weil's statement that "Mordell's conjecture" is a "question which an arithmetician can hardly fail to raise," I would ask when? It is quite a different matter to raise the question in 1921, as did Mordell, or decades later. Indeed, in his thesis [We 28] (see also *Collected Papers* Vol. I, p. 45) Weil wrote quite differently (my translation):

This conjecture, already stated by Mordell (loc. cit. note [4]) seems to be confirmed in some measure by an important result recently proved, and which I am happy to be able to cite here thanks to the kind permission of its author: "On every curve of genus $p > 0$, and for any number field k of rationality, there can be only a finite number of points whose coordinates are integers of k."

The above important result is of course Siegel's theorem on the finiteness of integral points. Weil made a similar evaluation in [We 36] (*Collected Papers* Vol. I, p. 126) without reference to Mordell:

On the other hand, Siegel's theorem, for curves of genus > 1, is only the first step in the direction of the following statement:

On every curve of genus > 1, there are only finitely many rational points.

This seems extremely plausible, but undoubtedly we are still far from a proof. Perhaps one will have to apply here the method of infinite descent directly to the curve itself rather than to associated algebraic varieties. But, first of all, it will be necessary to extend the theory of abelian functions to nonabelian extensions of fields of algebraic functions. As I hope to show, such an extension is indeed possible. In any case, we face here a series of important and difficult problems, whose solution will perhaps require the efforts of more than one generation.

In addition, in 1979 Weil made comments on page 525 of Vol. I of his collected works concerning his papers [1927c] and [1928], the latter being his thesis [We 28], and he makes similar comments on pages 528–529 about [1932c], showing clearly the influence Mordell had in making the conjecture.[3]

Second, the statements in 1974 and 1979 that there is no "shred of evidence" or "motif sérieux" for Mordell's conjecture went not only against Weil's own evaluations in earlier decades (1928, 1936), but they were made after Manin proved the function field analogue in 1963 [Man 63]; after Grauert gave his other proof in 1965 [Gr 65]; after Parshin gave his other proof in 1968 [Par 68], while indicating that Mordell's conjecture follows from Shafarevich's conjecture (which Shafarevich himself had proved for curves of genus 1); at the same time that Arakelov theory was being developed and that Zarhin was working actively on the net of conjectures in those directions; and within four years of Falting's proof of Mordell's conjecture. In addition to that, some mathematicians thought there was experimental evidence for Mordell's conjecture, as when Parshin wrote in [Pa 68]: "Finally when $g > 1$ numerous examples provide a basis for Mordell's conjecture..." Thus, as I stated in a letter dated 7 December 1985 (reproduced in the Taniyama-Shimura file), when Weil wrote that "one sees no reason to be for or against" Mordell's conjecture, all he showed was that he was completely out of it in 1979.

§4. The Spread of Improper Attributions

Weil's 1967 paper received considerable attention. At the time, Shimura's book including Theorems 7.14 and 7.15 was not yet available. Weil's formulation went beyond Shimura's in that when the differential of first kind on the elliptic curve corresponds to a modular form of level N, Weil made excplicit that this level is also the conductor of the elliptic curve, namely, a certain a priori definable integer divisible precisely by the primes dividing the discriminant, but usually to much smaller powers. This connection with the conductor suggested explicit computations to those working in

---

[3]I quote from these pages:

p. 525: Mon ambition avait été de prouver aussi que, sur une courbe de genre > 1, les points rationnels sont en nombre fini; c'est la "conjecture de Mordell". Je le dis à Hadamard. "Travaillez-y encore," me ditil: "vous vous devez de ne pas publier un demi-résultat." Apres quelques nouvelles tentatives, je décidai de ne pas suivre son conseil.

pp. 528–529: Je n'avais pas encore renoncé à démontrer un jour la "conjecture de Mordell"; je ne désespérais même pas de pouvoir me rapprocher de ce but (lointain encore aujourd'hui) par une analyse attentive et un approfondissement des moyens mis in oeuvre dans ma thèse...

Mon espoir secret, bien entendu, était qu'il permetrait d'avancer en direction de la conjecture de Mordell; il n'en a pas été ainsi, que je sache, jusqu'à present.

the field, and these computations (in addition to the more structural evidence provided by the conjectured functional equation) in turn led them to believe the conjecture. Modular elliptic curves over Q were then called "Weil curves." The idea that all elliptic curves over Q are modular was generally attributed to him, for example when Tate referred to it as "Weil's astounding idea" in [Ta 74]. I myself for a decade used the terminology "Weil curve" and "Taniyama-Weil conjecture," before I learned more information.

For instance, Barry Mazur told me in 1986 that he heard Weil give credit to Shimura verbally (offhand) in a colloquium talk in the early sixties. However, in 1986, when I discussed these matters with Serre (publicly, in front of others at a party in Berkeley), he claimed that Weil had reported a conversation with Shimura as follows:

Weil: Why did Taniyama think that all elliptic curves are modular?

Shimura: You told him that, and you have forgotten.

I am reporting here the gist of the exchange. Weil may have asked, "Why did Taniyama make the conjecture?" I immediately wrote to both Shimura and Weil to ask them whether such a conversation took place and to verify what Serre had attributed to them. In his letter to me of 13 August 1986, already cited in § 1, Shimura answered categorically:

Such a conversation never took place...

1. It doesn't fit, or rather contradicts, the actual conversation between Weil and myself: "Did you make the statement that every Q-rational elliptic curve is modular?" "Yes, don't you think it plausible?" etc.

2. It would have been stupid of Weil to have asked why Taniyama made the conjecture. Once the statement is given, it makes sense, and it shouldn't have occurred to Weil to pose such a question. Indeed, Serre never asked the reason behind my statement.

. . .

4. Knowing the above passage and Taniyama's problem, and having stated the conjecture in my own way, I couldn't and wouldn't have attributed the origin of the conjecture to Weil. Besides, there is one point which almost all people seem to have forgotten. In his paper [1967a], Weil views the statement problematic. In other words, he was not completely for it, and so he didn't have to attribute it to me. Thus there is nothing for which you can take him to task. Anyway, for these reasons I have consistently and consciously avoided speaking of the Weil conjecture. For example, in my Nagoya paper (vol. 43, 1971), I proved that the conjecture was true for every elliptic curve with complex multiplication. Other authors treating the same problem would very naturally have mentioned Weil. But I

didn't. Also I always thought the reader of my book (*Introduction to the Arithmetic Theory* [of automorphic functions]) would wonder why the conjecture was not mentioned. The fact is, I was unable, or rather did not try very hard, to find a presentation of the topic in a way agreeable to everybody, including myself.

I then wrote both to Serre and Weil to ask them to comment on Shimura's reply. Of course I also sent them the Taniyama-Shimura file as it developed. Serre wrote me back two letters. The first dated 16 August 1986 criticized my attempt to verify what he told me, and thus gave rise to a side exchange. (Among other things, in this exchange, I asked Serre to correct his false reporting of the conversation between Shimura and Weil and to stop spreading false stories.) Serre's second letter to me dated 11 September 1986 stated briefly, in full, "Merci pour tes lettres, ainsi que la copie de celle de Shimura. Je les ai trouvées très instructives."

§5. Weil's Letter

Somewhat later on 3 December 1986, Weil wrote me back a longer letter which also contained comments concerning other items in the file, such as those mentioned in § 2 and § 3 above.

Dear Lang,

I do not recall when and where your letter of August 9 first reached me. When it did, I had (and still have) far more serious matters to think about.

I cannot but resent strongly any suggestion that I ever sought to diminish the credit due to Taniyama and to Shimura. I am glad to see that you admire them. So do I.

Reports of conversations held long ago are open to misunderstandings. You choose to regard them as "history"; they are not. At best they are anecdotes. Concerning the controversy which you have found fit to raise, Shimura's letters seem to me to put an end to it, once and for all.

As to attaching names to concepts, theorems or (?) conjectures, I have often said: (a) that, when a proper name gets attached to (say) a concept, this should never be taken as a sign that the author in question had anything to do with the concept; more often than not, the opposite is true. Pythagoras had nothing to do with "his" theorem, nor Fuchs with the fonctions fuchsiennes, any more than Auguste Comte with rue Auguste-Comte; (b) proper names tend, quite properly, to get replaced by more appropriate ones; the Leray-Koszul sequence is now a spectral sequence (and, as Siegel once told Erdős, abelian is now written with a small a).

Why shouldn't I have made "stupid" remarks sometimes, as you are pleased to say? But indeed, I was "out of it" in 1979 when expressing some skepticism about Mordell's conjecture, since at that time I was totally ignorant of the work of the Russians (Parshin, etc.) in that direction. My excuse, if it is one, is that I had had long conversations with Shafarevich in 1972, and he never mentioned any of that work.

Sincerely,

A. Weil

AW:ig

P.S. Should you wish to run this letter through your Xerox machine, do feel free to do so. I wonder what the Xerox Co. would do without you and the like of you.

Da capo. I have no explanation why Serre's Bourbaki Seminar talk took no account of Serre's past letter where he found Shimura's and my letters "instructive," of Weil's own historical comments in his *Collected Papers* Vol. III, p. 450, or of Weil's warning about "when a proper name gets attached to (say) a concept," as well as Weil's clearcut statement: "Concerning the controversy which you have found fit to raise, Shimura's letters seem to me to put an end to it, once and for all."

Bibliography

[DDT 95]   H. DARMON, F. DIAMOND, and R. TAYLOR, *Fermat's Last Theorem*, Current Developments in Math. 1995, International Press, Cambridge MA, pp. 1–107.

[Fa 83]   G. FALTINGS, *Endlichkeitssätze für abelsche Varietäten über Zahlkörpern*, Invent. Math. 73 (1983) pp. 349–365.

[Fa 95]   G. FALTINGS, *The proof of Fermat's Last Theorem* by R. Taylor and A. Wiles, *Notices AMS* (July 1995), 743–746.

[Gr 65]   G. GRAUERT, *Mordell's Vermutung über rational Punkte auf algebraishchen Kurven and Funktionenköper*, Inst. Hautes Études Sci. Pub. Math. 25 (1965), pp. 131–149.

[Ha 54]   H. HASSE, *Zetafunktionen und L-Funktionen zu einem arithmetischen Funktionenkörper vom Fermatschen Typus*, Abh. S. Akad. Wiss. Berlin Math. Kl. (1954) 5–70; see also Hasse's *Collected papers* Math. Abh. Vol. II, p. 450.

[Ma 63]   I. MANIN, *Rational points of algebraic curves over function fields*, Izv Akad. Nauk SSSR Ser. Mat. 27 (1963), 1395–1440.

[Mo 21]   L. J. MORDELL, *On the rational solutions of the indeterminate equation of the third and fourth degrees*, Proc. Cambridge Philos. Soc. 21 (1922), 179–192.

[Pa 68]   A. N. PARSHIN, *Algebraic curves over function fields*, [zv. Akad. Nauk SSSR Ser. Mat. 32 (1968); English translation Math USSR Izv. 2 (1968), 1145–1170.

[Ri 95]   K. RIBET, *Galois representations and modular forms*, Bulletin AMS 32 (1995), 375–402.

[Se 95]   J.-P. Serre, *Travaux de Wiles (et Taylor. . .)*, Partie I, Séminaire Bourbaki 1994–95, June 1995, No. 803.

[Sh 58]   G. SHIMURA, *Correspondences modulaires et les fonctions zeta de courbes algébriques*, J. Math. Soc. Japan 10 (1958), 1–28.

[Sh 61]   G. SHIMURA, *On the zeta functions of the algebraic curves uniformized by certain automorphic functions*, J. Math. Soc. Japan 13 (1961), 275–331.

[Sh 67]   G. SHIMURA, *Class fields and zeta functions of algebraic curves*. Ann. of Math. 85 (1967), 58–159.

[Ta 74]   J. TATE, *The arithmetic of elliptic curves*, Invent. Math. 23 (1974), 179–206.

[We 28]   A. WEIL, *L'arithmétique sur les courbes algébriques*, Acta Math. 52 (1928), 281–315.

[We 36]   A. WEIL, *Arithmetic on algebraic varieties*, Uspekhi Mat. Nauk 3 (1936), 101–112; see also *Collected Papers* Vol. I, 1936b, p. 126.

[We 1950b]   A. WEIL, *Number theory and algebraic geometry*, Proc. Internat. Congr. 1950, Cambridge, Mass. Vol. II, pp. 90–100.

[We 1967a]   A. WEIL, *Über die Bestinnumg Dirichletscher Reihen Durch Funktionalgleichungen*, Math. Ann. (1967), 165–172.

[We 74]   A. WEIL, *Two lectures on number theory past and present*, Collected Papers, 1974a, p. 94.

[We 79]   A. WEIL, *Collected Papers*, Springer-Verlag, 1979 (three volumes).

janvier 1996 / n° 67

GAZETTE DES
MATHEMATICIENS

société   mathématique   de   france

Eric Legrand, GLCS, 80 Rue de Paris, 93100 Montreuil.

Lætitia Paoli, URA 740 CNRS Analyse Numérique, UER de Sciences, Université de Saint-Etienne, 23 rue Dr Paul Michelon, 42023 Saint-Etienne Cedex.

Michelle Schatzman, URA 740 CNRS Analyse Numérique, Université Lyon 1, 69622 Villeurbanne Cedex.

LA CONJECTURE DE BATEMAN-HORN

Serge LANG

Nn article récent d'Arnold sur le Congrès de Zurich traite entre autre de l'importance relative de certains problèmes en mathématiques. Sous le titre "Les mathématiques vont-elles survivre ? ", il écrit : «Des questions étranges, comme le problème de Fermat ou ceux sur les sommes de nombres premiers, ont été présentées comme centrales en mathématiques. "Pourquoi additionner les nombres premiers", se demandait perplexe le grand physicien Landau. "Les nombres premiers sont faits pour être multipliés et non additionnés" ». On peut déjà noter que Vinogradov en 1937 avait démontré le théorème que tout nombre impair suffisamment grand est la somme de trois nombres premiers [Vi 37], et que les mathématiques ont survécu à Vinogradov. Comme ça fait environ 60 ans, j'espère qu'Arnold ne perdra pas l'espoir pour le futur.

Je note en outre que le problème classique des nombres premiers jumeaux est un problème de la théorie additive des nombres premiers, et tombe donc sous la rubrique de "questions étranges" d'Arnold-Landau. Tous les goûts sont permis, mais si le "grand physicien" Landau avait un point de vue restreint partagé par Arnold, par contre des mathématiciens, petits, moyens, ou grands, un peu partout et un peu toujours, se sont occupés de nombres premiers dans des contextes où la structure additive se mélangeait avec la structure multiplicative. Mais ce qui est bien plus important, c'est que la conjecture des nombres premiers jumeaux n'est qu'un cas particulier de conjectures de Hardy-Littlewood [HaL 22] (voir plus bas), et de la conjecture de Bateman-Horn [BaH 62], qui n'est malheureusement pas bien connue (l'est elle par Arnold ?). Elle mérite de l'être, et je la considère comme une des grandes conjectures du siècle, d'où cette lettre.

Commençons par un cas particulier, considéré déja au siècle dernier par Bouniakovsky [Bo 1854]. Soit $f(X)$ un polynome à coefficients entiers, de degré au moins 1. On se pose la question de savoir si $f(n)$ représente une infinité de nombres premiers quand n parcourt les entiers positifs. On peut aussi aller plus avant, et demander un comportement asymptotique. Bien entendu, il y a des conditions nécéssaires, par exemple que les coefficients soit positif, et aussi que f soit irréductible. Ces conditions qui sont les premières auxquelles on pense sont elles suffisantes ? Eh bien non, comme l'a remarqué Bouniakowski. Par exemple, pour chaque nombre premier p, on a

$$X^p - X - p = 0 \mod p.$$

On peut naturellement remplacer le terme constant p par pk avec k entier quelconque, de sorte que le polynôme soit irréductible, par exemple avec $p = 2$, $p = 3$, etc. Il faut donc supposer au moins que l'ensemble des valeurs $f(n)$ pour n positif ne soient divisibles par aucun nombre premier. La conjecture de Bouniakowski est que ces conditions sont suffisantes. Un cas particulier de cette conjecture est que le polynome $X^2 + 1$ représente une infinité de nombres premiers. De même, le théorème de Dirichlet selon lequel il y a une infinité de nombres premiers dans une progression arithmétique en est un cas particulier, avec le polynôme $aX + b$, avec a, b entiers premiers entre eux, et a positif.

La conjecture de Bouniakowski fut redécouverte et généralisée à plusieurs polynômes par Schinzel [Sch 58], voir aussi les commentaires dans [HaR 74]. Mais Bateman-Horn conjecturent un comportement asymptotique comme suit. Soient $f_1, ..., f_r$ des polynômes à coefficients entiers, de degrés $d_1, ..., d_r$ positifs, irréductibles, et dont les coefficients de plus haut degré soient positifs. Soit

$$f = f_1 \cdots f_r$$

leur produit. On suppose la condition de Bouniakowski, qu'il n'y a pas de nombre premier p divisant toutes les valeurs $f(n)$ pour n entier positif. On définit

$\pi_f(x) = $ nombre d'entiers positifs $n \le x$ tels que $f_1(n), ..., f_r(n)$ sont tous premiers ;

$N_f(p) = $ nombre de solutions de la congruence $f(n) = 0 \mod p$ (pour p premier).

Soit $C(f)$ la constante (aux places finies !)

$$C(f) = \prod_p (1 - 1/p)^{-r} (1 - N_f(p)/p)$$

le produit étant pris sur tous les nombres premiers p. Alors la conjecture de Bateman-Horn dit que

$$\pi_f(x) \sim \frac{C(f)}{d_1 ... d_r} \int_2^x \frac{dt}{(logt)^r}$$

n° 67 – JANVIER 1996

La formule donne bien la densité de Dirichlet pour le polynomial $f(X) = aX + b$. Elle redonne, après des calculs, les comportements asymptotiques conjecturés par Hardy-Littlewood pour la représentation de nombres premiers par le polynôme $X^2 + 1$; et pour les nombres premiers jumeaux, quand on l'applique aux deux polynômes X et $X + 2$ [HarL 22]. Elle redonne aussi d'autres conjectures de Hardy-Littlewood dans ce même article, rempli de superbes conjectures, dont certaines sont des conséquences de l'hypothèse de Riemann généralisée. En particulier, Hardy-Littlewood donnent conjecturalement le nombre asymptotique de représentations d'un nombre entier comme somme de trois nombres premiers. Leur analyse heuristique traite en même temps un problème comme celui auquel s'est occupé Vinogradov, et un problème comme celui de Bouniakovski.

Bibliographie

[BaH 62] P. T. BATEMAN et R. HORN, *A heuristic asymptotic formula concerning the distribution of prime numbers*, Math. Comp. 16 (1962) pp. 363-367

[Bo 1854] V. BOUNIAKOWSKY, *Sur les diviseurs numériques invariables des fonctions rationnelles entières*, Mémoires sc. math. et phys. T. VI (1854-1855) pp. 307-329

[HalR 74] H. HALBERSTAM and H.-E. RICHERT, *Sieve methods*, Academic Press, (1974)

[HarL 22] G. H. HARDY and J. E. LITTLEWOOD, *Some problems of Partitio Numerorum III*, Acta Math. 44 (1922) pp. 1-70

[Sch 58] A. SCHINZEL and W. SIERPINSKI, *Sur certaines hypothèses concernant les nombres premiers*, Acta Arith. 4 (1958) pp. 185-208

[Vi 37] I. M. VINOGRADOV, *Representation of an odd number as a sum of three primes*, CR (Dokl) Acad. Sci. USSR (1937) pp. 169-172

Notices of the AMS
October 1996
pp. 1119–1123

Comments on Chow's Works

Serge Lang

Van der Waerden's pre-war series of article began an algebraization of Italian algebraic geometry. I was born into algebraic geometry in the immediate post war period. This period was mostly characterized by the work of Chevalley, Chow, Weil (starting with his *Foundations* and his books on correspondences and abelian varieties), and Zariski. In the fifties, there was a constant exchange of manuscripts among the main contributors of that period. I shall describe briefly some of Chow's contributions. I'll comment here mostly on some of Chow's works in algebraic geometry, which I know best.

§1. Chow Coordinates

One of Chow's most influential works was also his first, namely the construction of the Chow form, in a paper written jointly with van der Waerden [ChW 37]. To each projective variety, Chow saw how to associate a homogeneous polynomial in such a way that the association extends to a homomorphism from the additive monoid of effective cycles in projective space to the multiplicative monoid of homogeneous polynomials, and the association is compatible with the Zariski topology. In other words, if one cycle is a specialization of another, then the associated Chow form is also a specialization. Thus varieties of given degree in a given projective space decompose into a finite number of algebraic families, called Chow families. The coefficients of the Chow form are called the Chow coordinates of the cycle, or of the variety. Two decades later, he noted that the Chow coordinates can be used to generate the smallest field of definition of a divisor [Ch 50a]. He also applied the Chow form to a study of algebraic families when he gives a criterion for local analytic equivalence [Ch 50b]. He was to use them all his life, in various contexts dealing with algebraic families.

In Grothendieck's development of algebraic geometry, Chow coordinates were bypassed by Grothendieck's construction of Hilbert schemes, whereby two schemes are in the same family whenever they have the same Hilbert polynomial. The Hilbert schemes can be used more advantageously than the Chow families in some cases. However, as frequently happens in mathematics, neither is a substitute for the other in all cases. In recent times, say during the last decade, Chow forms and coordinates have made a reappearance due to a renewed emphasis on explicit constructions needed to make theorems effective (rather than having non-effective existence proofs, say), and for computational aspects of algebraic geometry whereby one wants not only theoretical effectiveness but good bounds for solutions of algebraic geometric problems as functions of bounds on the data. Projective constructions such as Chow's are very well suited for such purposes. Thus Chow coordinates reappeared both in general algebraic geometry, and also in Arakelov theory and in diophantine applications. The Chow coordinates can be used for example to define the height of a variety, and to compare it to other heights constructed by more intrinsic, non-projective methods as in [Ph 91], [Ph 94], [Ph 95]. They were used further in Arakelov theory by Bin Wang [Wa 96].

Chow coordinates were also used to prove a conjecture of Lie on a converse to Abel's theorem. See the papers by Wirtinger [Wi 38] and Chern [Che 83].

§2. Abelian Varieties and Group Varieties

(a) **Projective construction of the Jacobian variety.** In the fifties, Chow contributed in a major way to the general algebraic theory of abelian varieties due to Weil (who algebraicized the transcendental arguments of the Italian school, especially Castelnuovo). For one thing, Chow gave a construction of the Jacobian variety by projective methods, giving the projective embedding directly and also effectively [Ch 54]. The construction also shows that when a curve moves in an algebraic family, then the Jacobian also moves along in a corresponding family.

(b) **The Picard variety.** Chow complemented Igusa's transcendental construction of the Picard variety by showing how this variety behaves well in algebraic systems, using his "associated form" [Ch 52]. He announced an algebraic construction of the Picard variety in a "forthcoming paper." Indeed, such a paper circulated as an unpublished manuscript a few years later [Ch 55c], but was never published as far as I know.

(c) **Fixed part of an algebraic system.** Chow also developed a theory of algebraic systems of abelian varieties, defining the fixed part of such systems, i.e. that part which does not depend genuinely on the parameters [Ch 55a,b]. His notion of fixed part was used by others in an essential way, e.g. by Lang-Néron, who proved that for an abelian variety A defined over a function field K, the group of rational points of A in K modulo the group of points of the fixed part is finitely generated [LaN 59]. This is a relative version of the Mordell-Weil theorem.

(d) **Field of definition.** Chow gave conditions under which an abelian variety defined over an extension of a field k can actually be defined over k

itself [Ch 55a,b]. Chow's idea was extended by Lang [La 55] to give such a criterion for all varieties, not just abelian varieties, and Weil reformulated the criterion in terms of cohomology (splitting a cocycle) [We 56].

§3. Homogeneous Spaces

(a) **Projective embedding of homogeneous spaces.** Chow extended the Lefschetz-Weil proof of the projective embedding of abelian varieties to the case of homogeneous spaces over arbitrary group varieties, which may not be complete [Ch 57a]. Chow's proof has been overlooked in recent years, even though interest in projective constructions has been reawakened, but I expect Chow's proof to make it back to the front burner soon, just like his other contributions.

(b) **Algebraic properties.** Chow's paper [Ch 49b] dealt with the geometry of homogeneous spaces. The main aim of this paper is to characterize the group by geometric properties. The latter could refer to the lines in a space, as in projective geometry, or to certain kinds of matrices, such as symmetric matrices. For instance, a typical theorem says: Any bijective adjacence preserving transformation of the space of a polar system with itself is due to a transformation of the basic group, provided that the order of the space is greater than 1. Birational geometry is considered in this context.

§4. The Chow Ring

In topology, intersection theory holds for the homology ring. In 1956, Chow defined rational equivalence between cycles on an algebraic variety, he defined the intersection product for such classes, and thus obtained the Chow ring [Ch 56a], which proved to be just as fundamental in algebraic geometry as its topological counterpart.

§5. Algebraic Geometry over Rings

In the late fifties began the extension of algebraic geometry over fields to algebraic geometry over rings of various type, partly to deal with algebraic or analytic families, but partly because of the motivation from number theory, where one deals with local Dedekind rings, p-adic rings, and more generally complete Noetherian local rings. Chow contributed to this extension in several ways. Of course, in the sixties Grothendieck vastly and systematically went much further in this direction, but it is often forgotten that the process had begun earlier. I shall mention here some of Chow's contributions in this direction.

(a) **Connectedness theorem.** In 1951 Zariski had proved a general connectedness theorem for specializations of connected algebraic sets. Zariski based his proof on an algebraic theory of holomorphic functions which he developed for this purpose. In [Ch 57b] and [Ch 59], Chow gave a proof of a generalization over arbitrary complete Noetherian local domains, based on much simpler techniques of algebraic geometry, especially the Chow form.

(b) **Uniqueness of the integral model of a curve.** The paper [ChL 57c] proved the uniqueness of the model of a curve of genus ≥ 1 and abelian variety over a discrete valuation ring, in the case of nondegenerate reduction.

(c) **Cohomology.** Invoking the theory of deformations of complex analytic structures by Kodaira-Spencer, the connectedness theorem, and Igusa's work on moduli spaces of elliptic curves, Chow and Igusa proved the upper semicontinuity of the cohomology over a broad class of Noetherian local domains [ChI 58c]. Semicontinuity was proved subsequently in the complex analytic case by Grauert, and by Grothendieck in more general algebraic settings. However, Chow's and Igusa's contribution did not get the credit they deserved. Cf. [Ha 77], Chapter 3, §12, and the bibliographical references given there, referring to work in the sixties, but not to Chow-Igusa.

(d) **Bertini's theorem.** During that same period in the late fifties, Chow extended Bertini's theorem to local domains [Ch 58b].

(e) **Unmixedness theorem.** A homogeneous ideal defining a projective variety is said to be unmixed if it has no embedded prime divisors. Chow proved that the Segre product of two unmixed ideals is also unmixed, under fairly general conditions, in a ring setting [Ch 64].

§6. Algebraicity of Analytic Objects

Chow was concerned over many years with the algebraicity of certain complex analytic objects. We mention two important instances.

(a) **Meromorphic mappings and formal functions.** In 1949, Chow proved the fundamental fact, very frequently used from then on, that a complex analytic subvariety of projective space is actually algebraic [Ch 49a]. Twenty years later, he came back to similar questions, and proved in the context of homogeneous varieties that a meromorphic map is algebraic [Ch 69]. Remarkably, and wonderfully, almost twenty years after that, he came back once more to the subject and completed it in an important point [Ch 86]. I quote from the introduction to this paper, which shows how Chow was still lively mathematically: "Let X be a homogeneous algebraic variety on which a group G acts, and let Z be a subvariety of positive dimension. Assume that Z generates X [*in a sense which Chow makes precise*]... One asks whether a formal rational function on X along Z is the restriction along Z of an algebraic function (or even a rational function) on X. In a paper [Ch 69] some years ago, the author gave an affirmative answer to this question, under the assumption that the subvariety Z is complete, but only for the complex-analytic case with the formal function replaced by the usual analytic function defined in a neighborhood of Z. The question remains whether the result holds also for the formal functions in the abstract case over any ground field. We had then some thoughts on this question, but we did not pursue them any further as we did not see a way to reach the desired conclusion at the time. In a recent paper [3], Faltings raised this same question and gave a partial answer to it in a slightly different formulation. This result of Faltings led us to reconsider this question again, and this time we are more fortunate. In fact, we have been able not only to solve the problem, but also to do it by using essentially the same method we used in our original paper."

(b) **Analytic surfaces.** In a paper with Kodaira, it was proved that a Kähler surface with two algebraically independent meromorphic functions is a non-singular algebraic surface [ChK 52c].

§7. Other Works in Algebraic Geometry

Chow's papers in algebraic geometry include a number of others, which, as I already asserted, I am less well acquainted with, and won't comment upon, such as his paper on the braid group [Ch 48], on the fundamental group of a variety [Ch 52d], on rational dissections [Ch 56b], and on real traces of varieties [Ch 63].

§8. PDE

Chow's very early paper on systems of linear partial differential equations of first order [Ch 39c] gives a generalization of a theorem of Caratheodory on the foundations of thermodynamics. This paper had effects not well known to the present generation of mathematicians, including me. It was only just now brought to my attention. An anonymous colleague wrote to the editor of the present collection of articles on Chow's work: "This paper essentially asserts the identity of the integral submanifold of a set of vector fields and the integral submanifold of the Lie algebra generated by the set of vector fields. This is widely known as "Chow's theorem" in nonlinear control theory, and is the basis for the study of the controllability problem in nonlinear systems. Controllability refers to the existence of an input signal that drives the state of a system from a given initial state to a desired terminal state. A more detailed exposition of the role of Chow's theorem, with several references, is provided in the survey paper [Br 76]."

§9. Bibliography

Chow's Papers

[ChW 37a] (with van der Waerden) Zur algebraische Geometrie IX. *Math. Ann.* 113 (1937): pp. 692–704.

[Ch 37b] Die geometrische Theorie der algebraischen Funktionen für beliebige vollkommene Körper. *Math. Ann.* 114 (1937): pp. 655–682.

[Ch 39a] Einfacher topologischer Beweis des Fundamentalsatzes der Algebra. *Math. Ann.* 116 (1939): p. 463.

[Ch 39b] Über die Multiplizität der Schnittpunkte von Hyperflächen. *Math. Ann.* 116 (1939): pp. 598–601.

[Ch 39c] Über systemen von linearen partiellen Differentialgleichungen erster Ordnung. *Math. Ann.* 117 (1939): pp. 98–108.

[Ch 40] On electrical networks. *J. Chinese Math. Soc.* 2 (1940): pp. 3–160.

[Ch 48] On the algebraical braid group. *Ann. Math.* 49, no. 3 (1948): pp. 654–658.

[Ch 49a] On compact complex analytic varieties. *Amer. J. Math.* 71, no. 4 (1949): pp. 893–914.

[Ch 49b] On the geometry of algebraic homogeneous spaces. *Ann. Math.* 50, no. 1 (1949): pp. 32–67.

[Ch 49c] Über die Lösbarkeit gewisser algebraischer Gleichungssysteme. *Comment. Math. Helv.* 23, no. 1 (1949): pp. 76–79.

[Ch 49d] On the genus of curves of an algebraic system. *Trans. Am. Math. Soc.* 65 (1949): pp. 137–140.

[Ch 50a] On the defining field of a divisor in an algebraic variety. *Proc. Amer. Math. Soc.* 1, no. 6 (1950): pp. 797–799.

[Ch 50b] Algebraic systems of positive cycles in an algebraic variety. *Amer. J. Math.* 72, no. 2 (1950): pp. 247–283.

[Ch 52a] On the quotient variety of an abelian variety. *Proc. NAS* 38 (1952): pp. 1039–1044.

[Ch 52b] On Picard varieties. *Amer. J. Math.* 74, no. 4 (1952): pp. 895–909.

[ChK 52c] (with Kodaira) On analytic surfaces with two independent meromorphic functions. *Proc. NAS* 38, no. 4 (1952): pp. 319–325.

[Ch 52d] On the fundamental group of an algebraic variety. *Amer. J. Math.* 74 (1952): pp. 726–736.

[Ch 52e] On the quotient variety of an Abelian variety. *Proc. NAS* 38 (1952): pp. 1039–1044.

[Ch 54] The Jacobian variety of an algebraic curve. *Amer. J. Math.* 76, no. 2 (1954): pp. 453–476.

[Ch 55a] On Abelian varieties over function fields, *Proc. NAS* 41 (1955): pp. 582–586.

[Ch 55b] Abelian varieties over function fields, *Trans. Amer. Math. Soc.* 78 (1955): pp. 253–275.

[Ch 55c] Abstract theory of the Picard and Albanese varieties, unpublished manuscript.

[Ch 56a] On equivalence classes of cycles in an algebraic variety. *Ann. Math.* 64, no. 3 (1956): pp. 450–479.

[Ch 56b] Algebraic varieties with rational dissections. *Proc. NAS* 42 (1956): pp. 116–119.

[Ch 57a] On the projective embedding of homogeneous spaces, Lefschetz conference volume. *Algebraic Geometry and Topology.* Princeton University Press (1957).

[Ch 57b] On the principle of degeneration in algebraic geometry. *Ann. Math.* 66 (1957): pp. 70–79.

[ChL 57c] (With S. Lang) On the birational equivalence of curves under specialization. *Amer. J. Math.* 79 (1952): pp. 649–652.

[Ch 58a] The criterion for unit multiplicity and a generalization of Hensel's lemma. *Amer. J. Math.* 80, no. 2 (1958): pp. 539–552.

[Ch 58b] On the theorem of Bertini for local domains. *Proc. NAS* 44, no. 6 (1958): pp. 580–584.

[ChI 58c] (with Igusa) Cohomology theory of varieties over rings. *Proc. NAS* 44, no. 12 (1958): pp. 1244–1248.

[Ch 58d] Remarks on my paper "The Jacobian variety of an algebraic curve," *Amer. J. Math.* 80 (1958): pp. 238–240.

[Ch 59]   On the connectedness theorem in algebraic geometry. *Amer. J. Math.* 81, no. 4 (1959): pp. 1033–1074.

[Ch 63]   On the real traces of analytic varieties. *Am. J. Math.* 85, no. 4 (1963) pp. 723–733.

[Ch 64]   On the unmixedness theorem. *Am. J. Math.* 86 (1964): pp. 799–822.

[Ch 69]   On meromorphic maps of algebraic varieties. *Ann. Math.* 89, no. 2 (1969): pp. 391–403.

[Ch 86]   Formal functions on homogeneous spaces, *Invent. math.* 86 (1986): 115–130.

Papers by Others

[Br 76]   R.W. Brockett. Nonlinear systems and differential geometry. *Proc. IEEE* 64 (1976): pp. 61–71.

[Che 83]   S.S. Chern. Web Geometry. *AMS Proc. Symp. in Pure Math.* 39 (1983): pp. 3–10.

[Ha 77]   R. Hartshorne. *Algebraic Geometry*. Springer-Verlag, 1977.

[La 55]   S. Lang, Abelian varieties over finite fields. *Proc. Nat. Acad. Sci. USA* 41, no. 3 (1955): pp. 174–176.

[LaN 59]   S. Lang and A. Néron. Rational points of abelian varieties over function fields. *Amer. J. Math.* 81, no. 1 (1959): pp. 95–118.

[Ph 91]   P. Philippon. Sur des hauteurs alternatives I. *Math. Ann.* 289 (1991): pp. 255–283.

[Ph 94]   P. Philippon. Sur des hauteurs alternatives II. *Ann. Institut Fourier* 44, no. 4 (1994): pp. 1043–1065.

[Ph 95]   P. Philippon. Sur des hauteurs alternatives III. *J. Math. Pures Appl.* 74 (1995): pp. 345–365.

[Wa 96]   B. Wang. A note on Archimedean Height Pairing and Chow forms. Preprint. Brown University, 1996.

[We 56]   A. Weil. The field of definition of a variety. *Amer. J. Math.* 78, no. 3 (1956): 509–524.

[Wi 38]   W. Wirtinger. Lies Translationsmannigfaltigkeiten une Abelsche Integrale. *Monat. Math. u. Physik* 46 (1938): pp. 384–443.

Lecture Notes in Mathematics 1625

Editors:
A. Dold, Heidelberg
F. Takens, Groningen

Springer
Berlin
Heidelberg
New York
Barcelona
Budapest
Hong Kong
London
Milan
Paris
Santa Clara
Singapore
Tokyo

Serge Lang

Topics in
Cohomology of Groups

Springer

Author

Serge Lang
Mathematics Derpartment
Yale University, Box 208 283
10 Hillhouse Avenue
New Haven, CT 06520-8283, USA

Library of Congress Cataloging-in-Publication Data

Lang, Serge, 1927-
    [Rapport sur la cohomologie des groupes. English]
    Topics in cohomology of groups / Serge Lang.
        p.   cm. -- (Lecture notes in mathematics ; 1625)
    Includes bibliographical references (p.   -   ) and index.
    ISBN 3-540-61181-9 (alk. paper)
    1. Class field theory. 2. Group theory. 3. Homology theory.
I. Title.   II. Series: Lecture notes in mathematics (Springer
-Verlag) ; 1625.
QA247.L3513 1996
.512'.74--dc20   96-26607

The first part of this book was originally published in French with the title
"Rapport sur la cohomologie des groupes" by Benjamin Inc., New York, 1996.
It was translated into English by the author for this edition. The last part
(pp. 188–215) is new to this edition.

Mathematics Subject Classification (1991): 11S25, 11S31, 20J06, 12G05, 12G10

ISBN 3-540-61181-9 Springer-Verlag Berlin Heidelberg New York

Typesetting: Camera-ready TEX output by the author
SPIN: 10479722   46/3142-543210 - Printed on acid-free paper

Contents

Chapter I. Existence and Uniqueness

§1. The abstract uniqueness theorem 3
§2. Notations, and the uniqueness theorem in $\text{Mod}(G)$ 9
§3. Existence of the cohomological functor on $\text{Mod}(G)$ 20
§4. Explicit computations 29
§5. Cyclic groups 32

Chapter II. Relations with Subgroups

§1. Various morphisms 37
§2. Sylow subgroups 50
§3. Induced representations 52
§4. Double cosets 58

Chapter III. Cohomological Triviality

§1. The twins theorem 62
§2. The triplets theorem 68
§3. Splitting module and Tate's theorem 70

Chapter IV. Cup Products

§1. Erasability and uniqueness 73
§2. Existence 83
§3. Relations with subgroups 87
§4. The triplets theorem 88
§5. The cohomology ring and duality 89
§6. Periodicity 95
§7. The theorem of Tate-Nakayama 98
§8. Explicit Nakayama maps 101

Chapter V. Augmented Products

§1. Definitions 109
§2. Existence 112
§3. Some properties 113

Chapter VI. Spectral Sequences

§1. Definitions 116
§2. The Hochschild-Serre spectral sequence 118
§3. Spectral sequences and cup products 121

Chapter VII. Groups of Galois Type (Unpublished article of Tate)

§1. Definitions and elementary properties 123
§2. Cohomology 128
§3. Cohomological dimension 138
§4. Cohomological dimension ≤ 1 143
§5. The tower theorem 149
§6. Galois groups over a field 150

Chapter VIII. Group Extensions

§1. Morphisms of extensions 156
§2. Commutators and transfer in an extension 160
§3. The deflation 163

Chapter IX. Class formations

§1. Definitions 166
§2. The reciprocity homomorphism 171
§3. Weil groups 178

Chapter X. Applications of Galois Cohomology in Algebraic Geometry (from letters of Tate)

§1. Torsion-free modules 189
§2. Finite modules 191
§3. The Tate pairing 195
§4. $(0,1)$-duality for abelian varieties 199
§5. The full duality 201
§6. Brauer group 202
§7. Ideles and idele classes 210
§8. Idele class cohomology 212

Preface

The Benjamin notes which I published (in French) in 1966 on the cohomology of groups provided missing chapters to the Artin-Tate notes on class field theory, developed by cohomological methods. Both items were out of print for many years, but recently Addison-Wesley has again made available the Artin-Tate notes (which were in English). It seemed therefore appropriate to make my notes on cohomology again available, and I thank Springer-Verlag for publishing them (translated into English) in the Lecture Notes series.

The most basic necessary background on homological algebra is contained in the chapter devoted to this topic in my *Algebra* (derived functors and other material at this basic level). This material is partly based on what have now become routine constructions (Eilenberg-Cartan), and on Grothendieck's influential paper [Gr 59], which appropriately defined and emphasized δ-functors as such.

The main source for the present notes are Tate's private papers, and the unpublished first part of the Artin-Tate notes. The most significant exceptions are: Rim's proof of the Nakayama-Tate theorem, and the treatment of cup products, for which we have used the general notion of multilinear category due to Cartier.

The cohomological approach to class field theory was carried out in the late forties and early fifties, in Hochschild's papers [Ho 50a], [Ho 50b], [HoN 52], Nakayama [Na 41], [Na 52], Shafarevich [Sh 46], Weil's paper [We 51], giving rise to the Weil groups, and seminars of Artin-Tate in 1949-1951, published only years later [ArT 67].

As I stated in the preface to my *Algebraic Number Theory*, there are several approaches to class field theory. None of them makes any other obsolete, and each gives a different insight from the others.

The original Benjamin notes consisted of Chapters I through IX. Subsequently I wrote up Chapter X, which deals with applications to algebraic geometry. It is essentially a transcription of weekly installment letters which I received from Tate during 1958-1959. I take of course full responsibility for any errors which might have crept in, but I have made no effort to make the exposition anything more than a rough sketch of the material. Also the reader should not be surprised if some of the diagrams which have been qualified as being commutative actually have character -1.

The first nine chapters are basically elementary, depending only on standard homological algebra. The Artin-Tate axiomatization of class formations allows for an exposition of the basic properties of class field theory at this elementary level. Proofs that the axioms are satisfied are in the Artin-Tate notes, following Tate's article [Ta 52]. The material of Chapter X is of course at a different level, assuming some knowledge of algebraic geometry, especially some properties of abelian varieties.

I thank Springer Verlag for keeping all this material in print. I also thank Donna Belli and Mel Del Vecchio for setting the manuscript in AMSTeX, in a victory of person over machine.

Serge Lang
New Haven, 1995

CHAPTER I

Existence and Uniqueness

§1. The abstract uniqueness theorem

We suppose the reader is familiar with the terminology of abelian categories. However, we shall deal only with abelian categories which are categories of modules over some ring, or which are obtained from such in some standard ways, such as categories of complexes of modules. We also suppose that the reader is acquainted with the standard procedures constructing cohomological functors by means of resolutions with complexes, as done for instance in my *Algebra* (third edition, Chapter XX). In some cases, we shall summarize such constructions for the convenience of the reader.

Unless otherwise specified, all functors on abelian categories will be assumed additive. What we call a δ-**functor** (following Grothendieck) is sometimes called a **connected sequence of functors**. Such a functor is defined for a consecutive sequence of integers, and transforms an exact sequence

$$0 \to A \to B \to C \to 0$$

into an exact sequence

$$\cdots \to H^p(A) \to H^p(B) \to H^p(C) \xrightarrow{\delta} H^{p+1}(A) \to \cdots$$

functorially. If the functor is defined for all integers p with $-\infty < p < \infty$, then we say that this functor is **cohomological**.

Let H be a δ-functor on an abelian category $\mathfrak{A}$. We say that H is **erasable** by a subset $\mathfrak{M}$ of objects in A if for every A in $\mathfrak{A}$ there exists $M_A \in \mathfrak{M}$ and a monomorphism $\varepsilon_A : A \to M_A$ such that $H(M_A) = 0$. This definition is slightly more restrictive than the usual general definition (*Algebra*, Chapter XX, §7), but its conditions are those which are used in the forthcoming applications. An **erasing functor** for H consists of a functor

$$M : A \to M(A)$$

of $\mathfrak{A}$ into itself, and a monomorphism ε of the identity in M, i.e. for each object A we are given a monomorphism

$$\varepsilon_A : A \to M_A$$

such that, if $u : A \to B$ is a morphism in $\mathfrak{A}$, then there exists a morphism $M(u)$ and a commutative diagram

$$
\begin{array}{ccc}
0 \longrightarrow A & \xrightarrow{\;\varepsilon_A\;} & M(A) \\
\downarrow{\scriptstyle u} & & \downarrow{\scriptstyle M(u)} \\
0 \longrightarrow B & \xrightarrow[\;\varepsilon_B\;]{} & M(B)
\end{array}
$$

such that $M(uv) = M(u)M(v)$ for the composite of two morphisms u, v. In addition, one requires $H(M_A) = 0$ for all $A \in \mathfrak{A}$.

Let $X(A) = X_A$ be the cokernel of ε_A. For each u there is a morphism

$$X(u) : X_A \to X_B$$

such that the following diagram is commutative:

$$
\begin{array}{ccccccccc}
0 & \longrightarrow & A & \longrightarrow & M_A & \longrightarrow & X_A & \longrightarrow & 0 \\
& & \downarrow{\scriptstyle u} & & \downarrow{\scriptstyle M(u)} & & \downarrow{\scriptstyle X(u)} & & \\
0 & \longrightarrow & B & \longrightarrow & M_B & \longrightarrow & X_B & \longrightarrow & 0,
\end{array}
$$

and for the composite of two morphisms u, v we have $X(uv) = X(u)X(v)$. We then call X the **cofunctor** of M.

Let p_0 be an integer, and $H = (H^p)$ a δ-functor defined for some values of p. We say that M is an **erasing functor for H in dimension $> p_0$** if $H^p(M_A) = 0$ for all $A \in \mathfrak{A}$ and all $p > p_0$.

We have similar notions on the left. Let H be an exact δ-functor on $\mathfrak{A}$. We say that H is **coerasable** by a subset $\mathfrak{M}$ if for each object A there exists an epimorphism

$$\eta_A : M_A \to A$$

with $M_A \in \mathfrak{M}$, such that $H(M_A) = 0$. A **coerasing functor** M for H consists of an epimorphism of M with the identity. If η is such a functor, and $u : A \to B$ is a morphism, then we have a commutative diagram with exact horizontal sequences:

$$
\begin{array}{ccccccccc}
0 & \longrightarrow & Y_A & \longrightarrow & M_A & \overset{\eta_A}{\longrightarrow} & A & \longrightarrow & 0 \\
& & \downarrow{\scriptstyle Y(u)} & & \downarrow{\scriptstyle M(u)} & & \downarrow{\scriptstyle u} & & \\
0 & \longrightarrow & Y_B & \longrightarrow & M_B & \underset{\eta_B}{\longrightarrow} & B & \longrightarrow & 0
\end{array}
$$

and Y_A is functorial in A, i.e. $Y(uv) = Y(u)Y(v)$.

Remark. In what follows, erasing functors will have the additional property that the exact sequence associated with each object A will split over $\mathbf{Z}$, and therefore remains exact under tensor products or hom. An erasing functor into an abelian category of abelian groups having this property will be said to be **splitting**.

Theorem 1.1. First uniqueness theorem. *Let $\mathfrak{A}$ be an abelian category. Let H, F be two δ-functors defined in degrees $0, 1$ (resp. $0, -1$) with values in the same abelian category. Let (φ_0, φ_1) and $(\varphi_0, \bar\varphi_1)$ be δ-morphisms of H into F, coinciding in dimension 0 (resp. $(\varphi_{-1}, \varphi_0)$ and $(\bar\varphi_{-1}, \varphi_0)$). Suppose that H^1 is erasable (resp. H^{-1} is coerasable). Then we have $\varphi_1 = \bar\varphi_1$ (resp. $\varphi_{-1} = \bar\varphi_{-1}$).*

Proof. The proof being self dual, we give it only for the case of indices $(0, 1)$. For each object $A \in \mathfrak{A}$ we have an exact sequence

$$0 \to A \to M_A \to X_A \to 0$$

and $H^1(M_A) = 0$. There is a commutative diagram

$$
\begin{array}{ccccccc}
H^0(M_A) & \longrightarrow & H^0(X_A) & \overset{\delta_H}{\longrightarrow} & H^1(A) & \longrightarrow & 0 \\
\downarrow{\scriptstyle \varphi_0} & & \downarrow{\scriptstyle \varphi_0} & & \downarrow{\scriptstyle \varphi_1, \bar\varphi_1} & & \\
F^0(M_A) & \longrightarrow & F^0(X_A) & \overset{\delta_F}{\longrightarrow} & F^1(A) & \longrightarrow & 0
\end{array}
$$

with horizontal exact sequences, from which it follows that δ_H is surjective. It follows at once that $\varphi_1 = \bar{\varphi}_1$.

In the preceding theorem, φ_1 and $\bar{\varphi}_1$ are given. One can also prove a result which implies their existence.

Theorem 1.2. Second uniqueness theorem. *Let $\mathfrak{A}$ be an abelian category. Let H, F be δ-functors defined in degrees $(0, 1)$ (resp. $0, -1$) with values in the same abelian category. Let $\varphi_0 : H^0 \dashrightarrow F^0$ be a morphism. Suppose that H^1 is erasable by injectives (resp. H^{-1} is coerasable by projectives). Then there exists a unique morphism*

$$\varphi_1 : H^1 \to F^1 \ (\text{resp.} \ \varphi_{-1} : H^{-1} \to F^{-1})$$

such that (φ_0, φ_1) (resp. $(\varphi_0, \varphi_{-1})$) is also a δ-morphism. The association $\varphi_0 \mapsto \varphi_1$ is functorial in a sense made explicit below.

Proof. Again the proof is self dual and we give it only in the cases when the indices are $(0, 1)$. For each object $A \in \mathfrak{A}$ we have the exact sequence

$$0 \to A \to M_A \to X_A \to 0$$

and $H^1(M_A) = 0$. We have to define a morphism

$$\varphi_1(A) : H^1(A) \to F^1(A)$$

which commutes with the induced morphisms and with δ. We have a commutative diagram

$$
\begin{array}{ccccccc}
H^0(M_A) & \longrightarrow & H^0(X_A) & \xrightarrow{\delta_H} & H^1(A) & \longrightarrow & 0 \\
\varphi_0 \downarrow & & \varphi_0 \downarrow & & & & \\
F^0(M_A) & \longrightarrow & F^0(X_A) & \xrightarrow[\delta_F]{} & F^1(A) & &
\end{array}
$$

with exact horizontal sequences. The right surjectivity is just the erasing hypothesis. The left square commutativity shows that $\operatorname{Ker} \delta_H$ is contained in the kernel of $\delta_F \varphi_0(X_A)$. Hence there exists a unique morphism

$$\varphi_1(A) : H^1(A) \to F^1(A)$$

which makes the right square commutative. We shall prove that $\varphi_1(A)$ satisfies the desired conditions.

First, let $u : A \to B$ be a morphism. From the hypotheses, there exists a commutative diagram

$$
\begin{array}{ccccccccc}
0 & \longrightarrow & A & \longrightarrow & M_A & \longrightarrow & X_A & \longrightarrow & 0 \\
& & \downarrow{\scriptstyle u} & & \downarrow{\scriptstyle M(u)} & & \downarrow{\scriptstyle X(u)} & & \\
0 & \longrightarrow & B & \longrightarrow & M_B & \longrightarrow & X_B & \longrightarrow & 0
\end{array}
$$

the morphism $M(u)$ being defined because M_A is injective. The morphism $X(u)$ is then defined by making the right square commutative. To simplify notation, we shall write u instead of $M(u)$ and $X(u)$.

We consider the cube:

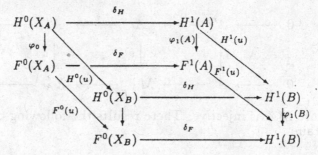

We have to show that the right face is commutative. We have:

$$
\begin{aligned}
\varphi_1(B)H^1(u)\delta_H &= \varphi_1(B)\delta_H H^0(u) \\
&= \delta_F \varphi_0 H^0(u) \\
&= \delta_F F^0(u)\varphi_0 \\
&= F^1(u)\delta_F \varphi_0 \\
&= F^1(u)\varphi_1(A)\delta_H.
\end{aligned}
$$

We have used the fact (implied by the hypotheses) that all the faces of the cube are commutative except possibly the right face. Since δ_H is surjective, one gets what we want, namely

$$
\varphi_1(B)H^1(u) = F^1(u)\varphi_1(A).
$$

The above argument may be expressed in the form of a useful general lemma.

If, in a cube, all the faces are commutative except possibly one, and one of the arrows as above is surjective, then this face is also commutative.

Next we have to show that φ_1 commutes with δ, that is (φ_0, φ_1) is a δ-morphism. Let

$$0 \to A' \to A \to A'' \to 0$$

be an exact sequence in $\mathfrak{A}$. Then there exist morphisms

$$v : A \to M_{A'} \qquad \text{and} \qquad w : A'' \to X_{A'}$$

making the following diagram commutative:

$$
\begin{array}{ccccccccc}
0 & \longrightarrow & A' & \longrightarrow & A & \longrightarrow & A'' & \longrightarrow & 0 \\
 & & \downarrow{\scriptstyle \text{id}} & & \downarrow{\scriptstyle v} & & \downarrow{\scriptstyle w} & & \\
0 & \longrightarrow & A' & \longrightarrow & M_{A'} & \longrightarrow & X_{A'} & \longrightarrow & 0
\end{array}
$$

because $M_{A'}$ is injective. There results the following commutative diagram:

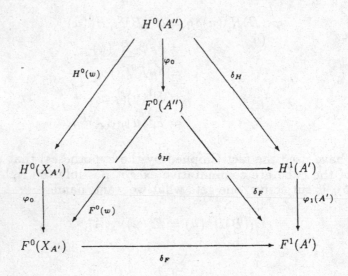

We have to show that the right square is commutative. Note that the top and bottom triangles are commutative by definition of a δ-functor. The left square is commutative by the hypothesis that φ_0 is a morphism of functors. The front square is commutative by definition of $\varphi_1(A')$. We thus find

$$
\begin{aligned}
\varphi_1(A')\delta_H &= \varphi_1(A')\delta_H H^0(w) \quad \text{(top triangle)} \\
&= \delta_F \varphi_0 H^0(w) \quad\quad\ \text{(front square)} \\
&= \delta_F F^0(w)\varphi_0 \quad\quad \text{(left square)} \\
&= \delta_F \varphi_0 \quad\quad\quad\quad\ \text{(bottom triangle)},
\end{aligned}
$$

which concludes the proof.

Finally, let us make explicit what we mean by saying that φ_1 depends functorially on φ_0. Suppose we have three functors H, F, E defined in degrees $0, 1$; and suppose given $\varphi_0 : H^0 \to F^0$ and $\psi_0 : F^0 \to E^0$. Suppose in addition that the erasing functor erases both H^1 and F^1. We can then construct φ_1 and ψ_1 by applying the theorem. On the other hand, the composite

$$
\psi_0 \varphi_0 = \theta_0 : H^0 \to E^0
$$

is also a morphism, and the theorem implies the existence of a morphism

$$
\theta_1 : H^1 \to E^1
$$

such that (θ_0, θ_1) is a δ-morphism. By uniqueness, we obtain

$$
\theta_1 = \psi_1 \circ \varphi_1.
$$

This is what we mean by the assertion that φ_1 depends functorially on φ_0.

§2. Notation, and the uniqueness theorem in Mod(G)

We now come to the cohomology of groups. Let G be a group. As usual, we let $\mathbf{Q}$ and $\mathbf{Z}$ denote the rational numbers and the integers respectively. Let $\mathbf{Z}[G]$ be the group ring over $\mathbf{Z}$. Then

$\mathbf{Z}[G]$ is a free module over $\mathbf{Z}$, the group elements forming a basis over $\mathbf{Z}$. Multiplicatively, we have

$$\left(\sum_{\sigma \in G} a_\sigma \sigma\right)\left(\sum_{\tau \in G} b_\tau \tau\right) = \sum_{\sigma,\tau} a_\sigma b_\tau \sigma\tau,$$

the sums being taken over all elements of G, but only a finite number of a_σ and b_τ being $\neq 0$. Similarly, one defines the group algebra $k[G]$ over an arbitrary commutative ring k.

The group ring is often denoted by $\Gamma = \Gamma_G$. It contains the ideal I_G which is the kernel of the **augmentation homomorphism**

$$\varepsilon : \mathbf{Z}[G] \to \mathbf{Z}$$

defined by $\varepsilon\left(\sum n_\sigma \sigma\right) = \sum n_\sigma$. One sees at once that I_G is $\mathbf{Z}$-free, with a basis consisting of all elements $\sigma - e$, with σ ranging over the elements of G not equal to the unit element. Indeed, if $\sum n_\sigma = 0$, then we may write

$$\sum n_\sigma \sigma = \sum n_\sigma(\sigma - e).$$

Thus we obtain an exact sequence

$$0 \to I_G \to \mathbf{Z}[G] \to \mathbf{Z} \to 0,$$

used constantly in the sequel. The sequence splits, because $\mathbf{Z}[G]$ is a direct sum of I_G and $\mathbf{Z} \cdot e_G$ (identified with $\mathbf{Z}$).

Abelian groups form an abelian category, equal to the category of $\mathbf{Z}$-modules, denoted by $\mathrm{Mod}(\mathbf{Z})$. Similarly, the category of modules over a ring R will be denoted by $\mathrm{Mod}(R)$.

An abelian group A is said to be a G-**module** if one is given an operation (or action) of G on A; in other words, one is given a map

$$G \times A \to A$$

satisfying

$$(\sigma\tau)a = \sigma(\tau a), \quad e \cdot a = a \quad \sigma(a + b) = \sigma a + \sigma b$$

for all $\sigma, \tau \in G$ and $a, b \in A$. We let $e = e_G$ be the unit element of G. One extends this operation by linearity to the group ring $\mathbf{Z}[G]$. Similarly, if k is a commutative ring and A is a k-module, one extends the operation of G on A to $k[G]$ whenever the operation of G commutes with the operation of k on A. Then the category of $k[G]$-modules is denoted by $\mathrm{Mod}_k(G)$ or $\mathrm{Mod}(k, G)$.

The G-modules form an abelian category, the morphisms being the G-homomorphisms. More precisely, if $f : A \to B$ is a morphism in $\mathrm{Mod}(\mathbf{Z})$, and if A, B are also G-modules, then G operates on $\mathrm{Hom}(A, B)$ by the formula

$$(\sigma f)(a) = \sigma(f(\sigma^{-1}a)) \quad \text{for} \quad a \in A \quad \text{and} \quad \sigma \in G.$$

If there is any danger of confusion one may write $[\sigma]f$ to denote this operation. If $[\sigma]f = f$, one says that f is a G-**homomorphism**, or a G-**morphism**. The set of G-morphisms from A into B is an abelian group denoted by $\mathrm{Hom}_G(A, B)$. The category consisting of G-modules and G-morphisms is denoted by $\mathrm{Mod}(G)$. It is the same as $\mathrm{Mod}(\Gamma_G)$.

Let $A \in \mathrm{Mod}(G)$. We let A^G denote the submodule of A consisting of all elements $a \in A$ such that $\sigma a = a$ for all $\sigma \in G$. In other words, it is the submodule of fixed elements by G. Then A^G is an abelian group, and the association

$$H_G^0 : A \mapsto A^G$$

is a functor from $\mathrm{Mod}(G)$ into the category of abelian groups, also denoted by **Grab**. This functor is left exact.

We let $\varkappa_G$ denote the canonical map (in the present case the identity) of an element $a \in A^G$ into $H_G^0(A)$.

Theorem 2.1. *Let H_G be a cohomological functor on $\mathrm{Mod}(G)$ with values in $\mathrm{Mod}(\mathbf{Z})$, and such that H_G^0 is defined as above. Assume that $H_G^r(M) = 0$ if M is injective and $r > 1$. Assume also that $H_G^r(A) = 0$ for $A \in \mathrm{Mod}(G)$ and $r < 0$. Then two such cohomological functors are isomorphic, by a unique morphism which is the identity on $H_G^0(A)$.*

This theorem is just a special case of the general uniqueness theorem.

Corollary 2.2. *If $G = \{e\}$ then $H_G^r(A) = 0$ for all $r > 0$.*

Proof. Define H_G by letting $H_G^0(A) = A^G$ and $H_G^r(A) = 0$ for $r \neq 0$. Then it is immediately verified that H_G is a cohomological functor, to which we can apply the uniqueness theorem.

Corollary 2.3. *Let $n \in \mathbf{Z}$ and let $n_A : A \to A$ be the morphism $a \mapsto na$ for $a \in A$. Then $H_G^r(n_A) = n_H$ (where H stands for $H_G^r(A)$).*

Proof. Since the coboundary δ is additive, it commutes with multiplication by n, and again we can apply the uniqueness theorem.

The existence of the functor H_G will be proved in the next section.

We say that G **operates trivially** on A if $A = A^G$, that is $\sigma a = a$ for all $a \in A$ and $\sigma \in G$. We always assume that G operates trivially on $\mathbf{Z}, \mathbf{Q}$, and $\mathbf{Q}/\mathbf{Z}$.

We define the abelian group

$$A_G = A/IA_G.$$

This is the factor group of A by the subgroup of elements of the form $(\sigma - e)a$ with $\sigma \in G$ and $a \in A$. The association

$$A \mapsto A_G$$

is a functor from $\mathrm{Mod}(G)$ into Grab.

Let U be a subgroup of finite index in G. We may then define the **trace**

$$S_G^U : A^U \to A^G \qquad \text{by the formula} \qquad S_G^U(a) = \sum_c \bar{c}a,$$

where $\{c\}$ is the set of left cosets of U in G, and $\bar{c}$ is a representative of c, so that

$$G = \bigcup_c \bar{c}U.$$

If $U = \{e\}$, then G is finite, and in that case the **trace** is written S_G, so

$$S_G(a) = \sum_{\sigma \in G} \sigma a.$$

For the record, we state the following useful lemma.

Lemma 2.4. *Let A, B, C be G-modules. Let U be a subgroup of finite index in G. Let*

$$A \xrightarrow{u} B \xrightarrow{v} C \xrightarrow{w} D$$

be morphisms in $\mathrm{Mod}(G)$, nad suppose that u, w are G-morphisms while v is a U-morphism. Then

$$\mathbf{S}_G^U(wvu) = w\mathbf{S}_G^U(v)u.$$

Proof. Immediate.

We shall now describe some embedding functors in $\mathrm{Mod}(G)$. These will turn out to erase some cohomological functors to be defined later. Indeed, injective or projective modules will not suffice to erase cohomology, for several reasons. First, when we change the group G, an injective does not necessarily remain injective. Second, an exact sequence

$$0 \to A \to J \to A'' \to 0$$

with an injective module J does not necessarily remain exact when we take its tensor product with an arbitrary module B. Hence we shall consider another class of modules which behave better in both respects.

Let G be a group and let B be an abelian group, i.e. a **Z**-module. We denote by $M_G(B)$ or $M(G, B)$ the set of functions from G into B, these forming an abelian group in the usual way (adding the values). We make $M_G(B)$ into a G-module by defining an operation of G by the formulas

$$(\sigma f)(x) = f(x\sigma), \qquad \text{for} \qquad x, \sigma \in G.$$

We have trivially $(\sigma\tau)(f) = \sigma(\tau f)$. Furthermore:

Proposition 2.6. *Let G' be a subgroup of G and let $G = \bigcup_\alpha x_\alpha G'$ be a coset decomposition. For $f \in M(G, B)$ let f_α be the function in $M(G', B)$ such that $f_\alpha(y) = f(x_\alpha y)$ for $y \in G'$. Then the map*

$$f \mapsto \prod_\alpha f_\alpha$$

is an isomorphism

$$M(G,B) \xrightarrow{\approx} \prod_\alpha M(G',B)$$

in the category of G'-modules.

The proof is immediate, and Proposition 2.5 is a special case with G' equal to the trivial subgroup.

Let $A \in \mathrm{Mod}(G)$, and define

$$\varepsilon_A : A \to M_G(A)$$

by the condition that $\varepsilon_A(a)$ is the function f_a such that $f_a(\sigma) = \sigma a$ for all $a \in A$ and $\sigma \in G$. We then obtain an exact sequence

$$(1) \qquad 0 \to A \xrightarrow{\varepsilon_A} M_G(A) \to X_A \to 0$$

in $\mathrm{Mod}(G)$. Furthermore, this sequence splits over $\mathbf{Z}$, because the map

$$M_G(A) \to A \qquad \text{given by} \qquad f \mapsto f(e)$$

splits the left arrow in this sequence, i.e. composed with ε_A it yields the identity on A. Consequently tensoring this sequence with an arbitrary G-module B preserves exactness.

We already know that M_G is an exact functor. In addition, if $f : A \to B$ is a morphism in $\mathrm{Mod}(G)$, then in the following diagram

$$(2) \quad
\begin{array}{ccccccccc}
0 & \longrightarrow & A & \xrightarrow{\varepsilon_A} & M_G(A) & \longrightarrow & X_A & \longrightarrow & 0 \\
 & & \downarrow{\scriptstyle f} & & \downarrow{\scriptstyle M_G(f)} & & \downarrow{\scriptstyle X(f)} & & \\
0 & \longrightarrow & B & \xrightarrow[\varepsilon_B]{} & M_G(B) & \longrightarrow & X_B & \longrightarrow & 0
\end{array}$$

the left square is commutative, and hence the right square is commutative. Therefore, we find:

Theorem 2.5. *Let G be a group. Notations as above, the pair (M_G, ε) is an embedding functor in $\mathrm{Mod}(G)$. The associated exact sequence (1) splits over $\mathbf{Z}$ for each $A \in \mathrm{Mod}(G)$.*

In the next section, we shall define a cohomological functor H_G on $\mathrm{Mod}(G)$ for which (M_G, ε) is an erasing functor. By Proposition 2.6, we shall then find:

Corollary 2.6. *Let G' be a subgroup of G, and consider $\mathrm{Mod}(G)$ as a subcategory of $\mathrm{Mod}(G')$. Then $H_{G'}$ is a cohomological functor on $\mathrm{Mod}(G)$, and (M_G, ε) is an erasing functor for $H_{G'}$.*

Thus we shall have achieved our objective of finding a serviceable erasing functor simultaneously for a group and its subgroups, behaving properly under tensor products. The erasing functor as above will be called the **ordinary erasing functor**.

Remark. Let U be a subgroup of finite index in G. Let $A, B \in \mathrm{Mod}(G)$. Let $f : A \to B$ be a U-morphism. We may take the trace

$$\mathrm{S}_G^U(f) : A \to B$$

which is a G-morphism. Furthermore, considering $\mathrm{Mod}(G)$ as a subcategory of $\mathrm{Mod}(U)$, we see that (M_G, ε) is an embedding functor relative to U, that is, there exist U-morphisms $M_G(f)$ and $X(f)$ such that the diagram (2) is commutative, but with vertical U-morphisms.

Applying the trace to these vertical morphisms, and using Lemma 2.4, we obtain a commutative diagram:

$$
(3) \qquad
\begin{array}{ccccccccc}
0 & \longrightarrow & A & \xrightarrow{\ \varepsilon_A\ } & M_G(A) & \longrightarrow & X_A & \longrightarrow & 0 \\
 & & \downarrow{\scriptstyle \mathrm{S}_G^U(f)} & & \downarrow{\scriptstyle \mathrm{S}_G^U M_G(f)} & & \downarrow{\scriptstyle \mathrm{S}_G^U X(f)} & & \\
0 & \longrightarrow & B & \xrightarrow[\ \varepsilon_B\ ]{} & M_G(B) & \longrightarrow & X_B & \longrightarrow & 0
\end{array}
$$

The case of finite groups

For the rest of this section, we assume that G is finite.

We define two functors from $\mathrm{Mod}(G)$ into $\mathrm{Mod}(\mathbf{Z})$ by

$$\mathrm{H}_G^0 : A \mapsto A^G / \mathrm{S}_G A$$
$$\mathrm{H}_G^{-1} : A \mapsto A_{\mathrm{S}_G} / I_G A.$$

We denote by A_{S_G} the kernel of S_G in A. This is a special case of the notation whereby if $f : A \to B$ is a homomorphism, we let A_f be its kernel.

We let

$$\varkappa_G : A^G \to \mathbf{H}^0_G(A) = A^G/\mathbf{S}_G A.$$

$$\varkappa_G : A_{\mathbf{S}_G} \to \mathbf{H}^{-1}_G(A) = A_{\mathbf{S}_G}/I_G A.$$

be the canonical maps. The proof of the following result is easy and straightforward, and will be left to the reader.

Theorem 2.7. *The functors* $\mathbf{H}^{-1}_G$ *and* $\mathbf{H}^0_G$ *form a* δ*-functor if one defines the coboundary as follows. Let*

$$0 \to A' \xrightarrow{u} A \xrightarrow{v} A'' \to 0$$

be an exact sequence in $\mathrm{Mod}(G)$*. For* $a'' \in A''_{\mathbf{S}_G}$ *we define*

$$\delta\varkappa_G(a'') = \varkappa_G(u^{-1}\mathbf{S}_G v^{-1} a'').$$

The inverse images in this last formula have the usual meaning. One chooses any element a such that $va = a''$, then one takes the trace $\mathbf{S}_G$. One shows that this is an element in the image of u, so we can take u^{-1} of this element to be an element of A'^G, whose class modulo $\mathbf{S}_G A'$ is well defined, i.e. is independent of the choices of a such that $va = a''$. The verification of these assertions is trivial, and left to the reader. (Cf. *Algebra*, Chapter III, §9.)

Theorem 2.8. *Let* $\mathbf{H}_G$ *be a cohomological functor on* $\mathrm{Mod}(G)$ *(with finite group* G*), with value in* $\mathrm{Mod}(\mathbf{Z})$*, and such that* $\mathbf{H}^0_G$ *is as above. Suppose that* $\mathbf{H}^r_G(M) = 0$ *if* M *is injective and* $r \leqq 0$*. If* $\mathbf{F}_G$ *is another cohomological functor having the same properties, then there exists a unique isomorphism of* $\mathbf{H}_G$ *with* $\mathbf{F}_G$ *which is the identity on* $\mathbf{H}^0_G$*.*

Proof. This is a particular case of the uniqueness theorem.

Corollary 1.9. *If* G *is trivial then* $\mathbf{H}^r_G(A) = 0$ *for all* $r \in \mathbf{Z}$*.*

Corollary 2.10. *Let* $n \in \mathbf{Z}$ *and suppose* G *finite. For* A *in* $\mathrm{Mod}(G)$ *we have* $\mathbf{H}^r_G(n_A) = n_\mathbf{H}$ *(abbreviating* $\mathbf{H}^r_G$ *by* $\mathbf{H}$*).*

Both corollaries are direct consequences of the uniqueness theorem, like their counterpart for the other functor, as in Corollaries 2.2 and 2.3.

Let K_1, K_2, K be commutative rings, and let T be a biadditive bifunctor

$$T : \mathrm{Mod}(K_1) \times \mathrm{Mod}(K_2) \rightarrow \mathrm{Mod}(K).$$

Suppose we are given an action of G on $A_1 \in \mathrm{Mod}(K_1)$ and on $A_2 \in \mathrm{Mod}(K_2)$, so $A_1 \in \mathrm{Mod}(K_1[G])$ and $A_2 \in \mathrm{Mod}(K_2[G])$. Then $T(A_1, A_2)$ is a $K[G]$-module, under the operation $T(\sigma, \sigma)$ if T is covariant in both variables, and $T(\sigma^{-1}, \sigma)$ if T is contravariant in the first variable and covariant in the second. This remark will be applied to the case when T is the tensor product or $T = \mathrm{Hom}$.

A $K[G]$-module A is called $K[G]$-**regular** if the identity 1_A is a trace, that is there exists a K-morphism $v : A \rightarrow A$ such that

$$1_A = \mathbf{S}_G(v).$$

When $K = \mathbf{Z}$, a $K[G]$-regular module is simply called G-**regular**.

Proposition 2.11. *Let K_1, K_2, K be commutative rings as above, and let T be as above. Let $A_i \in \mathrm{Mod}(K_i)$ $(i = 1, 2)$ and suppose A_i is $K_i[G]$-regular for $i = 1, 2$. Then $T(A_1, A_2)$ is $K[G]$-regular.*

Proof. Left to the reader.

Proposition 2.12. *Let G' be a subgroup of the finite group G. Let $A \in \mathrm{Mod}(K, G)$. If A is $K[G]$-regular then A is also $K[G']$-regular. If G' is normal in G, then $A^{G'}$ is $K[G/G']$-regular.*

Proof. Write $G = \bigcup G'x_i$ expressing G as a right coset decomposition of G'. Then by assumption, we can write 1_A in the form

$$1_A = \sum_{\tau \in G'} \sum_i \tau x_i v$$

with some K-morphism v. The two assertions of the proposition are then clear, according as we take the double sum in the given order, or reverse the order of the summation.

Proposition 2.13. *Let $A \in \mathrm{Mod}(K, G)$. Then A is $K[G]$-projective if and only if A is K-projective and $K[G]$-regular.*

Proof. We recall that a projective module is characterized by being a direct summand of a free module. Suppose that A is $K[G]$-projective. We may then write A as a direct summand of a free

$K[G]$-module $F = A \oplus B$, with the natural injection i and projection π in the sequence

$$A \xrightarrow{i} A \times B = F \xrightarrow{\pi} A$$

with $\pi i = 1_A$, and both i, π are $K[G]$-homomorphisms. Since F is $K[G]$-free, it follows that $1_F = \mathbf{S}_G(v)$ for some K-homomorphism v. Then by Lemma 2.4,

$$1_A = \pi 1_F i = \pi \mathbf{S}_G(v) i = \mathbf{S}_G(\pi v i),$$

whence A is $K[G]$-regular. Conversely, let A be K-projective and $K[G]$-regular. Let

$$F \xrightarrow{\pi} A \to 0$$

be an exact sequence in $\mathrm{Mod}(K, G)$, with F being $K[G]$-free. By hypothesis, there exists a K-morphism $i_K : A \to F$ such that $\pi i_K = 1_A$, and there exists a K-morphism $v : A \to A$ such that $1_A = \mathbf{S}_G(v)$. We then find

$$\pi \mathbf{S}_G(i_K v) = \mathbf{S}_G(\pi i_K v) = \mathbf{S}_G(v) = 1_A,$$

which shows that $\mathbf{S}_G(i_K v)$ splits π, whence A is a direct summand of a free module, and is therefore $K[G]$-projective. This proves the proposition.

With the same type of proof, taking the trace of a projection, one also obtains the following result.

Proposition 2.14. *In* $\mathrm{Mod}(G)$, *a direct summand of a G-regular module is also G-regular. In particular, every projective module in* $\mathrm{Mod}(G)$ *is also G-regular.*

Proof. The second assertion is obvious for free modules, whence it follows from the first assertion for projectives.

For finite groups we have a modification of the embedding functor defined previously for arbitrary groups, and this modification will enjoy stronger properties. We consider the following two exact sequences:

(3) $$0 \to I_G \to \mathbf{Z}[G] \xrightarrow{\varepsilon} \mathbf{Z} \to 0$$

(4) $$0 \to \mathbf{Z} \xrightarrow[\varepsilon']{} \mathbf{Z}[G] \to J_G \to 0.$$

The first one is just the one already considered, with the augmentation homomorphism ε. The second is defined as follows. We embed $\mathbf{Z}$ in $\mathbf{Z}[G]$ on the diagonal, that is

$$\varepsilon' : n \mapsto n \sum_{\sigma \in G} \sigma.$$

Since G acts trivially on $\mathbf{Z}$, it follows that ε' is a G-homomorphism. We denote its cokernel by J_G.

Proposition 2.15. *The exact sequences* (3) *and* (4) *split in* Mod($\mathbf{Z}$).

Proof. We already know this for (3). For (4), given $\xi = \sum n_\sigma \sigma$ in $\mathbf{Z}[G]$ we have a decomposition

$$\xi = n_e \left(\sum_{\sigma \in G} \sigma \right) + \sum_{\sigma \neq e} (n_\sigma - n_e)\sigma$$

$$\in \mathbf{Z}\left(\sum_{\sigma \in G} \sigma \right) + \sum_{\sigma \neq e} \mathbf{Z}\sigma,$$

which shows that $\mathbf{Z}[G]$ is a direct sum of $\varepsilon'(\mathbf{Z})$ and another module, as was to be shown.

Given any $A \in \text{Mod}(G)$, taking the tensor product (over $\mathbf{Z}$) of the split exact sequences (3) and (4) with A yields split exact sequences by a basic elementary property of the tensor product, with G-morphisms $\varepsilon_A = \varepsilon \otimes 1_A$ and $\varepsilon'_A = \varepsilon' \otimes 1_A$, as shown below:

(4A) $0 \to I_G \otimes A \to \mathbf{Z}[G] \otimes A \xrightarrow{\varepsilon_A} \mathbf{Z} \otimes A = A \to 0$

(5A) $0 \to A = \mathbf{Z} \otimes A \xrightarrow[\varepsilon'_A]{} \mathbf{Z}[G] \otimes A \to J_G \otimes A \to 0.$

As usual, we identify $\mathbf{Z} \otimes A$ with A. Let $\mathbf{M}_G$ be the functor given by

$$\mathbf{M}_G(A) = \mathbf{Z}[G] \otimes A.$$

We observe that $\mathbf{M}_G(A)$ is G-regular. In the next section, we shall define a cohomological functor on Mod(G) for which $\mathbf{M}_G$ will be an erasing functor.

Let $f : A \to B$ be a G-morphism, or more generally, suppose G' is a subgroup of G and $A, B \in \mathrm{Mod}(G)$, while f is a G'-morphism. Then

$$\mathbf{M}_G(f) = 1 \otimes f$$

is a G'-morphism.

§3. Existence of the cohomological functors on $\mathrm{Mod}(G)$

Although we reproduced the proofs of uniqueness, because they were short, we now assume that the reader is acquainted with standard facts of general homology theory. These are treated in *Algebra*, Chapter XX, of which we now use §8, especially Proposition 8.2 giving the existence of the derived functors. We apply this proposition to the bifunctor

$$T(A, B) = \mathrm{Hom}_G(A, B) \quad \text{for} \quad A, B \in \mathrm{Mod}(G),$$

with an arbitrary group G. We have

$$\mathrm{Hom}_G(\mathbf{Z}, A) = A^G.$$

We then find:

Theorem 3.1. *Let X be a projective resolution of $\mathbf{Z}$ in $\mathrm{Mod}(G)$. Let $H(A)$ be the homology of the complex $\mathrm{Hom}_G(X, A)$. Then $H = \{H^r\}$ is a cohomology functor on $\mathrm{Mod}(G)$, such that*

$$H^r(A) = 0 \text{ if } r < 0.$$

$$H^0(A) = A^G.$$

$$H^r(A) = 0 \text{ if } A \text{ is injective in } \mathrm{Mod}(G) \text{ and } r \geqq 1.$$

This cohomology functor is determined up to a unique isomorphism.

For the convenience of the reader, we write the first few terms of the sequences implicit in Theorem 3.1. From the resolution

$$\cdots \to X_1 \to X_0 \to \mathbf{Z} \to 0$$

we obtain the sequence

$$0 \to \mathrm{Hom}_G(\mathbf{Z}, A) \to \mathrm{Hom}_G(X_0, A) \to \mathrm{Hom}_G(X_1, A) \to$$

so the cohomology sequence arising from an exact sequence

$$0 \to A' \to A \to A'' \to 0$$

starts with an exact part

$$0 \to A'^G \to A^G \to A''^G \to H^1(A').$$

Dually, we work with the tensor product. We let I_G as before be the augmentation ideal. For $A \in \text{Mod}(G)$, $I_G A$ is the G-module generated by all elements $\sigma a - a$ with $a \in A$, and even consists of such elements. We consider the functor $A \mapsto A_G = A/I_G A$ from $\text{Mod}(G)$ into Grab. For $A, B \in \text{Mod}(G)$ we define

$$T_G(A, B) = A \otimes_G B = (A \otimes B)_G.$$

Then T_G is a bifunctor

$$T_G : \text{Mod}(G) \times \text{Mod}(G) \to \text{Grab},$$

covariant in both variables. From *Algebra*, Chapter XX, Proposition 3.2', we find:

Theorem 3.2. *Let X be a projective resolution of $\mathbf{Z}$ in $\text{Mod}(G)$. Let T_G be as above, and let $H = \{H_r\}$ be the homology of the complex $T_G(X, A)$. Then H is a homological functor such that:*

$H_r(A) = 0$ *if* $r > 0$.

$H_0(A) = A_G$.

$H_r(A) = 0$ *if A is projective in* $\text{Mod}(G)$ *and* $r \geq 1$.

The explicit determination of $H_0(A) = A_G$ comes from the fact that

$$X_1 \otimes_G A \to X_0 \otimes_G A \to \mathbf{Z} \otimes_G A \to 0$$

is exact, and that $\mathbf{Z} \otimes_G A$ is functorially isomorphic to A_G.

Given a short exact sequence $0 \to A' \to A \to A'' \to 0$, the long homology exact sequence starts

$$\cdots \to H_1(A'') \to A'_G \to A_G \to A''_G \to 0.$$

The previous two theorems fit a standard pattern of the derived functor. In some instances, we have to go back to the way these functors are constructed by means of complexes, say as in *Algebra*, Chapter XX, Theorem 2.1. We summarize this construction as follows for abelian categories.

Theorem 3.3. *Let $\mathfrak{A}, \mathfrak{B}$ be abelian categories. Let*

$$Y : \mathfrak{A} \to C(\mathfrak{B})$$

be an exact functor to the category of complexes in $\mathfrak{B}$. Then there exists a cohomological functor H on $\mathfrak{A}$ with values in $\mathfrak{B}$, such that $H^r(A) =$ homology of the complex $Y(A)$ in dimension r. Given a short exact sequence in $\mathfrak{A}$:

$$0 \to A' \xrightarrow{u} A \xrightarrow{v} A'' \to 0$$

and therefore the exact sequence

$$0 \to Y(A') \to Y(A) \to Y(A'') \to 0,$$

the coboundary is given by the usual formula $Y(u)^{-1} d Y(v)^{-1}$.

For the applications, readers may take $\mathfrak{B}$ to be the category of abelian groups, and $\mathfrak{A}$ is $\mathrm{Mod}(G)$ most of the time.

Corollary 3.4. *Let $\mathfrak{A}_1, \mathfrak{A}$ be abelian categories and F a bifunctor on $\mathfrak{A}_1 \times \mathfrak{A}$ with values in $\mathfrak{B}$, contravariant (resp. covariant) in the first variable and covariant in the second. Let X be a complex in $C(\mathfrak{A}_1)$ such that the functor $A \mapsto F(X, A)$ on $\mathfrak{A}$ is exact. Then there exists a cohomological functor (resp. homological functor) H on $\mathfrak{A}$ with values in $\mathfrak{B}$, obtained as in Theorem 3.3, with $F(X, A) = Y(A)$.*

Next we deal with finite groups, for which we obtain a non-trivial cohomological functor in all dimensions, using constructions with complexes as in the above two theorems.

Finite groups. Suppose now that G is finite, so we have the trace homomorphism

$$\mathbf{S} = \mathbf{S}_G : A \to A$$

for every $A \in \mathrm{Mod}(G)$. We omit the index G for simplicity, so the kernel of the trace in A is denoted by $A_\mathbf{S}$. We also write I instead of I_G as long as G is the only group under consideration. It is clear that IA is contained in $A_\mathbf{S}$ and the association

$$A \mapsto A_\mathbf{S}/IA$$

is a functor from $\mathrm{Mod}(G)$ into Grab. We then have **Tate's theorem**.

Theorem 3.5. *Let G be a finite group. There is a cohomological functor $\mathbf{H}$ on $mod(G)$ with values in* Grab *such that:*

$\mathbf{H}^0$ *is the functor $A \mapsto A^G/\mathbf{S}_G A$.*

$\mathbf{H}^r(A) = 0$ *if A is injective and $r \geqq 1$.*

$\mathbf{H}^r(A) = 0$ *if A is projective and r is arbitrary.*

$\mathbf{H}$ *is erased by G-regular modules, and thus is erased by $\mathbf{M}_G$.*

Proof. Fix the projective resolution X of $\mathbf{Z}$ and apply the two bifunctors $\otimes_G$ and Hom_G to obtain a diagram:

$$
\begin{array}{ccccccc}
\to X_1 \otimes_G A & \to & X_0 \otimes_G A & & \mathrm{Hom}_G(X_0,A) & \to & \mathrm{Hom}_G(X_1,A) \to \\
 & & \downarrow & & \downarrow & & \\
 & & \mathbf{Z} \otimes_G A & & \mathrm{Hom}_G(\mathbf{Z},A) & & \\
 & & \downarrow & & \downarrow & & \\
 & & 0 & & 0 & &
\end{array}
$$

We have $A_G = \mathbf{Z} \otimes_G A$ and $\mathrm{Hom}_G(\mathbf{Z},A) = A^G$. The right side with Hom_G comes from Theorem 3.1 and the left side comes from Theorem 3.2. We shall splice these two sides together. The trace maps $A_G \to A^G$ and yields a morphism of the functor $A \mapsto A_G$ to the functor $A \mapsto A^G$. Hence there exists a unique homomorphism δ which makes the following diagram commutative.

$$
\begin{array}{ccccccc}
\to X_1 \otimes_G A & \to & X_0 \otimes_G^A & \overset{\delta}{\to} & \mathrm{Hom}_G(X_0,A) & \to & \mathrm{Hom}_G(X_1,A) \to \\
 & & \downarrow & & \uparrow & & \\
 & & \downarrow & & \uparrow & & \\
 & & A_G & \underset{\mathbf{s}_G}{\to} & A^G & & \\
 & & 0 & & 0 & &
\end{array}
$$

The upper horizontal line is then a complex. Each $X_r \otimes_G A$ may be considered as a functor in A, and similarly for $\mathrm{Hom}_G(X_r,A)$. These functors are exact since X_r is projective. Furthermore, δ is a morphism of the functor $X_0 \otimes_G$ into the functor $\mathrm{Hom}_G(X_0, \cdot)$. We let

$$
Y_r(A) = \begin{cases} \mathrm{Hom}_G(X_r, A) & \text{for } r \geqq 0 \\ X_{-r-1} \otimes_G A & \text{for } r < 0. \end{cases}
$$

Then $Y(A)$ is a complex, and $A \mapsto Y(A)$ is exact, meaning that if $0 \to A' \to A \to A'' \to 0$ is a short exact sequence, then

$$
0 \to Y(A') \to Y(A) \to Y(A'') \to 0
$$

is exact. We are thus in the standard situation of constructing a homology functor, say as in *Algebra*, Chapter XX, Theorem 2.1, whereby $\mathbf{H}^r(A)$ is the homology in dimension r of the complex $Y(A)$. In dimensions 0 and -1, we find the functors of Theorem 3.1 and 3.2, thus proving all but the last statement of the theorem, concerning the erasability.

To show that G-regular modules erase the cohomology, we do it first in dimension $r > 0$. There exists a homotopy (Cf. *Algebra*, Chapter XX, §5), i.e. a family of $\mathbf{Z}$-morphisms

$$D_r : X_r \to X_{r+1}$$

such that

$$\mathrm{id}_r = \mathrm{id}_{X_r} = \partial_{r+1} D_r + D_{r-1} \partial_r.$$

(Cf. the Remark at the end of §5, loc. cit.) Let $f : X_r \to A$ be a cocycle. By definition, $f\partial_{r+1} = 0$ and hence

$$f = f \circ \mathrm{id}_r = f D_{r-1} \partial_r.$$

On the other hand, by hypothesis there exists a $\mathbf{Z}$-morphism $v : A \to A$ such that $1_A = \mathbf{S}_G(v)$. Thus we find

$$\begin{aligned} f = 1_A f = \mathbf{S}_G(v)f = \mathbf{S}_G(vf) &= \mathbf{S}_G(vf D_{r-1}\partial_r) \\ &= \mathbf{S}_G(vf D_{r-1})\partial_r, \end{aligned}$$

which shows that f is a coboundary, in other words, the cohomology group is trivial.

For $r = 0$ we obviously have $\mathbf{H}^0(A) = 0$ if A is G-regular. For $r = -1$, the reader will check it directly. For $r < -1$, one repeats the above argument for $r \geq 1$ with the tensor product, essentially dualizing the argument (reversing the arrows). This concludes the proof of Theorem 3.3.

Alternatively, the splicing used to prove Theorem 3.3 could also be done as follows, using a complete resolution of $\mathbf{Z}$.

Let $X = (X_r)_{r \geq 0}$ be a G-free resolution of $\mathbf{Z}$, with X_r finitely generated for all r, acyclic, with augmentation ε. Define

$$X_{-r-1} = \mathrm{Hom}(X_r, \mathbf{Z}) \text{ for } r \geq 0.$$

Thus we have defined G-modules X_s for negative dimensions s. One sees immediately that these modules are G-free. If (e_i) is a basis of X_r over $\mathbf{Z}[G]$ for $r \geq 0$, then we have the dual basis $(e_i^\vee)$ for the dual module. By duality, we thus obtain G-free modules in negative dimensions. We may splice these two complexes around 0. For simplicity, say X_0 has dimension 1 over $\mathbf{Z}[G]$ (which is the case in the complexes we select for the applications). Thus let

$$X_0 = \mathbf{Z}[G] \cdot \xi, \text{ with } \varepsilon(\xi) = 1.$$

Let $\xi^\vee$ be the dual basis of $X_{-1} = \mathrm{Hom}(X_0, \mathbf{Z})$. We define ∂_0 by

$$\partial_0 \xi = S_G(\xi^\vee) = \sum_{\sigma \in G} \sigma \xi^\vee.$$

We can illustrate the relevant maps by the diagram:

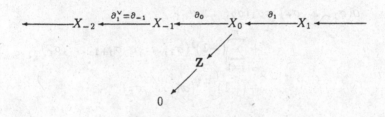

The boundaries ∂_{-r-1} for $r \geq 0$ are defined by duality. One verifies easily that the complex we have just obtained is acyclic (for instance by using a homotopy in positive dimension, and by dualising). We can then consider $\mathrm{Hom}_G(X, A)$ for A variable in $\mathrm{Mod}(G)$, and we obtain an exact functor

$$A \mapsto \mathrm{Hom}_G(X, A)$$

from $\mathrm{Mod}(G)$ into the category of complexes of abelian groups. In dimension 0, the homology of this complex is obviously

$$H^0(A) = A^G / S_G A,$$

and the uniqueness theorem applies.

The cohomological functor of Theorem 3.5 for finite groups will be called the **special** cohomology, to distinguish it from the cohomology defined for arbitrary groups, differing in dimension 0 and the negative dimensions. We write it as $\mathbf{H}_G$ if we need to specify G in the notation, especially when we shall deal with different groups and subgroups. It is uniquely determined, up to a unique isomorphism. When G is fixed throughout a discussion, we continue to denote it by $\mathbf{H}$. Thus depending on the context, we may write

$$\mathbf{H}(A) = \mathbf{H}_G(A) = \mathbf{H}(G, A) \text{ for } A \in \mathrm{Mod}(G).$$

The standard complex

The complex which we now describe allows for explicit computations of the cohomology groups. Let X_r be the free $\mathbf{Z}[G]$-module having for basis r-tuples $(\sigma_1, \ldots, \sigma_r)$ of elements of G. For $r = 0$ we take X_0 to be free over $\mathbf{Z}[G]$, of dimension 1, with basis element denoted by $(\cdot)$. We define the boundary maps by

$$\partial(\sigma_1, \ldots, \sigma_r) = \sigma_1(\sigma_2, \ldots, \sigma_r)$$
$$+ \sum_{j=1}^{r} (-1)^j (\sigma_1, \ldots, \sigma_j \sigma_{j+1}, \ldots, \sigma_r)$$
$$+ (-1)^{r+1} (\sigma_1, \ldots, \sigma_r).$$

We leave it to the reader to verify that $dd = 0$, i.e. we have a complex called the **standard complex**.

The above complex is the non-homogeneous form of another standard complex having nothing to do with groups. Indeed, let S be a set. For $r = 0, 1, 2, \ldots$ let E_r be the free $\mathbf{Z}$-module generated by $(r+1)$-tuples $(x_0, \ldots, x_r)$ with $x_0, \ldots, x_r \in S$. Thus such $(r+1)$-tuples form a basis of E_r over $\mathbf{Z}$. There is a unique homomorphism

$$d_{r+1} : E_{r+1} \to E_r$$

such that

$$d_{r+1}(x_0, \ldots, x_r) = \sum_{j=0}^{r+1} (-1)^j (x_0, \ldots, \hat{x}_j, \ldots, x_{r+1}),$$

where the symbol $\hat{x}_j$ means that this term is to be omitted. For $r = 0$ we define $\varepsilon : E_0 \to \mathbf{Z}$ to be the unique homomorphism such that $\varepsilon(x_0) = 1$. The map ε is also called the **augmentation**. Then we obtain a resolution of $\mathbf{Z}$ by the complex

$$\to E_{r+1} \to E_r \to \cdots \to E_0 \to \mathbf{Z} \to 0.$$

The above standard complex for an arbitrary set is called the **homogeneous standard complex**. It is exact, as one sees by using a homotopy as follows. Let $z \in S$ and define

$$h : E_r \to E_{r+1} \qquad \text{by} \qquad h(x_0, \ldots, x_r) = (z, x_0, \ldots, x_r).$$

Then it is routinely verified that

$$dh + hd = \text{id} \qquad \text{and} \qquad dd = 0.$$

Exactness follows at once.

Suppose now that the set S is the group G. Then we may define an action of G on the homogeneous complex E by letting

$$\sigma(\sigma_0, \ldots, \sigma_r) = (\sigma\sigma_0, \ldots, \sigma\sigma_r).$$

It is then routinely verified that each E_r is $\mathbf{Z}[G]$-free. We take $z = e$. Thus the homogeneous complex gives a $\mathbf{Z}[G]$-free resolution of $\mathbf{Z}$.

In addition, we have a $\mathbf{Z}[G]$-isomorphism $X \xrightarrow{\approx} E$ between the non-homogeneous and the homogeneous complex uniquely determined by the value on basis elements such that

$$(\sigma_1, \ldots, \sigma_r) \mapsto (e, \sigma_1, \sigma_1\sigma_2, \ldots, \sigma_1\sigma_2 \ldots \sigma_r).$$

The reader will immediately verify that the boundary operator ∂ given for X corresponds to the boundary operator as given on E under this isomorphism.

If G is finite, then each X_r is finitely generated. We may then proceed as was done in general to define the standard modules X_s in negative dimensions. The dual basis of $\{(\sigma_1, \ldots, \sigma_r)\}$ will be denoted by $\{[\sigma_1, \ldots, \sigma_r]\}$ for $r \geq 1$. The dual basis of $(\cdot)$ in dimension 0 will be denoted by $[\cdot]$. For finite groups, we thus obtain:

Theorem 3.6. *Let G be a finite group. Let $X = \{X_r\}(r \in \mathbf{Z})$ be the standard complex. Then X is $\mathbf{Z}[G]$-free, acyclic, and such that the association*

$$A \mapsto \operatorname{Hom}_G(X, A)$$

is an exact functor of $\operatorname{Mod}(G)$ into the category of complexes of abelian groups. The corresponding cohomological functor $\mathbf{H}$ is such that $\mathbf{H}^0(A) = A^G/\mathbf{S}_G A$.

Examples. In the standard complex, the group of 1-cocycles consists of maps $f : G \to A$ such that

$$f(\sigma) + \sigma f(\tau) = f(\sigma\tau) \text{ for all } \sigma, \tau \in G.$$

The 1-coboundaries consist of maps f of the form $f(\sigma) = \sigma a - a$ for some $a \in A$. Observe that if G has trivial action on A, then by the above formulas,

$$H^1(A) = \operatorname{Hom}(G, A).$$

In particular, $H^1(\mathbf{Q}/\mathbf{Z}) = \hat{G}$ is the character group of G.

The 2-cocycles have also been known as **factor sets**, and are maps $f(\sigma, \tau)$ of two variables in G satisfying

$$f(\sigma, \tau) + f(\sigma\tau, \rho) = \sigma f(\tau, \rho) + f(\sigma, \tau\rho).$$

In Theorem 3.5, we showed that for finite groups, H is erased by M_G. The analogous statement in Theorem 3.1 has been left open. We can now settle it by using the standard complex.

Theorem 3.7. *Let G be any group. Let $B \in \operatorname{Mod}(\mathbf{Z})$. Then for all subgroups G' of G we have*

$$H^r(G', M_G(B)) = 0 \text{ for } r > 0.$$

Proof. By Proposition 2.6 it suffices to prove the theorem when $G' = G$. Define a map h on the chains of the standard complex by

$$h : C^r(G, M_G(B)) \to C^{r-1}(G, M_G(B))$$

by letting

$$(hf)_{\sigma_2,\ldots,\sigma_r}(\sigma) = f_{\sigma,\sigma_2,\ldots,\sigma_r}(e).$$

One verifies at once that

$$f = hdf + dhf,$$

whence the theorem follows. (Cf. *Algebra*, Chapter XX, §5.)

§4. Explicit computations

In this section we compute some low dimensional cohomology groups with some special coefficients.

We recall the exact sequences

$$0 \to I_G \to \mathbf{Z}[G] \to \mathbf{Z} \to 0$$

$$0 \to \mathbf{Z} \to \mathbf{Q} \to \mathbf{Q}/\mathbf{Z} \to 0.$$

We suppose G finite of order n. We write $I = I_G$ and $\mathbf{H} = \mathbf{H}_G$ for simplicity. We find:

$$
\begin{array}{llll}
\mathbf{H}^{-3}(\mathbf{Q}/\mathbf{Z}) & \approx \mathbf{H}^{-2}(\mathbf{Z}) & \approx \mathbf{H}^{-1}(I) & = I/I^2 \\
\mathbf{H}^{-2}(\mathbf{Q}/\mathbf{Z}) & \approx \mathbf{H}^{-1}(\mathbf{Z}) & \approx \mathbf{H}^{0}(I) & = 0 \\
\mathbf{H}^{-1}(\mathbf{Q}/\mathbf{Z}) & \approx \mathbf{H}^{0}(\mathbf{Z}) & \approx \mathbf{H}^{1}(I) & = \mathbf{Z}/n\mathbf{Z} \\
\mathbf{H}^{0}(\mathbf{Q}/\mathbf{Z}) & \approx \mathbf{H}^{1}(\mathbf{Z}) & \approx \mathbf{H}^{2}(I) & = 0 \\
\mathbf{H}^{1}(\mathbf{Q}/\mathbf{Z}) & \approx \mathbf{H}^{2}(\mathbf{Z}) & \approx \mathbf{H}^{3}(I) & = \hat{G}.
\end{array}
$$

The proof of these formulas arises as follows. Each middle term in the above exact sequences annuls the cohomological functor because $\mathbf{Z}[G]$ is G-regular in the first case, and $\mathbf{Q}$ is uniquely divisible in the second case. The stated isomorphisms other than those furthest to the right are then those induced by the coboundary in the cohomology exact sequence.

As for the values furthest to the right, they are proved as follows.

For the first one with I/I^2, we note that every element of I has trace 0, and hence $\mathbf{H}^{-1}(I) = I/I^2$ directly from its definition as in Theorem 2.7.

For the second line, we immediately have $\mathbf{H}^{-1}(\mathbf{Z}) = 0$ since an element of $\mathbf{Z}$ with trace 0 can only be equal to 0.

For the third line, we have obviously $H^0(\mathbf{Z}) = \mathbf{Z}/n\mathbf{Z}$ from the definition.

For the fourth line, $H^1(\mathbf{Z}) = 0$ because from the standard complex, this group consists of homomorphism from G into $\mathbf{Z}$, and G is finite.

For the fifth line, we find $\hat{G}$ because $H^1(\mathbf{Q}/\mathbf{Z})$ is the dual group $\text{Hom}(G, \mathbf{Q}/\mathbf{Z})$.

Remark. Let U be a subgroup of G. Then

$$\mathbf{H}^r(U, \mathbf{Z}[G]) = \mathbf{H}^r(U, \mathbf{Q}) = 0 \text{ for } r \in \mathbf{Z}.$$

Hence the above table applies also to a subgroup U, if we replace $\mathbf{H}_G$ by $\mathbf{H}_U$ and replace $n_G = \#(G)$ by $n_U = \#(U)$ in the third line, as well as G by U in the last line.

As far as I/I^2 is concerned, we have another characterization.

Proposition 4.1. *Let G be a group (possible infinite). Let G^c be its commutator group. Let I_G be the ideal of $\mathbf{Z}[G]$ generated by all elements of the form $\sigma - e$. Then there is a functorial isomorphism (covariant on the category of groups)*

$$G/G^c \approx I_G/I_G^2, \text{ given by } \sigma G^c \mapsto (\sigma - e) + I_G^2.$$

Proof. We can define a map $G \to I/I^2$ by $\sigma \mapsto (\sigma - e) + I^2$. One verifies at once that this map is a homomorphism. Since I/I^2 is commutative, G^c is contained in the kernel of the homomorphism, whence we obtain a homomorphism $G/G^c \to I/I^2$. Conversely, I is $\mathbf{Z}$-free, and the elements $(\sigma - e)$ with $\sigma \in G$, $\sigma \neq e$ form a basis over $\mathbf{Z}$. Hence there exists a homomorphism $I \mapsto G/G^c$ defined by the formula $(\sigma - e) \mapsto \sigma G^c, \sigma \neq e$. In addition, this homomorphism is trivial on I^2, as one verifies at once. Thus we obtain a homomorphism of I/I^2 into G/G^c, which is visibly inverse of the previous homomorphism of G/G^c into I/I^2. This proves the proposition.

In particular, from the first line of the table, we find the isomorphism

$$\mathbf{H}^{-2}(\mathbf{Z}) \approx G/G^c,$$

obtained from the coboundary and the isomorphism of Proposition 4.1. This isomorphism is important in class field theory.

We end our explicit computations with one more result on H^1.

Proposition 4.2. *Let G be a group, $A \in \text{Mod}(G)$, and let $\alpha \in H^1(G, A)$. Let $\{a(\sigma)\}$ be a 1-cocycle representing α. There exists a G-morphism*

$$f : I_G \to A$$

such that $f(\sigma - e) = a(\sigma)$, i.e. one has $f \in (\text{Hom}(I_G, A))^G$. Then the sequence

$$0 \to A = \text{Hom}(\mathbf{Z}, A) \to \text{Hom}(\mathbf{Z}[G], A) \to \text{Hom}(I_G, A) \to 0$$

is exact, and taking the coboundary with respect to this short exact sequence, one has

$$\delta(\varkappa_G f) = -\alpha.$$

Proof. Since the elements $(\sigma - e)$ form a basis of I_G over $\mathbf{Z}$, one can define a $\mathbf{Z}$-morphism f satisfying $f(\sigma - e) = a(\sigma)$ for $\sigma \neq e$. The formula is even valid for $\sigma = e$, because putting $\sigma = \tau = e$ in the formula for the coboundary

$$a(\sigma\tau) = a(\sigma) + \sigma a(\tau),$$

we find $a(e) = 0$. We claim that f is a G-morphism. Indeed, for $\sigma, \tau \in G$ we find:

$$\begin{aligned}
f(\sigma(\tau - e)) = f(\sigma\tau - \sigma) &= f((\sigma\tau - e) - (\sigma - e)) \\
&= f(\sigma\tau - e) - f(\sigma - e) \\
&= a(\sigma\tau) - a(\sigma) \\
&= \sigma a(\tau) \\
&= \sigma f(\tau - e).
\end{aligned}$$

To compute $\varkappa_G f$ we first have to find a standard cochain of $\text{Hom}(\mathbf{Z}[G], A)$ in dimension 0, mapping on f, that is an element $f' \in \text{Hom}(\mathbf{Z}[G], A)$ whose restriction to I_G is f. Since

$$\mathbf{Z}[G] = I_G + \mathbf{Z}e$$

is a direct sum, we can define f' by prescribing that $f'(e) = 0$ and f' is equal to f on I_G. One then sees that $g_\sigma = \sigma f' - f'$ is a cocycle of dimension 1, in $\text{Hom}(\mathbf{Z}, A)$, representing $\varkappa_G f$ by definition. I claim that under the identification of $\text{Hom}(\mathbf{Z}, A)$ with A, the map g_σ corresponds to $-a(\sigma)$. In other words, we have to verify that $g_\sigma(e) = -a(\sigma)$. Here goes:

$$
\begin{aligned}
g_\sigma(e) &= (\sigma f')(e) - f'(e) \\
&= \sigma f'(\sigma^{-1}) \\
&= \sigma f(\sigma^{-1} - e) \\
&= f(e - \sigma) \\
&= -a(\sigma),
\end{aligned}
$$

thus proving our assertion and concluding the proof of Proposition 4.2.

§5. Cyclic groups

Throughout this section we let G be a finite cyclic group, and we let σ be a generator of G

The main feature of the cohomology of such a cyclic group is that the cohomology is periodic of period 2, as we shall now prove.

We start with the δ-functor in two dimensions

$$\mathbf{H}_G^{-1} \text{ and } \mathbf{H}_G^0.$$

Recall that $\varkappa : A^G \to \mathbf{H}^0(G, A) = A^G / S_G A$ and

$$\mathbf{x} : A_{S_G} \to \mathbf{H}^{-1}(G, A) = A_{S_G} / I_G A$$

are the canonical homomorphisms. We are going to define a cohomological functor directly from these maps. For $r \in \mathbf{Z}$ we let:

$$
\mathbf{H}^r(G, A) = \begin{cases}
\mathbf{H}^{-1}(G, A) & \text{if } r \text{ is odd} \\
\mathbf{H}^0(G, A) & \text{if } r \text{ is even.}
\end{cases}
$$

We then have to define the coboundary. Let

$$0 \to A' \xrightarrow{u} A \xrightarrow{v} A'' \to 0$$

be a short exact sequence in $\text{Mod}(G)$. For each $r \in \mathbf{Z}$ we define $\varkappa_r$ and $\varkappa_r$ in the natural way given the above definition, and for $a'' \in A''^G$ we pick any element $a \in A$ such that $va = a''$ and define

$$\delta_\sigma \varkappa_r(a'') = \varkappa_r(\sigma a - a).$$

Similarly, in odd dimensions, for $a'' \in A''_{\mathbf{S}_G}$ we pick $a \in A$ such that $va = a''$ and we define

$$\delta_\sigma \varkappa_r(a'') = \varkappa_r(\mathbf{S}_G a).$$

It is immediately verified that δ_σ is well-defined (depending on the choice of generator σ), that is independent of the choice of a such that $va = a''$. It is then also routinely and easily verified that the sequence $\{\mathbf{H}^r\}$ $(r \in \mathbf{Z})$ with the coboundary δ_σ is a cohomological functor. Since it vanishes on G-regular modules, and is given as before in dimension < 0, it follows that it is isomorphic to the special functor defined previously, by the uniqueness theorem. Directly from this new definition, we now see that for all $r \in \mathbf{Z}$ and $A \in \text{Mod}(G)$ we have the periodicity

$$\mathbf{H}^{r+2}(G, A) = \mathbf{H}^r(G, A).$$

Of course, by truncating on the left we can define a similar functor in the ordinary case. We put $H^0(A) = A^G$ as before, and for $r \geq 1$ we let:

$$H^r(A) = \begin{cases} H^{-1}(G, A) & \text{if } r \text{ is odd} \\ H^0(G, A) & \text{if } r \text{ is even.} \end{cases}$$

We define the coboundary as before, so we find a cohomological functor which is periodic for $r \geq 0$, and 0 in negative dimensions. Again the uniqueness theorem shows that it coincides with the functor defined in the previous sections. The beginning of the cohomology sequence reads:

$$0 \to A'^G \to A^G \to A''^G \to H^{-1}(G, A)$$

and it continues as for the special functor.

The cohomology sequence for the special functor can be conveniently written as in the next theorem.

Theorem 5.1. *Let G be finite cyclic and let σ be a generator. Let*

$$0 \rightarrow A' \rightarrow A \rightarrow A'' \rightarrow 0$$

be a short exact sequence in $\text{Mod}(G)$. *Then the following hexagon is exact:*

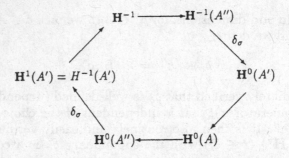

Suppose given an exact hexagon of finite abelian groups as shown:

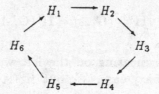

and let h_i be the order of H_i, that is $h_i = (H_i : 0)$. Let

$$f_i : H_i \rightarrow H_{i+1} \quad \text{with} \quad i \mod 6$$

be the corresponding homomorphism in the diagram. Then

$$h_i = (H_i : f_{i-1}H_{i-1})(f_{i-1}H_{i-1} : 0) = m_i m_{i-1}.$$

Hence

$$1 = \frac{m_0 m_1 m_2 m_3 m_4 m_5}{m_1 m_2 m_3 m_4 m_5 m_6} = \frac{h_1 h_3 h_5}{h_2 h_4 h_6}.$$

We may apply this formula to the exact sequence of Theorem 5.1. Assume that each group $\mathbf{H}^i(A)$ is finite, and let $h_i(A) = $ order of $\mathbf{H}^i(G, A)$. Then

$$1 = \frac{h_1(A')h_1(A'')h_2(A)}{h_1(A)h_2(A')h_2(A'')}.$$

Now let $A \in \text{Mod}(G)$ be arbitrary. If $h_1(A)$ and $h_2(A)$ are finite, we define the **Herbrand quotient** $h_{2/1}(A)$ to be

$$h_{2/1}(A) = \frac{h_2(A)}{h_1(A)} = \frac{(A_{\sigma-e} : S_G A)}{A_{S_G} : (\sigma - e)A}.$$

If $h_1(A)$ or $h_2(A)$ is not finite, we say that the Herbrand quotient is **not defined**. This Herbrand quotient is in fact a Euler characteristic, cf. for instance *Algebra*, Chapter XX, §3. If A is a finite abelian group in $\text{Mod}(G)$, then the Herbrand quotient is defined. The main properties of the Herbrand quotient are contained in the next theorems.

Theorem 5.2. Herbrand's Lemma. *Let G be a finite cyclic group, and let*

$$0 \to A' \to A \to A'' \to 0$$

be a short exact sequence in $\text{Mod}(G)$. *If two out of three Herbrand quotients $h_{2/1}(A'), h_{2/1}(A), h_{2/1}(A'')$ are defined so is the third, and one has*

$$h_{2/1}(A) = h_{2/1}(A')h_{2/1}(A'').$$

Proof. This follows at once from the discussion preceding the theorem. It is also a special case of *Algebra*, Chapter XX, Theorem 3.3.

Theorem 5.3. *Let G be finite cyclic and suppose $A \in \text{Mod}(G)$ is finite. Then $h_{2/1}(A) = 0$.*

Proof. We have a lattice of subgroups:

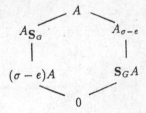

The factor groups A/A_{S_G} and $S_G A$ are isomorphic, and so are $A/A_{\sigma-e}$ and $(\sigma - e)A$. Computing the order $(A : 0)$ going around both sides of the hexagon, we find that the factor groups of the two vertical sides have the same order, that is

$$(A_{S_G} : (\sigma - e)A) = (A_{\sigma-e} : S_G A).$$

That $h_{2/1}(A) = 1$ now follows from the definitions.

Finally we have a result concerning the Herrbrand quotient for trivial action.

Theorem 5.4. *Let G be a finite cyclic group of prime order p. Let $A \in \mathrm{Mod}(G)$. Let $t(A)$ be the Herbrand quotient relative to the trivial action of G on the abelian group A, so that*

$$t(A) = \frac{(A_p : 0)}{(A/pA : 0)}.$$

Suppose this quotient is defined. Then $t(A^G), t(A_G)$, and $h_{2/1}(A)$ are defined, and one has

$$h_{2/1}(A)^{p-1} = t(A^G)^p/t(A) = t(A_G)^p/t(A).$$

Proof. We leave the proof as an exercise (not completely trivial).

CHAPTER II

Relations with Subgroups

This chapter tabulates systematically a number of relations between the cohomology of a group and that of its subgroups and factor groups.

§1. Various morphisms

(a) **Changing the group** G. Let $\lambda : G' \to G$ be a group homomorphism. Then λ gives rise to an exact functor

$$\Phi_\lambda : \mathrm{Mod}(G) \to \mathrm{Mod}(G')$$

because every G-module can be considered as a G'-module if we define the action of an element $\sigma' \in G'$ on an element $a \in A$ by

$$\sigma' a = \lambda(\sigma') a.$$

We may therefore consider the cohomological functor $H_{G'} \circ \Phi_\lambda$ (or the special functor $\mathbf{H}_{G'} \circ \Phi_\lambda$ if G' is finite) on $\mathrm{Mod}(G)$.

In dimension 0, we have a morphism of functors

$$H_G^0 \to H_{G'}^0 \circ \Phi_\lambda$$

given by the inclusion $A^G \hookrightarrow A^{G'} = (\Phi_\lambda(A))^{G'}$. If in addition G and G' are finite, then we have a morphism of functors

$$\mathbf{H}_G^0 \to \mathbf{H}_{G'}^0 \circ \Phi_\lambda$$

267

given by the homomorphism $A^G/S_G A \to A^{G'}/S_{G'} A$, with G' acting on A in the manner prescribed above, via λ.

By the uniqueness theorem, there exists a unique morphism of cohomological functor (δ-morphism)

$$\lambda^* : H_G \to H_{G'} \circ \Phi_\lambda \quad \text{or} \quad \mathbf{H}_G \to \mathbf{H}_{G'} \circ \Phi_\lambda,$$

the second possibility arising when G and G' are finite. We shall now make this map λ^* explicit in various special cases.

Suppose λ is surjective. Then we call λ^* the **lifting** morphism, and we denote it by $\text{lif}_{G'}^G$. In this case, G may be viewed as a factor group of G' and the lifting goes from the factor group to the group. On the other hand, when G' is a subgroup of G, then λ^* will be called the **restriction**, and will be studied in detail below.

Let $A \in \text{Mod}(G)$ and $B \in \text{Mod}(G')$. We may consider A as a G'-module as above (via the given λ). Let $v : A \to B$ be a G'-morphism. Then we say that the pair (λ, v) is a **morphism** of (G, A) to (G', B). One can define formally a category whose objects are pairs (G, A) for which the morphisms are precisely the pairs (λ, v). Every morphism (λ, v) induces a homomorphism

$$(\lambda, v)_* : H^r(G, A) \to H^r(G', B),$$

and similarly replacing H by the special $\mathbf{H}$ if G and G' are finite, by taking the composite

$$\mathbf{H}^r(G, A) \xrightarrow{\lambda^*} \mathbf{H}^r(G', A) \xrightarrow{\mathbf{H}_{G'}(v)} \mathbf{H}^r(G', B).$$

Of course, we should write more correctly $\mathbf{H}^r(G', \Phi_\lambda(A))$, but usually we delete the explicit reference to Φ_λ when the reference is clear from the context.

Proposition 1.1. *Let (λ, v) be a morphism of (G, A) to $G', B)$, and let (φ, w) be a morphism of (G', B) to (G'', C). Then $(\lambda\varphi, wv)$ is a morphism of (G, A) to (G'', C), and one has*

$$(\varphi\lambda, wv)_* = (\varphi, w)_*(\lambda, v)_*.$$

Proof. Since φ^* is a morphism of functors, the following diagram is commutative:

$$
\begin{array}{ccc}
H^r(G', \Phi_\lambda(A)) & \xrightarrow{H_{G'}(v)} & H^r(G', B) \\
{\scriptstyle \varphi^*}\Big\downarrow & & \Big\downarrow{\scriptstyle \varphi^*} \\
H^r(G'', \Phi_\varphi \Phi_\lambda(A)) & \xrightarrow[H_{G''}(v)]{} & H^r(G'', \Phi_\varphi(B)).
\end{array}
$$

Consequently we find

$$
\begin{aligned}
(\lambda\varphi, wv)_* &= H_{G''}(wv) \circ (\lambda\varphi)^* \\
&= H_{G''}(w) \circ H_{G''}(v) \circ \varphi^* \circ \lambda^* \\
&= H_{G''}(w) \circ \varphi^* H_{G'}(v) \circ \lambda^* \\
&= (\varphi, w)_*(\lambda, v)_*,
\end{aligned}
$$

thus proving the proposition.

(b) Restriction. This is the case when λ is an injection, so that we may consider G' as a subgroup of G. We therefore have for $A \in \mathrm{Mod}(G)$:

$$
\mathrm{res}^G_{G'} : H^r(G, A) \to H^r(G', A),
$$

and similarly for the special functor $\mathbf{H}^r$ when G is finite. One verifies at once that for $r > 0$ the restriction homomorphism is obtained from the standard complex by restricting a cochain $\{f(\sigma_1, \ldots, \sigma_r)\}$ as a function of r-tuples of elements of G to r-tuples of elements in G', because the restriction is a morphism of cohomological functors to which we can apply the uniqueness theorem. In dimension 0, the restriction is induced by the inclusion mapping.

We have transitivity:

Proposition 1.2. *Let $G' \supset G''$ be subgroups of G. Then on H_G, or $\mathbf{H}_G$ if G is finite, we have*

$$
\mathrm{res}^{G'}_{G''} \circ \mathrm{res}^G_{G'} = \mathrm{res}^G_{G''}.
$$

Proof. Immediate from Proposition 1.1.

(c) **Inflation.** Let $\lambda : G \to G/G'$ be a surjective homomorphism. Let $A \in \operatorname{Mod}(G)$. Then $A^{G'}$ is a G/G'-module for the obvious action induced by the action of G, trivial on G', and of course $A^{G'}$ is also a G-module for this operation. We have a morphism of inclusion

$$u : A^{G'} \to A$$

in $\operatorname{Mod}(G)$, which induces a homomorphism

$$H_G^r(u) = u_r : H^r(G, A^{G'}) \to H^r(G, A) \text{ for } r \geqq 0.$$

We define inflation

$$\operatorname{inf}_G^{G/G'} : H^r(G/G', A^{G'}) \to H^r(G, A)$$

to be the composite of the functorial morphism

$$H^r(G/G', A^{G'}) \to H^r(G, A^{G'})$$

followed by the induced homomorphism u_r for $r \geqq 0$. Note that inflation is NOT defined for the special cohomology functor when G is finite.

In dimension 0, the inflation therefore gives the identity map

$$(A^{G'})^{G/G'} \to A^G.$$

In dimension $r > 0$, it is induced by the cochain homomorphism in the standard complex, which to each cochain $\{f(\bar\sigma_1, \ldots, \bar\sigma_r)\}$ with $\bar\sigma_i \in G/G'$ associates the cochain $\{f(\sigma_1, \ldots, \sigma_r)\}$ whose values are constant on cosets of G'.

We have already observed that if G acts trivially on A, then $H^1(G, A)$ is simply $\operatorname{Hom}(G, A)$. Therefore we obtain:

Proposition 1.3. *Let G' be a normal subgroup of G and suppose G acts trivially on A. Then the inflation*

$$\operatorname{inf}_G^{G/G'} : H^1(G/G', A^{G'}) \to H^1(G, A)$$

induces the inflation of a homomorphism $\bar\chi : G/G' \to A$ to a homomorphism $\chi : G \to A$.

Let G' be a normal subgroup of G. We may consider the association

$$F_G : A \mapsto A^{G'}$$

as a functor, not exact, from $\text{Mod}(G)$ to $\text{Mod}(G/G')$. Inflation is then a morphism of functors (but not a cohomological morphism)

$$H_{G/G'} \circ F_{G'} \to H_G$$

on the category $\text{Mod}(G)$. Even though we are not dealing with a cohomological morphism, we can still use the uniqueness theorem to prove certain commutativity formulas, by decomposing the inflation into two pieces.

As another special case of Proposition 1.1, we have:

Proposition 1.4. *Let G', N be subgroups of G with N normal in G, and N contained in G'. Then on $H^r(G/N, A^N)$ we have*

$$\inf_{G'}^{G'/N} \circ \operatorname{res}_{G'/N}^{G/N} = \operatorname{res}_{G'}^{G} \circ \inf_{G}^{G/N}.$$

We also have transitivity, also as a special case of Proposition 1.1.

Proposition 1.4. *Let $G \to G_1 \to G_2$ be surjective group homomorphisms. Then*

$$\inf_{G}^{G_1} \circ \inf_{G_1}^{G_2} = \inf_{G}^{G_2}.$$

(d) Conjugation. Let U be a subgroup of G. For $\sigma \in G$ we have the conjugate subgroup

$$U^\sigma = \sigma^{-1} U \sigma = U[\sigma] = [\sigma^{-1}]U.$$

The notation is such that $U^{(\sigma\tau)} = (U^\sigma)^\tau$. On $\text{Mod}(G)$ we have two cohomological functors, H_U and H_{U^σ}. In dimension 0, we have an isomorphism of functors

$$A^U \to A^{U^\sigma} \quad \text{given by} \quad a \mapsto \sigma^{-1} a.$$

We may therefore extend this isomorphism uniquely to an isomorphism of H_U with H_{U^σ} which we denote by σ_* and which we call **conjugation**.

Similarly if U is finite, we have conjugation σ_* on the special functor $\mathbf{H}_U \to \mathbf{H}_{U^\sigma}$.

Proposition 1.5. *If $\sigma \in U$ then σ_* is the identity on H_U (resp. $\mathbf{H}_U$ if U is finite).*

Proof. The assertion is true in dimension 0, whence in all dimensions.

Let $f : A \to B$ be a U-morphism with $A, B \in \text{Mod}(G)$. Then

$$f^\sigma = [\sigma^{-1}]f = f[\sigma] : A \to B$$

is a U^σ-morphism. The fact that σ_* is a morphism of functors shows that

$$H_{U^\sigma}(f^\sigma) \circ \sigma_* = \sigma_* \circ H_U(f)$$

as morphisms on $H(U, A)$ (and similarly for H replaced by $\mathbf{H}$ if U is finite).

If U is a normal subgroup of G, then σ_* is an automorphism of H_U (resp. $\mathbf{H}_U$ if U is finite). In other words, G acts on H_U (or $\mathbf{H}_U$). Since we have seen that σ_* is trivial if $\sigma \in U$ it follows that actually G/U acts on H_U (resp. $\mathbf{H}_U$).

Proposition 1.7. *Let $V \subset U$ be subgroups of G, and let $\sigma \in G$. Then*

$$\sigma_* \circ \text{res}^U_V = \text{res}^{U^\sigma}_{V^\sigma} \circ \sigma_*$$

on H_U (resp. $\mathbf{H}_U$ if U is finite).

Proposition 1.8. *Let $V \subset U$ be subgroups of G of finite index, and let $\sigma \in G$. Suppose V normal in U. Then*

$$\text{inf}^{U^\sigma/V^\sigma}_{U^\sigma} \circ \sigma_* = \sigma_* \circ \text{inf}^{U/V}_U$$

on $H(U/V, A^V)$, with $A \in \text{Mod}(G)$.

Both the above propositions are special cases of Proposition 1.1.

(e) **The transfer.** Let U be a subgroup of G, of finite index. The trace gives a morphism of functors $H^0_U \to U^0_G$ by the formula

$$S^U_G : A^U \to A^G,$$

and similarly in the special case when G is finite, $\mathbf{H}^0_U \to \mathbf{H}^0_G$ by

$$S^U_G : A^U/S_U A \to A^G/S_G A.$$

The unique extension to the cohomological functors will be denoted by tr_G^U, and will be called the **transfer**. The following proposition is proved by verifying the asserted commutativity in dimension 0, and then applying the uniqueness theorem. In the case of inflation, we decompose this map in its two components.

Proposition 1.9. *Let $V \subset U \subset G$ be subgroups of finite index in G. Then on H_V (resp. $\mathbf{H}_V$) we have*

(1) $\operatorname{tr}_G^U \circ \operatorname{tr}_U^V = \operatorname{tr}_G^V$.

(2) $\sigma_* \circ \operatorname{tr}_U^V = \operatorname{tr}_U^V \circ \sigma_*$ *for $\sigma \in G$.*

(3) *If V is normal in G, then on $H^r(U/V, A^V)$ with $r \geq 0$ we have*
$$\operatorname{inf}_G^{G/V} \circ \operatorname{tr}_{G/V}^{U/V} = \operatorname{tr}_G^U \circ \operatorname{inf}_U^{U/V}.$$

The next result is particularly important.

Proposition 1.10. *Let U be a subgroup of finite index in G. Then on H_G (resp. $\mathbf{H}_G$) we have*
$$\operatorname{tr}_G^U \circ \operatorname{res}_U^G = (G : U),$$

where $(G : U)$ on the right abbreviates $(G : U)_H$, i.e. multiplication by the index on the cohomology functor.

Proof Again the formula is immediate in dimension 0, since restriction is just inclusion, and so the trace simply multiplies elements by $(G : U)$. Then the proposition follows in general by applying the uniqueness theorem.

Corollary 1.11. *Suppose G finite of order n. Then for all $r \in \mathbf{Z}$ and $A \in \operatorname{Mod}(G)$ we have $n\mathbf{H}^r(G, A) = 0$.*

Proof. Take $U = \{e\}$ in the proposition and use the fact that $H^r(e, A) = 0$.

Corollary 1.12. *Suppose G finite, and $A \in \operatorname{Mod}(G)$ finitely generated over $\mathbf{Z}$. Then $\mathbf{H}^r(G, A)$ is a finite group for all $r \in \mathbf{Z}$.*

Proof. First $\mathbf{H}^r(G, A)$ is finitely generated, because in the standard complex, the cochains are determined by their values on the finite number of generators of the complex in each idmension. Since $\mathbf{H}^r(G, A)$ is a torsion group by the preceding corollary, it follows that it is finite.

Corollary 1.13. *Suppose G finite and $A \in \mathrm{Mod}(G)$ is uniquely divisible by every integer $m \in \mathbf{Z}, m \neq 0$. Then $\mathbf{H}^r(G, A) = 0$ for all $r \in \mathbf{Z}$.*

Proposition 1.14. *Let $U \subset G$ be a subgroup of finite index. Let $A, B \in \mathrm{Mod}(G)$ and let $f : A \to B$ be a U-morphism. Then*

$$H_G(\mathbf{S}_G^U(f)) = \mathrm{tr}_G^U \circ H_U(f) \circ \mathrm{res}_U^G,$$

and similarly with $\mathbf{H}$ instead of H when G is finite.

Proof. We use the fact that the assertion is immediate in dimension 0, together with the technique of dimension shifting. We also use Chapter I, Lemma 2.4, that we can take a G-morphism in and out of a trace, so we find a commutative diagram

$$
\begin{array}{ccccccccc}
0 & \longrightarrow & A & \longrightarrow & M_G(A) & \longrightarrow & X_A & \longrightarrow & 0 \\
 & & \downarrow & & \downarrow & & \downarrow & & \\
0 & \longrightarrow & B & \longrightarrow & M_G(B) & \longrightarrow & X_B & \longrightarrow & 0
\end{array}
$$

the three vertical maps being $\mathbf{S}_G^U(f), \mathbf{S}_G^U(M(f))$ and $\mathbf{S}_G^U(X(f))$ respectively. In the hypothesis of the proposition, we replace f by $X(f) : X_A \to X_B$, and we suppose the proposition proved for $X(f)$. We then have two squares which form the faces of a cube as shown:

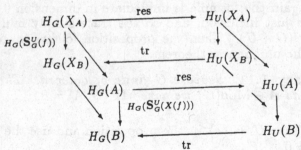

The maps going forward are the coboundary homomorphisms, and are surjective since M_G erases cohomology. Thus the diagram allows an induction on the dimension to conclude the proof. In the case of the special functor H, we use the dual diagram going to the left for the induction.

Corollary 1.15. *Suppose G finite and $A, B \in \mathrm{Mod}(G)$. Let $f : A \to B$ be a $\mathbf{Z}$-morphism. Then $\mathbf{S}_G(f) : A \to B$ induces 0 on all the cohomology groups.*

Proof. We can take $U = \{e\}$ in the preceding proposition.

Explicit formulas

We shall use systematically the following notation. We let $\{c\}$ be the set of right cosets of a subgroup U of G (not necessarily finite). We choose a set of coset representatives denoted by $\bar{c}$. If $\sigma \in G$, we denote by $\bar{\sigma}$ the representative of $U\sigma$. We may then write

$$G = \bigcup_c U\bar{c} = \bigcup_c \bar{c}^{-1} U$$

since $\{\bar{c}^{-1}\}$ is a system of representatives for the left cosets of U in G. By definition, we have $U\bar{c} = Uc$, whence for all $\sigma \in G$, we have

$$\bar{c}\sigma\overline{c\sigma}^{-1} \in U.$$

We now give the explicit formula for the transfer on standard cochains. It is induced by the cochain map $f \mapsto \mathrm{tr}_G^U(f)$ given by

$$\mathrm{tr}_G^U(f)(\sigma_0, \ldots, \sigma_r) = \sum_c \bar{c}^{-1} f(\bar{c}\sigma_0\overline{c\sigma_0}^{-1}, \ldots, \bar{c}\sigma_r\overline{c\sigma_r}^{-1}).$$

For a non-homogeneous cochain, we have the formula

$$\mathrm{tr}_G^U(f)(\sigma_1, \ldots, \sigma_r) =$$
$$\sum_c \bar{c}^{-1} f(\bar{c}\sigma_1\overline{c\sigma_1}^{-1}, \ldots, \overline{c\sigma_1}\sigma_2\overline{c\sigma_1\sigma_2}^{-1}, \ldots, \overline{c\sigma_1\cdots\sigma_{r-1}}\sigma_r\overline{c\sigma_1\cdots\sigma_r}^{-1}).$$

(f) Translation. Let G be a group, U a subgroup and N a normal subgroup of G. Let $A \in \mathrm{Mod}(G)$. Then we have a lattice of submodules of A:

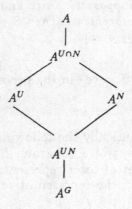

We have $UN/N \approx U/(U \cap N)$, and U acts on A^N since G acts on A^N. Furthermore $U \cap N$ leaves A^N fixed, and so we have a homomorphism called **translation**

$$\text{tsl}_* : H^r(UN/N, A^N) \to H^r(U/(U \cap N), A^{U \cap N})$$

for $r \geq 0$. The isomorphism $UN/N \approx U/(U \cap N)$ is compatible with the inclusion of A^N in $A^{U \cap N}$. Similarly, if G is finite, we get the translation for the special cohomology $\mathbf{H}$ instead of H, with $r \geq 0$.

Taking G arbitrary and $r \geq 0$, we have a commutative diagram:

$$
\begin{array}{ccccc}
H^r(G/N, A^N) & \xrightarrow{\text{res}} & H^r(UN/N, A^N) & \xrightarrow{\text{inf}} & H^r(UN, A) \\
 & & \downarrow{\text{tsl}} & & \\
\text{inf}\downarrow & & H^r(U/(U \cap N), A^{U \cap N}) & & \downarrow{\text{res}} \\
 & & \downarrow{\text{inf}} & & \\
H^r(G, A) & \xrightarrow{\text{res}} & H^r(U, A) & \longrightarrow & H^r(U, A)
\end{array}
$$

The composition, which one can achieve in three ways,

$$\text{tsl}_* : H^r(G/N, A^N) \to H^r(U, A)$$

will also be called **translation**, and is denoted tsl_*.

In dimension -1, we have the following explicit determination of cohomology.

Proposition 1.16. *Let G be a finite group and U a subgroup. Let $A \in \mathrm{Mod}(G)$.*

(1) *For $a \in A_{S_G}$ we have $\mathbf{S}_G^U(a) \in A_{S_U}$ and*

$$\mathrm{res}_U^G \divideontimes_G(a) = \divideontimes(\mathbf{S}_G^U(a)).$$

(2) *For $a \in A_{S_U}$ we have $a \in A_{S_G}$ and*

$$\mathrm{tr}_G^U \divideontimes_U(a) = \divideontimes_G(a).$$

(3) *Let $a \in A_{S_U}$ and $\sigma \in G$. Then $\sigma^{-1}a \in A_{S_{U^\sigma}}$ and*

$$\sigma_* \divideontimes_U(a) = \divideontimes_{U^\sigma}(\sigma^{-1}a).$$

Proof. In each case, one verifies explicitly that the morphism of the functor $\mathbf{H}^{-1}$ given by the expression on the right of the formulas is a δ-morphism, for the pair of functors $(\mathbf{H}^{-1}, \mathbf{H}^0)$. We can then apply the uniqueness theorem. The verification is done routinely, and is left to the reader, who will use the explicit determination of δ in Chapter I, Theorem 2.7.

Roughly speaking, Proposition 1.16 asserts that the restriction and transfer correspond respectively to the trace and the inclusion (so the order is reversed with respect to dimension 0). Conjugation just consists in applying σ^{-1} to a cochain representing a cohomology class.

In Chapter I, §4 we gave an explicit determination of $\mathbf{H}^{-2}(G, \mathbf{Z})$. We shall now indicate how the transfer, restriction and conjugation behave with respect to this determination.

We recall the isomorphisms:

$$\mathbf{H}^{-2}(G, \mathbf{Z}) \underset{\approx}{\xrightarrow{\delta}} \mathbf{H}^{-1}(G, I_G) = I_G/I_G^2 \xrightarrow{\approx} G/G^c.$$

If $\tau \in G$, we denote by ζ_τ the element of $\mathbf{H}^{-2}(G, \mathbf{Z})$ which corresponds to τG^c under the above isomorphism.

Directly in terms of groups, we have some natural homomorphisms as follows. If $\lambda : G \to \bar{G}$ is a homomorphism, then we have an induced homomorphism

$$\lambda^c : G/G^c \to \bar{G}/\bar{G}^c.$$

In particular, if U is a subgroup of G, we have the canonical homomorphism

$$\text{inc}_* : U/U^c \to G/G^c$$

induced by the inclusion.

If U is of finite index in G then we have the transfer from group theory

$$\text{Tr}_U^G = \text{Tr}_{U/U^c}^{G/G^c} : G/G^c \to U/U^c$$

defined by the product

$$\text{Tr}_U^G(\sigma G^c) = \prod_c (\bar{c}\sigma\overline{c\sigma}^{-1}U^c).$$

Cf. for instance Artin-Tate, Chapter XIII, §4.

We shall now see that the transfer and restriction on $\mathbf{H}^{-2}$ correspond to the inclusion and transfer on the groups (so the order is reversed).

Theorem 1.17. *Let G be a finite group and U a subgroup. Then:*

1. *The transfer $\text{tr}_G^U : \mathbf{H}^{-2}(U, \mathbf{Z}) \to \mathbf{H}^{-2}(G, \mathbf{Z})$ corresponds to the natural map $U/U^c \to G/G^c$ induced by the inclusion of U in G. Thus we may write*

$$\text{tr}(\zeta_\tau) = \zeta_\tau.$$

2. *The restriction $\text{res}_U^G : \mathbf{H}^{-2}(G, \mathbf{Z}) \to \mathbf{H}^{-2}(U, \mathbf{Z})$ corresponds to the transfer of group theory. Thus we may write*

$$\text{res}_U^G(\zeta_\sigma) = \zeta_{\text{tr}(\sigma)}.$$

3. *Conjugation $\sigma_* : \mathbf{H}^{-2}(U, \mathbf{Z}) \to \mathbf{H}^{-2}(U^\sigma, \mathbf{Z})$ corresponds to the map of U/U^c into $U^\sigma/(U^\sigma)^c$ induced by conjugation with $\sigma \in G$, so we may write*

$$\sigma_*(\zeta_\tau) = \zeta_{\sigma\tau\sigma^{-1}}.$$

Proof. Since $\mathbf{Z}[U]$ is naturally contained in $\mathbf{Z}[G]$ we obtain a commutative diagram

$$
\begin{array}{ccccccccc}
0 & \longrightarrow & I_U & \longrightarrow & \mathbf{Z}[U] & \longrightarrow & \mathbf{Z} & \longrightarrow & 0 \\
& & \text{inc} \downarrow & & \text{inc} \downarrow & & \text{id} \downarrow & & \\
0 & \longrightarrow & I_G & \longrightarrow & \mathbf{Z}[G] & \longrightarrow & \mathbf{Z} & \longrightarrow & 0
\end{array}
$$

the vertical maps being inclusions, and the map on $\mathbf{Z}$ being the identity. The horizontal sequences are exact. Consequently we obtain a commutative diagram

$$
\begin{array}{ccc}
\mathbf{H}^{-2}(U, \mathbf{Z}) & \xrightarrow{\;\delta\;} & \mathbf{H}^{-1}(U, I_U) \approx I_U / I_U^2 \\
id \downarrow & & \downarrow \text{inc}_* \\
\mathbf{H}^{-2}(U, \mathbf{Z}) & \xrightarrow{\;\delta\;} & \mathbf{H}^{-1}(U, I_G) \approx (I_G)_{\mathbf{s}_U} / I_U I_G .
\end{array}
$$

The coboundaries are isomorphisms, and hence inc_* is also an isomorphism. Thus we may write, as we have done, $(I_G)_{\mathbf{s}_U} / I_U I_G$ instead of I_U / I_U^2. In dimension -1 we may then use the explicit determination of $\mathbf{H}^{-2}$ from the preceding proposition, which we do case by case.

Let $\tau \in U$. Then $\tau - e$ is in $(I_G)_{\mathbf{s}_U}$, and we have

$$
\text{tr}_G^U \varkappa_U(\tau - e) = \varkappa_G(\tau - e),
$$

which proves the first formula.

Next let $\sigma \in G$. We have

$$
\text{res}_U^G \varkappa_U(\sigma - e) = \varkappa_U \left(\sum_c \bar{c}(\sigma - e) \right)
$$

where c ranges over the right cosets of U, and $\bar{c}$ is a representative of c. Furthermore

$$
\sum_c \bar{c}(\sigma - e) = \sum_c (\bar{c}\sigma - \overline{c\sigma})
$$

because $c \mapsto c\sigma$ permutes the cosets. Since $\bar{c}\sigma\overline{c\sigma}^{-1}$ is in U, we may rewrite this equality in the form

$$\sum_c (\bar{c}\sigma\overline{c\sigma}^{-1} - e)(\overline{c\sigma} - e) + \sum_c (\bar{c}\sigma\overline{c\sigma}^{-1} - e)$$

$$\equiv \sum_c (\bar{c}\sigma\overline{c\sigma}^{-1} - e) \mod I_U I_G.$$

If we apply ж to both sides, one sees that the second formula is proved, taking into account the formula for the transfer in group theory, which gives

$$\mathrm{Tr}_U^G(\sigma G^c) = \prod_c (\bar{c}\sigma\overline{c\sigma}^{-1} U^c).$$

As to the third formula, it is proved similarly, using the equalities

$$\sigma^{-1}(\sigma\tau\sigma^{-1} - e) = \tau\sigma^{-1} - \sigma^{-1}$$

$$= (\tau - e)(\sigma^{-1} - e) + (\tau - e)$$

$$\equiv (\tau - e) \mod I_U I_G.$$

This concludes the proof of Theorem 1.17.

§2. Sylow subgroups

Let G be a finite group of order N. For each prime $p \mid N$ there exists a Sylow subgroup G_p, i.e. a subgroup of order a power of p such that the index $(G : G_p)$ is prime to p. Furthermore, two Sylow subgroups are conjugate.

In particular, if $A \in \mathrm{Mod}(G)$ then $\mathbf{H}^r(G_p, A)$ is well defined, up to a conjugation isomorphism.

By Corollary 1.11 we know that $\mathbf{H}^r(G, A)$ is a torsion group. Therefore

$$\mathbf{H}^r(G, A) = \bigoplus_{p \mid N} \mathbf{H}^r(G, A, p),$$

where $\mathbf{H}^r(G, A, p)$ is the p-primary subgroup of $\mathbf{H}^r(G, A)$, i.e. consists of those elements whose period is a power of p. In particular, if $G = G_p$ is a p-group, then

$$\mathbf{H}^r(G, A) = \mathbf{H}^r(G, A, p).$$

Theorem 2.1. *Let G_p be a p-Sylow subgroup of G. Then for all $r \in \mathbf{Z}$, the restriction*

$$\mathrm{res}_{G_p}^G : \mathbf{H}^r(G, A, p) \to \mathbf{H}^r(G, A)$$

is injective, and the transfer

$$\mathrm{tr}_G^{G_p} : \mathbf{H}^r(G_p, A) \to \mathbf{H}^r(G, A, p)$$

is surjective. We have a direct sum decomposition

$$\mathbf{H}^r(G_p, A) = \mathrm{Im}\,\mathrm{res}_{G_p}^G + \mathrm{Ker}\,\mathrm{tr}_G^{G_p}.$$

Proof. Let $q = (G_p : e)$ be the order of G_p and $m = (G : G_p)$. These integers are relatively prime, and so there exists an integer m' such that $m'm \equiv 1 \mod q$. For all $\alpha \in \mathbf{H}^r(G_p, A)$ we have

$$\alpha = m'm\alpha = m' \cdot \mathrm{tr}_G^{G_p}\mathrm{res}_{G_p}^G(\alpha) = \mathrm{tr}_G^{G_p}m' \cdot \mathrm{res}_{G_p}^G(\alpha),$$

whence the injectivity and surjectivity follow as asserted. For the third, we have for $\beta \in \mathbf{H}^r(G_p, A)$,

$$\beta = \mathrm{res}\ m'\mathrm{tr}(\beta) + (\beta - \mathrm{res}\ m'\mathrm{tr}(\beta)),$$

the restriction and transfer being taken as above. One sees immediately that the first term on the right is the image of the restriction, and the second term is the kernel of the transfer. The sum is direct, because if $\beta = \mathrm{res}(\alpha), \mathrm{tr}(\beta) = 0$, then

$$\mathrm{tr}(\mathrm{res}(\alpha)) = m\alpha = 0,$$

whence $m'm\alpha = \alpha = 0$ and so $\beta = 0$. This concludes the proof.

Corollary 2.2. *Given $r \in \mathbf{Z}$, and $A \in \mathrm{Mod}(G)$, the map*

$$\alpha \mapsto \prod_{p|N} \mathrm{res}_{G_p}^G(\alpha)$$

gives an injective homomorphism

$$\mathbf{H}^r(G, A) \to \prod_{p|N} H^r(G_p, A).$$

Corollary 2.3. *If $H^r(G_p, A)$ is of finite order for all $p \mid N$, then so is $H^r(G, A)$, and the order of this latter group divides the product of the orders of $H^r(G_p, A)$ for all $p \mid N$.*

Corollary 2.4. *If $H^r(G_p, A) = 0$ for all p, then $H^r(G, A) = 0$.*

§3. Induced representations

Let G be a group and U a subgroup. We are going to define a functor
$$M_G^U : \text{Mod}(U) \to \text{Mod}(G).$$

Let $B \in \text{Mod}(U)$. We let $M_G^U(B)$ be the set of mappings from G into B satisfying

$$\sigma f(x) = f(\sigma x) \quad \text{for} \quad \sigma \in U \quad \text{and} \quad x \in G.$$

One can also write

$$M_G^U(B) = (M_G(B))^U.$$

The sum of two mappings is taken as usual summing their values, so $M_G^U(B)$ is an abelian group. We can define an action of G on $M_G^U(B)$ by the formula

$$(\sigma f)(x) = f(x\sigma) \quad \text{for} \quad \sigma, x \in G.$$

We then have transitivity.

Proposition 3.1. *Let $V \subset U$ be subgroups of G. Then the functors*
$$M_G^U \circ M_U^V \quad \text{and} \quad M_G^V$$
are isomorphic in a natural way.

We leave the proof to the reader.

We use the same notation as in §1 with a right coset decomposition $\{c\}$ of U in G, and chosen representatives $\bar{c}$. We continue to use $B \in \text{Mod}(U)$.

Proposition 3.2. *Let G be a group and U a subgroup with right cosets $\{c\}$. Then the map $f \mapsto \mathrm{res} f$, which to an element $f \in M_G^U(B)$ associates its restriction to the coset representatives $\{\bar{c}\}$, is a $\mathbf{Z}$-isomorphism*

$$M_G^U(B) \to M(G/U, B)$$

where $M(G/U, B)$ is the additive group of maps from the coset space G/U into B.

Proof. The formula $f(\sigma\tau) = \sigma f(\tau)$ for $\sigma \in U$ and $\tau \in G$ shows that the values $f(\bar{c})$ of f on coset representatives determine f, so the restriction map above is injective. Furthermore, given a map $f_0 : G/U \to B$, if we define $f_0(\bar{c}) = f_0(c)$, then we may extend f_0 to a map $f : G \to B$ by the same formula, so the proposition is clear.

Proposition 3.3. *Let G be a group and U a subgroup. Then M_G^U is an additive, covariant, exact functor of $\mathrm{Mod}(U)$ to $\mathrm{Mod}(G)$.*

Proof. Let $h : B \to B'$ be a surjective morphism in $\mathrm{Mod}(U)$, and suppose $f' : G/U \to B'$ is a given map. For each value $f'(\bar{c})$ there exists an element $b \in B$ such that $h(b) = f'(\bar{c})$. We may then define a map $f : G/U \to B$ such that $f \circ h = f'$. From this one sees that $M_G^U(B) \to M_G^U(B')$ is surjective. The rest of the proposition is even more routine.

Theorem 3.4. *Let G be a group and U a subgroup. The bifunctors*

$$\mathrm{Hom}_G(A, M_G^U(B)) \quad and \quad \mathrm{Hom}_U(A, B)$$

from $\mathrm{Mod}(G) \times \mathrm{Mod}(U)$ to $\mathrm{Mod}(\mathbf{Z})$ are isomorphic under the following associations. Given $f \in \mathrm{Hom}_G(A, M_G^U(B))$, we let f_1 be the map $A \to B$ such that $f_1(a) = f(a)(e)$. Then f_1 is in $\mathrm{Hom}_U(A, B)$. Conversely, given $h \in \mathrm{Hom}_U(A, B)$ and $a \in A$, let g_a be defined by $g_a(\sigma) = h(\sigma a)$. Then $a \mapsto g_a$ is in $\mathrm{Hom}_G(A, M_G^U(B))$. The maps $f \mapsto f_1, a \mapsto (a \mapsto g_a)$ are inverse to each other.

Proof. Routine verification left to the reader.

The above theorem is fundamental, and is one version of the basic formalism of induced representations. Cf. *Algebra*, Chapter XVIII, §7.

Corollary 3.5. *We have* $(M_G^U(B))^G = B^U$.

Proof. Take $A = \mathbf{Z}$ in the theorem.

Corollary 3.6. *If* B *is injective in* $\mathrm{Mod}(U)$ *then* $M_G^U(B)$ *is injective in* $\mathrm{Mod}(G)$.

Proof. Immediate from the definition of injectivity.

Theorem 3.7. *Let* G *be a group and* U *a subgroup. The map*

$$H^r(G, M_G^U(B)) \to H^r(U, B)$$

obtained by composing the restriction res_U^G *followed by the* U-*morphism* $g \mapsto g(e)$, *is an isomorphism for* $r \geq 0$.

Proof. We have two cohomological functors $H_G \circ M_G^U$ and H_U on $\mathrm{Mod}(U)$, because M_G^U is exact. By the two above corollaries, they are both equal to 0 on injective modules, and are isomorphic in dimension 0. By the uniqueness theorem, they are isomorphic in all dimensions. This isomorphism is the one given in the statement of the theorem, because if we denote by $\pi : M_G^U(B) \to B$ the U-morphism such that $\pi g = g(e)$, then

$$H_U(\pi) \circ \mathrm{res}_U^G : H_U \circ M_G^U \to H_U$$

is clearly a δ-morphism which, in dimension 0, induces the prescribed isomorphism $(M_G^U(B))^G$ on B^U. This proves the theorem.

Suppose now that U is of finite index in G. Let $A = M_G^U(B)$. For each coset c, we may define a U-endomorphism $\pi_c : A \to A$ of A into itself by the formula:

$$\pi_c(f)(\sigma) = \begin{cases} 0. & \text{if } \sigma \notin U \\ f(\sigma \bar{c}) & \text{if } \sigma \in U. \end{cases}$$

Indeed, π_c is additive, and if $\tau \in U$, then

$$\tau(\pi_c f)(\sigma) = (\pi_c f)(\tau \sigma) \quad \text{for all} \quad \sigma \in G.$$

Indeed, if $\sigma \notin U$ then both sides are equal to 0; and if $\sigma \in U$, then we use the fact that $f \in M_G^U(B)$ to conclude that they are equal.

Let us denote by A_1 the set of elements $f \in M_G^U(B)$ such that $f(\sigma) = 0$ if $\sigma \notin U$. Then A_1 is a U-module, as one verifies at once.

Theorem 3.8. *Let U be of finite index in G. Then:*

(i) *Let A_1 be the U-submodule of elements $f \in M_G^U(B)$ such that $f(\sigma) = 0$ if $\sigma \notin U$. Then*

$$M_G^U(B) = \bigoplus_c \bar{c}^{-1} A_1,$$

and every such f can be written uniquely in the form

$$f = \sum \bar{c}^{-1}(\pi_c f).$$

(ii) *The map $f \mapsto f(e)$ gives a U-isomorphism $A_1 \xrightarrow{\approx} B$.*

Proof. For the first assertion, let $\sigma = \tau \bar{c}_0$ with $\tau \in U$. Then

$$\sum \bar{c}^{-1}(\pi_c f)(\sigma) = \sum \bar{c}^{-1}(\pi_c f)(\tau \bar{c}_0)$$
$$= \sum (\pi_c f)(\tau \bar{c}_0 \bar{c}^{-1}).$$

If $c \neq c_0$ then the corresponding term is 0. Hence in the above sum, there will be only one term $\neq 0$, with $c = c_0$. In this case, we find the value $f(\tau \bar{c}_0) = f(\sigma)$. This shows that f can be written as asserted, and it is clear that the sum is direct. Finally A_1 is U-isomorphic to B because each $f \mid A_1$ is uniquely determined by its value $f(e)$, taking into account that $\tau f(e) = f(\tau)$ for $\tau \in U$. This same fact shows that we can define $f \mid A_1$ by prescribing $f(e) = b \in B$ and $f(\tau) = \tau b$. This proves the theorem.

We continue to consider the case when U is of finite index in G. Let $A \in \text{Mod}(G)$. We say that A is **semilocal for** U, or **relative to** U, if there exists a U-submodule A_1 of A such that A is equal to the direct sum

$$A = \bigoplus_c \bar{c}^{-1} A_1.$$

We then say that the U-module A_1 is the **local component**. It is clear that A is uniquely determined by its local component, up to an isomorphism. More precisely:

Proposition 3.9. *Let* $A_1, A_1' \in \text{Mod}(U)$.

(i) *Let* $f_1 : A_1 \to A_1'$ *be a* U*-isomorphism, and let* A, A' *be* G*-modules, which are semilocal for* U*, with local components* A_1 *and* A_1' *respectively. Then there exists a unique* G*-isomorphism* $f : A \to A'$ *which extends* f_1.

(ii) *Let* $A \in \text{Mod}(G)$ *and let* A_1 *be a* $\mathbf{Z}$*-submodule of* A*. Suppose that* A *is direct sum of a finite number of* σA_1 *(for* $\sigma \in G$*). Then* A *is semilocal for the subgroup of elements* $\tau \in G$ *such that* $\tau A_1 = A_1$.

Proof. Immediate.

Theorem 3.8 and Proposition 3.9 express the fact that to each U-module B there exists a unique G-module semilocal for U, with local component (U, B).

Proposition 3.10. *Let* $A \in \text{Mod}(G)$*,* U *a subgroup of finite index in* G*, and* A *semilocal for* U *with local component* A_1*. Let* $\pi_1 : A \to A_1$ *be the projection, and* π *its composition with the inclusion of* A_1 *in* A*. Then*

$$1_A = \mathbf{S}_G^U(\pi),$$

in other words, the identity on A *is the trace of the projection.*

Proof. Every element $a \in A$ can be written uniquely

$$a = \sum_c \bar{c}^{-1} a_c$$

with $a_c \in A_1$. By definition,

$$\mathbf{S}_G^U(\pi)(a) = \sum \bar{c}^{-1} \pi \bar{c} a.$$

The proposition is then clear from the definitions, taking into account the fact that if c, c' are two distinct cosets, then $\pi \bar{c}\bar{c}' a_{c'} = 0$.

If G is finite, one can make the trace more explicit.

Proposition 3.11. *Let G be a group, and U a subgroup of finite index. Let $A \in \mathrm{Mod}(G)$ be semilocal for U with local component A_1. Then an element $a \in A$ is in A^G if and only if*

$$a = \sum_c \bar{c}^{-1} a_1 \quad \text{with some} \quad a_1 \in A_1^U.$$

If G is finite, then $a \in \mathbf{S}_G A$ if and only if $a_1 \in \mathbf{S}_U A_1$ in the above formula. The functors

$$\mathbf{H}_G^0(M_G^U(B)) \quad \text{and} \quad \mathbf{H}_U^0(B)$$

(with variable $B \in \mathrm{Mod}(U)$) are isomorphic.

Proof. One verifies at once that for the first assertion, if an element a is expressed as the indicated sum with $a_1 \in A_1$ and $a \in A^G$ then the projection maps A^G into A_1^U. Since we already know that the projection gives an isomorphism between A^G and A_1^U it follows that all elements of A^G are expressed as stated, with $a_1 \in A_1^U$. If G is finite, then for $b \in A$ we have

$$\mathbf{S}_G(b) = \sum_c \bar{c}^{-1} \left(\sum_{\tau \in U} \tau b \right),$$

and the second assertion follows directly from this formula.

For finite groups, M_G^U maps U-regular modules to G-regular modules. This is important because such modules erase cohomology.

Proposition 3.12. *Let G be finite with subgroup U. If A is semilocal for U with local component (U, A_1) and if A_1 is U-regular, then A is G-regular.*

Proof. If one can write $1_{A_1} = \mathbf{S}_U(f)$ with some $\mathbf{Z}$-morphism f, then

$$1_A = \mathbf{S}_G^U(\pi \mathbf{S}_U(f)) = \mathbf{S}_G^U(\mathbf{S}_U(\pi f)) = \mathbf{S}_G(\pi f),$$

which proves that A is G-regular.

From the present view point, we recover a result already found previously.

Corollary 3.13. *Let G be a finite group, with subgroup U and B in $\mathrm{Mod}(U)$. If B is U-regular then $M_G^U(B)$ is G-regular.*

Theorem 3.14. *Let U be a subgroup of finite index in a group G. Suppose $A \in \mathrm{Mod}(G)$ is semilocal for U, with local component A_1. Let $\pi_1 : A \to A_1$ be the projection and $\mathrm{inc} : A_1 \to A$ the inclusion. Then the maps*

$$H_U(\pi_1) \circ \mathrm{res}_U^G \qquad and \qquad \mathrm{tr}_G^U \circ H_U(\mathrm{inc})$$

are inverse isomorphisms

$$H^r(G, A) \xleftarrow{\approx} H^r(U, A_1).$$

If G is finite, the same holds for the special functor $\mathbf{H}^r$ instead of H^r.

Proof. The composite

$$A \xrightarrow{\pi_1} A_1 \xrightarrow{\mathrm{inc}} A$$

is a U-morphism of A into itself, which we denoted by π. We know that the identity 1_A is the trace of this morphism. We can then apply Proposition 1.15 to prove the theorem for H. When G is finite, and we deal with the special functor $\mathbf{H}$, we use Corollary 3.13, the uniqueness theorem on cohomological functors vanishing on U-regular modules, the two functors being

$$\mathbf{H}_G \circ M_G^U \qquad and \qquad \mathbf{H}_U.$$

We thus obtain inverse isomorphisms of $\mathbf{H}^r(G, A)$ and $\mathbf{H}^r(U, A_1)$. This concludes the proof.

Remark. Theorem 3.14 is one of the most fundamental of the theory, and is used constantly in algebraic number theory when considering objects associated to a finite Galois extension of a number field.

§4. Double cosets

Let G be a group and U a subgroup of finite index. Let S be an arbitrary subgroup of G. Then there is a disjoint decomposition of G into double cosets

$$G = \bigcup_\gamma U\gamma S = \bigcup_\gamma S\gamma^{-1}U,$$

with $\{\gamma\}$ in some finite subset of G (because U is assumed of finite index), representing the double cosets. For each γ there is a decomposition into simple cosets:

$$S = \bigcup_{\tau_\gamma}(S \cap U[\gamma])\tau_\gamma = \bigcup_{\tau_\gamma}\tau_\gamma^{-1}(S \cap U[\gamma]),$$

where τ_γ ranges over a finite subset of S, depending on γ. Then we claim that the elements $\{\gamma\tau_\gamma\}$ form a family of right coset representatives for U in G, so that

$$G = \bigcup_{\gamma,\tau_\gamma} U\gamma\tau_\gamma = \bigcup_{\gamma,\tau_\gamma} \tau_\gamma^{-1}\gamma_U^{-1}$$

is a decomposition of G into cosets of U. The proof is easy. First, by hypothesis, we have

$$G = \bigcup_{\tau_\gamma}\bigcup_{\gamma} U\gamma(S \cap U[\gamma])\tau_\gamma,$$

and every element of G can be written in the form

$$u_1\gamma\gamma^{-1}u_2\gamma\tau_\gamma = u\gamma\tau_\gamma, \quad \text{with} \quad u_1, u_2 \in U.$$

Second, one sees that the elements $\{\gamma\tau_\gamma\}$ represent different cosets, for if

$$U\gamma\tau_\gamma = U\gamma'\tau_{\gamma'},$$

then $\gamma = \gamma'$ since the γ represent distinct double cosets, whence τ_γ and $\tau\gamma'$ represent the same coset of $\gamma^{-1}U\gamma$, and are consequently equal.

For the rest of this section, we preserve the above notation.

Proposition 4.1. *On the cohomological functor H_U (resp. $\mathbf{H}_U$ if G is finite) on $\mathrm{Mod}(G)$, the following morphisms are equal:*

$$\mathrm{res}_S^G \circ \mathrm{tr}_G^U = \sum_\gamma \mathrm{tr}_S^{S \cap U[\gamma]} \circ \mathrm{res}_{S \cap U[\gamma]}^{U[\gamma]} \circ \gamma_*.$$

Proof. As usual, it suffices to verify this formula in dimension 0. Thus let $a \in A^U$. The operation on the left consists in first taking the trace $S_G^U(a)$, and then applying the restriction which is just the inclusion. The operation on the right consists in taking first $\gamma^{-1}a$, and then applying the restriction which is the inclusion, followed by the trace

$$\operatorname{tr}_S^{S \cap U[\gamma]}(\gamma^{-1}a) = \sum_{\tau_\gamma} \tau_\gamma^{-1}\gamma^{-1}a$$

according to the coset decomposition which has been worked out above. Finally taking the sum over γ, one finds tr_G^U, which proves the proposition.

Corollary 4.2. *If U is normal, then for $A \in \operatorname{Mod}(G)$ and $\alpha \in H^r(U, A)$ we have*

$$\operatorname{res}_U^G \operatorname{tr}_G^U(\alpha) = S_{G/U}(\alpha),$$

and similarly for $\mathbf{H}$ if G is finite.

Proof. Clear.

Let again $A \in \operatorname{Mod}(G)$ and $\alpha \in H(U, A)$. We say that A is **stable** if for every $\sigma \in G$ we have

$$\operatorname{res}_{U \cap U[\sigma]}^{U[\sigma]} \sigma_*(\alpha) = \operatorname{res}_{U \cap U[\sigma]}^{U}(\alpha).$$

If U is normal in G, then α is stable if and only if $\sigma_* \alpha = \alpha$ for all $\sigma \in G$.

Proposition 4.3. *Let $\alpha \in H^r(U, A)$ for $A \in \operatorname{Mod}(G)$. If $\alpha = \operatorname{res}_U^G(\beta)$ for some $\beta \in H^r(G, A)$ then α is stable.*

Proof. By Proposition 1.5 we know that $\sigma_*\beta = \beta$. Hence we find

$$\operatorname{res}_{U \cap U[\sigma]}^{U[\sigma]} \circ \sigma_*(\alpha) = \operatorname{res}_{U \cap U[\sigma]}^{U[\sigma]} \circ \sigma_* \circ \operatorname{res}_U^G(\beta)$$

$$= \operatorname{res}_{U \cap U[\sigma]}^{U[\sigma]} \operatorname{res}_{U[\sigma]}^G \sigma_*(\beta)$$

$$= \operatorname{res}_{U \cap U[\sigma]}^G(\beta).$$

If we unwind this formula via the intermediate subgroup U instead of $U[\sigma]$, we find what we want to prove the proposition.

Proposition 4.4. *If $\alpha \in H^r(U, A)$ is stable, then*

$$\text{res}_U^G \circ \text{tr}_G^U(\alpha) = (G : U)\alpha.$$

Proof. We apply the general formula to the case when $U = S$. The verification is immediate.

Suppose finally that A is semilocal for U, with local component A_1, projection $\pi_1 : A \to A_1$ as before. Then elements of S possibly do not permute the submodules A_c transitively. However, we have:

Proposition 4.5. *Let $\sigma \in G$. Then $\sigma \in S\gamma^{-1}U$ if and only if σA_1 is in the same orbit of S as $\gamma^{-1}A_1$. For each γ, the sum*

$$\sum_{\tau_\gamma} \tau_\gamma^{-1}\gamma^{-1}A_1$$

is an S-module, semilocal for $S \cap U[\gamma]$, with local component $\tau_\gamma^{-1}\gamma^{-1}A_1$.

Proof. The assertion is immediate from the decomposition of G into cosets $\tau_\gamma^{-1}\gamma^{-1}U$.

If S and U are both of finite index in G, then the reader will observe a symmetry in the above formulas, in particular in the double coset decomposition. In particular, we can rewrite the formula of Proposition 4.3 in the form

$$(*) \qquad \text{res}_U^G \circ \text{tr}_G^S = \sum_\gamma \gamma_*^{-1} \circ \text{tr}_{U[\gamma]}^{U[\gamma] \cap S} \circ \text{res}_{U[\gamma] \cap S}^S.$$

We just take into account the commutativity of γ_* and the other maps, replacing U by S, S by U and γ by γ^{-1}.

CHAPTER III

Cohomological Triviality

In this chapter we consider only finite groups, and the special functor $\mathbf{H}_G$, such that $\mathbf{H}_G(A) = A^G/\mathbf{S}_G A$. The main result is Theorem 1.7.

§1. The twins theorem

We begin by auxiliary results. We let $\mathbf{F}_p = \mathbf{Z}/p\mathbf{Z}$ for a prime p. We always assume G has trivial action on $\mathbf{F}_p$.

Proposition 1.1. *Let G be a p-group, and $A \in \mathrm{Mod}(G)$ finite of order equal to a p-power. Then $A^G = 0$ implies $A = 0$.*

Proof. We express A as a disjoint union of orbits of G. For each $x \in A$ we let G_x be the isotropy group, i.e. the subgroup of elements $\sigma \in G$ such that $\sigma x = x$. Then the number of elements in the orbit Gx is the index $(G : G_x)$. Since G leaves 0 fixed, and

$$(A : 0) = \sum m_i p^i$$

where m_i is the number of orbits having p^i elements, it follows that either $A = 0$ or there is an element $x \neq 0$ whose orbit also has only one element, i.e. x is fixed by G, as was to be shown.

292

Corollary 1.2. *Let G be a p-group. If A is a simple p-torsion G-module then $A \approx \mathbf{F}_p$.*

Proof. Immediate.

Corollary 1.3. *The radical of $\mathbf{F}_p[G]$ is equal to the ideal I_p generated by all elements $(\sigma - e)$ over $\mathbf{F}_p$.*

Proof. A simple G-module over $\mathbf{F}_p[G]$ is finite, of order a power of p, so isomorphic to $\mathbf{F}_p$, annihilated by I_p which is therefore contained in the radical. The reverse inclusion is immediate since $\mathbf{F}_p[G]/I_p \approx \mathbf{F}_p$.

Proposition 1.4. *Let G be a p-group and A an $\mathbf{F}_p[G]$-module. The following conditions are equivalent.*

1. $\mathbf{H}^i(G, A) = 0$ *for some i with $-\infty < i < \infty$.*

2. *A is G-regular.*

3. *A is G-free.*

Proof. Since $\mathbf{F}_p$ is a field every $\mathbf{F}_p$-module is free in $\mathrm{Mod}(\mathbf{F}_p)$. If A is G-free then A is obviously G-regular. Conversely, if A is G-regular, with local component A_1 for the unit element of G, then A_1 is a direct sum of $\mathbf{F}_p$ a certain number of times, and the G-orbit of a factor $\mathbf{F}_p$ is G-isomorphic to $\mathbf{F}_p[G]$, so A itself is a G-direct sum of such G-modules isomorphic to $\mathbf{F}_p[G]$, thus proving the equivalence of the last two conditions.

It suffices now to prove the equivalence between the first and third conditions. Abbreviate $I = I_p$ for the augmentation ideal of $\mathbf{F}_p[G]$. Then A/IA is a vector space over $\mathbf{F}_p$. Let $\{a_j\}$ be representatives in A for a basis over $\mathbf{F}_p$, so that A is generated over $\mathbf{F}_p[G]$ by these elements a_j and IA. Let E be the $\mathbf{F}_p[G]$-free module with free generators $\bar{a}_j$. There is a G-morphism $E \to A$ such that $\bar{a}_j \mapsto a_j$. Let B be the image of E in A, so that $A = B + IA$. Since I is nilpotent, $I^n = 0$ for some positive integer n, and we find

$$A = B + IA + B + IB + I^2A = \cdots = B + IB + \cdots + I^nA = B$$

by iteration. Hence the map $E \to A$ is surjective. Let A' be its kernel. We have

$$E/IE = A/IA$$

and so $A' \subset IE$. Hence $\mathbf{S}_G A' = 0$. Since the following sequence is exact,

$$0 \to A' \to E \to A \to 0,$$

and E is G-regular, one finds $\mathbf{H}^r(G, A) = \mathbf{H}^{r+1}(G, A')$ for all $r \in \mathbf{Z}$.

Suppose that the index i in the hypothesis is -2. Then $\mathbf{H}^{-1}(G, A') = 0$, so $A'_{\mathbf{S}_G} = IA'$. Since $A' = A'_{\mathbf{S}_G}$, we find $A' = IA'$ and hence $A' = 0$ so $E \approx A$.

On the other hand, if $i \neq -2$, we know by dimension shifting that there exists a G-module C such that $\mathbf{H}^r(U, C) \approx \mathbf{H}^{r+1}(U, A)$ for some integer d, all $r \in \mathbf{Z}$ and all subgroups U of G, and also $\mathbf{H}^{-2}(G, C) = 0$. Hence C is $\mathbf{F}_p[G]$-free, so cohomologically trivial, and therefore similarly for A, so in particular $\mathbf{H}^{-2}(G, A) = 0$. This proves the proposition.

Proposition 1.5. *Let G be a p-group, $A \in \mathrm{Mod}(G)$. Suppose there exists an integer i such that $\mathbf{H}^i(G, A) = \mathbf{H}^{i+1}(G, A) = 0$. Suppose in addition that A is $\mathbf{Z}$-free. Then A is G-regular.*

Proof. We have an exact sequence

$$0 \to A \xrightarrow{p} A \to A/pA \to 0$$

and hence

$$0 = \mathbf{H}^i(A) \to \mathbf{H}^i(A/pA) \to \mathbf{H}^{i+1}(A) = 0,$$

the functor $\mathbf{H}$ being $\mathbf{H}_G$. Therefore $\mathbf{H}^i(A/pA) = 0$. Since A/pA is an $\mathbf{F}_p[G]$-module, we conclude from Proposition 1.3 that A/pA is $\mathbf{F}_p[G]$-free, and therefore G-regular.

Since we supposed A is $\mathbf{Z}$-free, we see immediately that the sequence

$$0 \to \mathrm{Hom}_A(A, A) \xrightarrow{p} \mathrm{Hom}_A(A, A) \to \mathrm{Hom}_{\mathbf{Z}}(A, A/pA) \to 0$$

is exact. But $\mathrm{Hom}_{\mathbf{Z}}(A, A/pA)$ is G-regular, so

$$p = p_* : \mathbf{H}^i(\mathrm{Hom}_{\mathbf{Z}}(A, A)) \to \mathbf{H}^i(\mathrm{Hom}_{\mathbf{Z}}(A, A))$$

is an automorphism. Hence so is its iteration, and hence so is multiplication by the order $(G : e)$ since G is assumed to be a p-group. But $(G : e)_* = 0$, whence all the cohomology groups $\mathbf{H}^r(\mathrm{Hom}_{\mathbf{Z}}(A, A)) = 0$ for $r \in \mathbf{Z}$. In particular, $\mathbf{H}^0(\mathrm{Hom}_A(A, A)) = 0$. From the definitions, we conclude that the identity 1_A is a trace, and so A is G-regular, thus proving Proposition 1.5.

Corollary 1.6. *Hypotheses being as in the proposition, then A is projective in* Mod(G).

Proof. Immediate consequence of Chapter I, Proposition 2.13.

Let G be a finite group and $A \in$ Mod(G). We define A to be **cohomologically trivial** if $\mathbf{H}^r(U, A) = 0$ for all subgroups U of G and all $r \in \mathbf{Z}$.

Theorem 1.7. Twin theorem. *Let G be a finite group and $A \in$ Mod(G). Then A is cohomologically trivial if and only if for each $p \mid (G : e)$ there exists an integer i_p such that*

$$\mathbf{H}^{i_p}(G_p, A) = \mathbf{H}^{i_p+1}(G_p, A) = 0$$

for a p-Sylow subgroup G_p of G.

Proof. Let $E \in \mathbf{Z}[G]$ be G-free such that

$$0 \to A' \to E \to A \to 0$$

is exact. Then

$$\mathbf{H}^{i_p+1}(G_p, A) = \mathbf{H}^{i_p+2}(G_p, A') = 0.$$

Since A' is $\mathbf{Z}$-free, it is also G_p-regular by Proposition 1.5. Hence for all subgroups G'_p of G_p we have $\mathbf{H}^r(G'_p, A) = 0$ for all $r \in \mathbf{Z}$. Since there is an injection

$$0 \to \mathbf{H}^r(G', A) \to \prod_p \mathbf{H}^r(G'_p, A)$$

for all subgroups G' of G by Chapter II, Corollary 2.2. It follows that A is cohomologically trivial. The converse is obvious.

Corollary 1.8. *Let G be finite and $A \in$ Mod(G). The following conditions are equivalent:*

1. *A is cohomologically trivial.*

2. *The projective dimension of A is $\leqq 2$.*

3. *The projective dimension of A is finite.*

Proof. Recall from *Algebra* that A has projective dimension $\leq s < \infty$ if one can find an exact sequence

$$0 \to P_1 \to P_2 \to \cdots \to P_s \to A \to 0$$

with projectives P_j. One can complete this sequence by introducing the kernels and cokernels as shown, the arches being exact.

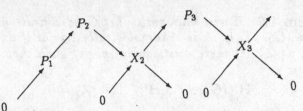

Therefore

$$\mathbf{H}^r(G', A) = \mathbf{H}^{r+1}(G', X_{s-1}) = \cdots = \mathbf{H}^{r+s-1}(G', P_1) = 0.$$

It is clear that a G-module of finite projective dimension is cohomologically trivial.

Conversely, let us write an exact sequence

$$0 \to A' \to P \to A \to 0$$

where P is $\mathbf{Z}[G]$-free. Then A' is $\mathbf{Z}$-free, and by Proposition 1.5 it is also G_p-regular for all p. We now need a lemma.

Lemma 1.9. *Suppose $M \in \mathrm{Mod}(G)$ is $\mathbf{Z}$-free and G_p-regular for all primes p. Then M is G-regular, and so projective in $\mathrm{Mod}(G)$.*

Proof. We view $\mathbf{H}^0(G, \mathrm{Hom}_{\mathbf{Z}}(M, M))$ as being injected in the product

$$\prod_p \mathbf{H}^0(G_p, \mathrm{Hom}_{\mathbf{Z}}(M, M)),$$

and we apply the definition. We conclude that M is G-regular. Since M is $\mathbf{Z}$-free it is $\mathbf{Z}[G]$-projective by Corollary 1.6.

We apply the lemma to $M = A'$ to conclude that the projective dimension of A is ≤ 2. This proves Corollary 1.8.

Corollary 1.10. *Let $A \in \mathrm{Mod}(G)$ be cohomologically trivial, and let $M \in \mathrm{Mod}(G)$ be without torsion. Then $A \otimes M$ is cohomologically trivial.*

Proof. There is an exact sequence

$$0 \to P_1 \to P_2 \to A \to 0$$

with P_1, P_2 projective in $\mathrm{Mod}(G)$. Since M has no torsion, the sequence

$$0 \to P_1 \otimes M \to P_2 \otimes M \to A \otimes M \to 0$$

is exact. But P_1, P_2 are G-regular (direct summands in free modules, so regular), whence $P_i \otimes M$ is cohomologically trivial for $i = 1, 2$, whence $A \otimes M$ is cohomologically trivial.

More generally, we have one more result, which we won't use in the sequel.

Suppose A is cohomologically trivial, and that we have an exact sequence

$$0 \to P_1 \to P_2 \to A \to 0$$

with projectives P_1, P_2. Then we have an exact sequence

$$\to \mathrm{Tor}^1(P_2, M) \to \mathrm{Tor}^1(A, M) \to P_1 \otimes M \to P_2 \otimes M \to A \otimes M \to 0$$

for an arbitrary $M \in \mathrm{Mod}(G)$. Furthermore $\mathrm{Tor}^1(P_2, M) = 0$ since P_2 has no torsion (because $\mathbf{Z}[G]$ is $\mathbf{Z}$-free). By dimension shifting, and similar reasoning for Hom, we find:

Theorem 1.11. *Let G be a finite group, $A, B \in \mathrm{Mod}(G)$, and suppose A or B is cohomologically trivial. Then for all $r \in \mathbf{Z}$ and all subgroups G' of G, we have*

$$\mathbf{H}^r(G', A \otimes B) \approx \mathbf{H}^{r+2}(G', \mathrm{Tor}_1^{\mathbf{Z}}(A, B))$$

$$\mathbf{H}^r(G', \mathrm{Hom}(A, B)) \approx \mathbf{H}^{r-2}(G', \mathrm{Ext}_{\mathbf{Z}}^1(A, B)).$$

Corollary 1.12. *Let G, A, B be as in the theorem. Then $A \otimes B$ is cohomologically trivial if and only if $\mathrm{Tor}_1^{\mathbf{Z}}(A, B)$ is cohomologically trivial; and $\mathrm{Hom}(A, B)$ is cohomologically trivial if and only if $\mathrm{Ext}_{\mathbf{Z}}^1(A, B)$ is cohomologically trivial.*

Corollary 1.13. *Let G, A, B be as in the theorem. Then $A \otimes B$ is cohomologically trivial if A or B is without p-torsion for each prime p dividing $(G : e)$.*

§2. The triplets theorem

Let $f : A \to B$ be a morphism in $\mathrm{Mod}(G)$. Let U be a subgroup of G, and let

$$f_r : \mathbf{H}^r(U, A) \to \mathbf{H}^r(U, B)$$

be the homomorphisms induced on cohomology. Actually we should write $f_{r,U}$ but we omit the index U for simplicity. We say that f is a **cohomology isomorphism** if f_r is an isomorphism for all r and all subgroups U. We say that A and B are **cohomologically equivalent** if there exists a cohomology isomorphism f as above.

Theorem 2.1. *Let $f : A \to B$ be a morphism in $\mathrm{Mod}(G)$, and suppose there exists some $i \in \mathbf{Z}$ such that f_{i-1} is surjective, f_i is an isomorphism, and f_{i+1} is injective, for all subgroups U of G. Then f is a cohomology isomorphism.*

Proof. Suppose first that f is injective. We shall reduce the general case to this special case later. We therefore have an exact sequence

$$0 \to A \xrightarrow{f} B \xrightarrow{g} C \to 0,$$

with $C = B/fA$, and the corresponding cohomology sequence

$$\longrightarrow \mathbf{H}^{i-1}(U,A) \xrightarrow{f_{i-1}} \mathbf{H}^{i-1}(U,B) \xrightarrow{g_{i-1}} \mathbf{H}^{i-1}(U,C) \longrightarrow$$

$$\xrightarrow{\delta_i} \mathbf{H}^i(U,A) \xrightarrow{f_i} \mathbf{H}^i(U,B) \xrightarrow{g_i} \mathbf{H}^i(U,C) \longrightarrow$$

$$\xrightarrow{\delta_{i+1}} \mathbf{H}^{i+1}(U,A) \xrightarrow{f_{i+1}} \mathbf{H}^i(U,B) \longrightarrow$$

We shall see that $\mathbf{H}^{i-1}(U,C) = \mathbf{H}^i(U,C) = 0$. As to $\mathbf{H}^{i-1}(U,C) = 0$, it comes from the fact that f_{i-1} surjective implies $g_{i-1} = 0$, and f_i being an isomorphism implies $\delta_{i-1} = 0$. As to $\mathbf{H}^i(U,C) = 0$, it comes from the fact that f_i surjective implies $g_i = 0$, and f_{i+1} injective implies $\delta_{i+1} = 0$. By the twin theorem, we conclude that $\mathbf{H}^r(U,C) = 0$ for all $r \in \mathbf{Z}$, whence f_r is an isomorphism for all r.

We now reduce the theorem to the preceding case, by the method of the mapping cylinder. Let us put $\mathbf{M}_G(A) = \bar{A}$ and $\varepsilon_A = \varepsilon$. We have an injection

$$\varepsilon : A \to \bar{A}.$$

We map A into the direct sum $B \oplus \bar{A}$ by

$$\bar{f} : A \to B \oplus \bar{A} \quad \text{such that} \quad \bar{f}(a) = f(a) + \varepsilon(a).$$

One sees at once that $\bar{f}$ is a morphism in $\text{Mod}(G)$. We have an exact sequence

$$0 \to A \xrightarrow{\bar{f}} B \oplus \bar{A} \xrightarrow{h} C \to 0$$

where C is the cokernel of $\bar{f}$.

We also have the projection morphism

$$p : B \oplus \bar{A} \to B \quad \text{defined by} \quad p(b + \bar{a}) = b.$$

Its kernel is $\bar{A}$, and we have $f = p\bar{f}$, whence the commutative diagram:

$$
\begin{array}{ccccccccc}
& & & & 0 & & & & \\
& & & & \downarrow & & & & \\
0 & \longrightarrow & A & \xrightarrow{\bar{f}} & B \oplus \bar{A} & \longrightarrow & C & \longrightarrow & 0 \\
& & & \searrow & \downarrow & & & & \\
& & & & B & & & &
\end{array}
$$

We then obtain the diagram

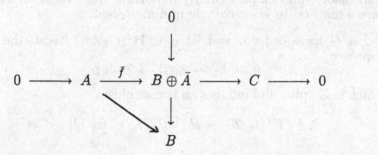

$$\mathbf{H}^i(\bar{A}) = 0$$

$$\mathbf{H}^{i-1}(C) \longrightarrow \mathbf{H}^i(A) \xrightarrow{f_i} \mathbf{H}^i(B \oplus \bar{A}) \xrightarrow{h_i} \mathbf{H}^i(C) \longrightarrow \mathbf{H}^{i+1}(A)$$

$$f_i \qquad \qquad \downarrow p_i \qquad \qquad g_i$$

$$\mathbf{H}^i(B)$$

$$\mathbf{H}^{i+1}(\bar{A}) = 0$$

the cohomology groups $\mathbf{H}$ being $\mathbf{H}_U$ for any subgroup U of G. The triangles are commutative.

The extreme vertical maps in the middle are 0 because $\bar{A} = \mathbf{M}_G(A)$, and consequently p_i is an isomorphism, which has an inverse p_i^{-1}. We have put

$$g_i = h_i p_i^{-1}$$

From the formula $f = p\bar{f}$ we obtain $f_i = p_i \bar{f_i}$. We can therefore replace $\mathbf{H}^i(B \oplus \bar{A})$ by $\mathbf{H}^i(B)$ in the horizontal sequence, and we obtain an exact sequence which is the same as the one obtained in the first part of the proof. Thus the theorem is reduced to this first part, thus concluding the proof.

§3. The splitting module and Tate's theorem

The second cohomology group in many cases, especially class field theory, plays a particularly important role. We shall describe here a method to kill a cocycle in dimension 2.

Let G have order n and let $\alpha \in \mathbf{H}^2(G, A)$. Recall the exact sequence
$$0 \to I_G \to \mathbf{Z}[G] \to \mathbf{Z} \to 0,$$
which is $\mathbf{Z}$-split, and induces an isomorphism
$$\delta : H^r(G, \mathbf{Z}) \to H^{r+1}(G, I_G) \quad \text{for all} \quad r.$$

Theorem 3.1. *Let $A \in \operatorname{Mod}(G)$ and $\alpha \in H^2(G, A)$. There exists $A' \in \operatorname{Mod}(G)$ and an exact sequence*
$$0 \to A \xrightarrow{u} A' \to I_G \to 0,$$
splitting over $\mathbf{Z}$, such that $\alpha = \delta\delta\zeta$, where ζ is the generator of $\mathbf{H}^0(G, \mathbf{Z})$ corresponding to the class of 1 in $H^0(G, \mathbf{Z}) = \mathbf{Z}/n\mathbf{Z}$, and
$$u_*\alpha = 0,$$
in other words, α splits in A'.

Proof. We define A' to be the direct sum of A and a free abelian group on elements $x_\sigma(\sigma \in G, \sigma \neq e)$. We define an action of G on

A' by means of a cocycle $\{a_{\sigma,\tau}\}$ representing α. We put $x_e = a_{e,e}$ for convenience, and let

$$\sigma x_\tau = x_{\sigma\tau} - x_\sigma + a_{\sigma,\tau}.$$

One verifies by brute force that this definition is consistent by using the coboundary relation satisfied by the cocycle $\{a_{\sigma,\tau}\}$, namely

$$\lambda a_{\sigma,\tau} - a_{\lambda\sigma,\tau} + a_{\lambda,\sigma\tau} - a_{\sigma,\tau} = 0.$$

One sees trivially that α splits in A'. Indeed, $(a_{\sigma,\tau})$ is the coboundary of the cochain (x_σ).

We define a morphism $v : A' \to I_G$ by letting

$$v(a) = 0 \text{ for } a \in A \text{ and } v(x_\sigma) = \sigma - e \text{ for } \sigma \neq e.$$

Here we identify A as a direct summand of A'. The map v is a G-morphism in light of the definition of the action of G on A'. It is obviously surjective, and the kernel of v is equal to A.

There remains to verify that $\alpha = \delta\delta\zeta$. The coboundary $\delta\zeta$ is represented by the 1-cocycle $b_\sigma = \sigma - e$ in I_G, representing an element β of $\mathbf{H}^1(G, I_G)$. We find $\delta\beta$ by selecting a cochain of G in A', for instance (x_σ), such that $v(x_\sigma) = b_\sigma$. The coboundary of (x_σ) represents $\delta\beta$, and one then sees that this gives α, thus proving the theorem.

We define an element $A \in \mathrm{Mod}(G)$ to be a **class module** if for every subgroup U of G, we have $\mathbf{H}^1(U, A) = 0$, and if $\mathbf{H}^2(G, A)$ is cyclic of order $(G : e)$, generated by an element α such that $\mathrm{res}_U^G(\alpha)$ generates $\mathbf{H}^2(U, A)$ and is of order $(U : e)$. An element α as in this definition will be called **fundamental**. The terminology comes from class field theory, where one meets such modules. See also Chapter IX.

Theorem 3.2. *Let $G, A, \alpha, u : A \to A'$ be as in Theorem 3.1. Then A is a class module and α is fundamental if and only if A' is cohomologically trivial.*

Proof. Suppose A is a class module and α fundamental. We have an exact sequence for all subgroups U of G:

$$0 \to \mathbf{H}^1(A) \to \mathbf{H}^1(A') \to \mathbf{H}^1(I) \to \mathbf{H}^2(A) \to \mathbf{H}^2(A') \to 0.$$

The 0 furthest to the right is due to the fact that

$$\mathbf{H}^2(I_G) = \mathbf{H}^1(\mathbf{Z}) = 0.$$

Since
$$\alpha = \delta\beta \text{ and } \beta = \delta\zeta,$$

and $\mathbf{H}^1(U, I)$ is cyclic of order $(U : e)$ generated by β, it follows that
$$\mathbf{H}^1(I) \to \mathbf{H}^2(A)$$

is an isomorphism. We conclude that $\mathbf{H}^1(U, A') = 0$ for all U.

Since we also have the exact sequence

$$\mathbf{H}^2(A) \to \mathbf{H}^2(A') \to \mathbf{H}^2(I) = 0,$$

and α splits in A', we conclude that $\mathbf{H}^2(U, A') = 0$ for all U. Hence A' is cohomologically trivial by the twin theorem.

Conversely, suppose A' cohomologically trivial. Then we have isomorphisms

$$\mathbf{H}^1(I) \xrightarrow{\delta} \mathbf{H}^2(A) \quad \text{and} \quad \mathbf{H}^0(\mathbf{Z}) \xrightarrow{\delta} \mathbf{H}^1(I),$$

for all subgroups U of G. This shows that $\mathbf{H}^2(U, A)$ is indeed cyclic of order $(U : e)$, generated by $\delta\delta\zeta$. This concludes the proof.

CHAPTER IV
Cup Products

§1. Erasability and uniqueness

To treat cup products, we have to start with the general notion of multilinear categories, due to Cartier.

Let $\mathfrak{A}$ be an abelian category. A structure of **multilinear category** on $\mathfrak{A}$ consists in being given, for each $(n+1)$-tuple $A_1, \ldots, A_n, B$ of objects in $\mathfrak{A}$, an abelian group $L(A_1, \ldots, A_n, B)$ satisfying the following conditions.

MUL 1. For $n = 1, L(A, B) = \operatorname{Hom}(A, B)$.

MUL 2. Let

$$f_1 : \quad A_{11} \times \cdots \times A_{ln_1} \quad \rightarrow \quad B_1$$
$$\cdots$$
$$f_r : \quad A_{r1} \times \cdots \times A_{rn_r} \quad \rightarrow \quad B_r$$
$$g : \quad B_1 \times \cdots \times B_r \quad \rightarrow \quad C$$

be multilinear. Then we may compose $g(f_1, \ldots, f_r)$ in $L(A_{11}, \ldots, A_{rn_r}, C)$, and this composition is multilinear in $g, f_1, \ldots, f_r$.

MUL 3. With the same notation, we have $g(\operatorname{id}, \ldots, \operatorname{id}) = g$.

303

MUL 4. The composition is associative, in the sense that (with obvious notation)

$$g(f_1(h\ldots), f_2(h\ldots), \ldots, f_r(h\ldots)) = g(f_1, \ldots, f_r)(h\ldots).$$

As usual the reader may think in terms of ordinary multilinear maps on abelian groups. These define a multilinear category, from which others can be defined by placing suitable conditions.

Example. Let G be a group. Then $\mathrm{Mod}(G)$ is a multilinear category if we define $L(A_1, \ldots, A_n, B)$ to consist of those **Z**-multilinear maps θ satisfying

$$\theta(\sigma a_1, \ldots, \sigma a_n) = \sigma\theta(a_1, \ldots, a_n)$$

for all $\sigma \in G$ and $a_i \in A_i$.

We can extend in the obvious way the notion of functor to multilinear categories. Explicitly, a functor $F : \mathfrak{A} \to \mathfrak{B}$ of such a category into another is given by a map $T : f \mapsto T(f) = f_*$, which to each multilinear map f in $\mathfrak{A}$ associates a multilinear map in B, satisfying the following condition. Let

$$f_1 : A_1 \quad \times \cdots \times \quad A_{N_1} \longrightarrow B_1$$
$$\cdots$$
$$f_p : A_{n_{p-1}} \times \cdots \times \quad A_{n_p} \longrightarrow B_p$$
$$g : B_1 \quad \times \cdots \times \quad B_p \longrightarrow C$$

be multilinear in $\mathfrak{A}$. Then we can compose $g(f_1, \ldots, f_p)$ and $Tg(Tf_1, \ldots, Tf_p)$. The condition is that

$$T(g(f_1, \ldots, f_p)) = Tg(Tf_1, \ldots, Tf_p) \quad \text{and} \quad T(\mathrm{id}) = \mathrm{id}.$$

We could also define the notion of tensor product in a multilinear abelian category. It is a bifunctor, bilinear, on $\mathfrak{A} \times \mathfrak{A}$, satisfying the universal mapping property just as for the ordinary tensor product. In the applications, it will be made explicit in each case how such a tensor product arises from the usual one.

Furthermore, in the specific cases of multilinear abelian categories to be considered, the category will be closed under taking tensor products, i.e. if A, B are objects of $\mathfrak{A}$, then the linear factorization of a multilinear map is also in $\mathfrak{A}$.

Let now $E_1 = (E_1^{p_1}), \ldots, E_n = (E_n^{p_n})$ and $H = (H^r)$ be δ-functors on the abelian category $\mathfrak{A}$, which we suppose multilinear, and the functors have values in a multilinear abelian category $\mathfrak{B}$. We assume that for each value of $p_1, \ldots, p_n$ taken on by $E_1, \ldots, E_n$ respectively, the sum $p_1 + \cdots + p_n$ is among the values taken on by r. By a **cup product**, or **cupping**, of $E_1, \ldots, E_n$ into H we mean that to each multilinear map $\theta \in L(A_1, \ldots, A_n, B)$ we have associated a multilinear map

$$\theta_{(p)} = \theta_* = \theta_{p_1, \ldots, p_n} : E_1^{p_1}(A_1) \times \cdots \times E_n^{p_n}(A_n) \longrightarrow H^p(B)$$

where $p = p_1 + \cdots + p_n$, satisfying the following conditions.

Cup 1. The association $\theta \mapsto \theta_{(p)}$ is a functor from the multilinear category $\mathfrak{A}$ into $\mathfrak{B}$ for each $(p) = (p_1, \ldots, p_n)$.

Cup 2. Given exact sequences in $\mathfrak{A}$,

$$0 \to A_i' \to A_i \to A_i'' \to 0$$

$$0 \to B' \to B \to B'' \to 0$$

and multilinear maps f', f', f'' in $\mathfrak{A}$ making the following diagram commutative:

$$A_1 \times \cdots \times A_i' \times \cdots \times A_n \longrightarrow A_1 \times \cdots \times A_i \times \cdots \times A_n \longrightarrow A_1 \times \cdots \times A_i'' \times \cdots \times A_n$$
$$\downarrow f' \qquad\qquad\qquad \downarrow f \qquad\qquad\qquad \downarrow f''$$
$$B' \longrightarrow \qquad\qquad B \qquad\qquad \longrightarrow \qquad B''$$

then the diagram

$$E_1^{p_1}(A_1) \times \cdots \times E_i^{p_i}(A_i'') \times \cdots \times E_n^{p_n}(A_n) \xrightarrow{f_*''} H^p(B'')$$
$$\text{id} \downarrow \qquad\qquad \delta \downarrow \qquad\qquad \text{id} \downarrow \qquad\qquad \delta \downarrow$$
$$E_1^{p_1}(A_1) \times \cdots \times E_i^{p_i+1}(A_i') \times \cdots \times E_n^{p_n}(A_n) \xrightarrow{f_*'} H^{p+1}(B')$$

has character $(-1)^{p_1 + \cdots + p_{i-1}}$, which means

$$f_*'(\text{id}, \ldots, \delta, \ldots, \text{id}) = (-1)^{p_1 + \cdots + p_{i-1}} \delta \circ f_*''.$$

The accumulation of indices is inevitable if one wants to take all possibilities into account. In practice, we mostly have to deal with the following cases.

First, let H be a cohomological functor. For each $n \geq 1$ suppose given a cupping

$$H \times \cdots \times H \to H$$

(the product on the left occurring n times) such that for $n = 1$ the cupping is the identity. Then we say that H is a **cohomological cup functor**.

Next, suppose we have only two factors, i.e. a cupping

$$E \times F \to H$$

from two δ-functors into another. Most of the time, instead of indexing the induced maps by their degrees, we simply index them by a $*$.

If $\mathfrak{A}$ is closed under the tensor product, then by **Cup 1** a cohomological cup functor is uniquely determined by its values on the canonical bilinear maps

$$A \times B \to A \otimes B.$$

Indeed, if $f : A \times B \to C$ is bilinear, one can factorize f through $A \otimes B$,

$$A \times B \xrightarrow{\theta} A \otimes B \xrightarrow{\varphi} C$$

where θ is bilinear and φ is a morphism in $\mathfrak{A}$. Thus certain theorems will be reduced to the study of the cupping on tensor products.

Let $E = (E^p)$ and $F = (F^q)$ be δ-functors with a cupping into the δ-functor $H = (H^r)$. Given a bilinear map

$$A \times B \to C$$

and two exact sequences

$$0 \longrightarrow A' \longrightarrow A \longrightarrow A'' \longrightarrow 0$$

$$0 \longrightarrow C' \longrightarrow C \longrightarrow C'' \longrightarrow 0$$

such that the bilinear map $A \times B \to C$ induces bilinear maps

$$A' \times B \to C' \quad \text{and} \quad A'' \times B \to C'',$$

then we obtain a commutative diagram

$$
\begin{array}{ccc}
E^p(A'') \times F^q(B) & \longrightarrow & H^{p+q}(C'') \\
\Big\downarrow \delta & \Big\downarrow & \Big\downarrow \delta \\
E^{p+1}(A') \times F^q(B) & \longrightarrow & H^{p+q+1}(C')
\end{array}
$$

and therefore we get the formula

$$\delta(\alpha'' \beta) = (\delta \alpha'')\beta \quad \text{with} \quad \alpha'' \in E^p(A'') \quad \text{and} \quad \beta \in F^q(B).$$

Proposition 1.1. *Let H be a cohomological cup functor on a multilinear category $\mathfrak{A}$. Then the product*

$$(\alpha, \beta) \mapsto (-1)^{pq}\beta\alpha \qquad for \qquad \alpha \in H^p(A), \beta \in H^q(B)$$

also defines a cupping $H \times H \to H$, making H into another cohomological cup functor, equal to the first one in dimension 0.

Proof. Clear.

Remark. Since we shall prove a uniqueness theorem below, the preceding proposition will show that we have

$$\alpha\beta = (-1)^{pq}\beta\alpha.$$

Now we come to the question of uniqueness for cup products. It will be applied to the uniqueness of a cupping on a cohomological functor H given in one dimension. More precisely, let H be a cohomological cup functor on a multilinear category $\mathfrak{A}$. When we speak of the cup functor in dimension 0, we mean the cupping

$$H^0 \times H^0 \times \cdots \times H^0 \to H^0,$$

which to each multilinear map $\theta : A_1 \times \cdots \times A_n \to B$ associates the multilinear map

$$\theta_0 : H^0(A_1) \times \cdots \times H^0(A_n) \to H^0(B).$$

We note that to prove a uniqueness theorem, we only need to deal with two factors (i.e. bilinear morphisms), because a cupping of several functors can be expressed in terms of cuppings of two functors, by associativity.

Theorem 1.2. *Let $\mathfrak{A}$ be an abelian multilinear category. Let $E = (E^0, E^1)$ and $H = (H^0, H^1)$ be two exact δ-functors, and let F be a functor of $\mathfrak{A}$ into a multilinear category $\mathfrak{B}$. Suppose E^1 is erasable by (M, ε), that $\mathfrak{A}, \mathfrak{B}$ are closed under tensor products, and that for all $A, B \in \mathfrak{A}$ the sequence*

$$0 \to A \otimes B \xrightarrow{\varepsilon_A \otimes 1} M_A \otimes B \to X_A \otimes B \to 0$$

is exact. Suppose given a cupping $E \times F \to H$. Then we have a commutative diagram

$$
\begin{array}{ccccc}
E^0(X_A) & \times & F(B) & \to & H^0(A \otimes B) \\
\downarrow{\scriptstyle \delta} & & \downarrow{\scriptstyle \mathrm{id}} & & \downarrow{\scriptstyle \delta} \\
E^1(A) & \times & F(B) & \to & H^1(A \otimes B)
\end{array}
$$

and the coboundary on the left is surjective.

Proof. Clear.

Corollary 1.3. *Hypotheses being as in the theorem, if two cuppings $E \times F \to H$ coincide in dimension 0, then they coincide in dimension 1.*

Proof. For all $\alpha \in E^1(A)$ there exists $\xi \in E^0(X_A)$ such that $\alpha = \delta \xi$, and by hypothesis we have for $\beta \in F(B)$,

$$\alpha \beta = (\delta \xi)\beta = \delta(\xi \beta),$$

whence the corollary follows.

Theorem 1.4. *Let G be a group and let H be the cohomological functor H_G from Chapter I, such that $H^0(A) = A^G$ for A in $\mathrm{Mod}(G)$. Then a cupping*

$$H \times H \to H$$

such that in dimension 0, the cupping is induced by the bilinear map

$$(a, b) \mapsto \theta(a, b) \quad \text{for} \quad \theta : A \times \to C \quad \text{and} \quad a \in A^G, b \in B^G,$$

is uniquely determined by this condition.

Proof. This is just a special case of Theorem 1.2, since in Chapter I we proved the existence of the erasing functor (M, ε) necessary to prove uniqueness.

We shall always consider the functor H_G as having its structure of cup functor as we have just defined it in Theorem 1.4. Its existence will be proved in the next section.

If G is finite, then in the category $\mathrm{Mod}(G)$ we have an erasing functor $\mathbf{M}_G$ for the special cohomology functor $\mathbf{H}_G$. So we now formulate the general situation.

Let $E = (E^p), F = (F^q)$, and $H = (H^r)$ be three exact δ-functors on a multilinear category $\mathfrak{A}$. We suppose given a cupping $E \times F \to H$. As in Chapter I, suppose given an erasing functor M for E in dimensions $> p_0$. We say that $\mathbf{M}$ is **special** if for each bilinear map

$$\theta : A \times B \to C$$

in $\mathfrak{A}$, there exists bilinear maps

$$\mathbf{M}(\theta) : \mathbf{M}_A \times B \to \mathbf{M}_C \quad \text{and} \quad X(\theta) : X_A \times B \to X_C$$

such that the following diagram is commutative:

$$
\begin{array}{ccccc}
A \times B & \longrightarrow & \mathbf{M}_A \times B & \longrightarrow & X_A \times B \\
\theta \downarrow & & M(\theta) \downarrow & & M(\theta) \downarrow \\
C & \longrightarrow & \mathbf{M}_C & \longrightarrow & X_C
\end{array}
$$

Theorem 1.5 (right). *Let $E = (E^p), F = (F^q)$ and $H = (H^r)$ be three exact δ-functors on a multilinear category $\mathfrak{A}$. Suppose given a cupping $E \times F \to H$. Let $\mathbf{M}$ be a special erasing functor for E and for H in all dimensions. Then there is a commutative diagram associated with each bilinear map $\theta : A \times B \to C$:*

$$
\begin{array}{ccc}
E^p(X_A) \times F^q(B) & \longrightarrow & H^{p+q}(C) \\
\delta \downarrow & \quad \mathrm{id} \downarrow \quad & \delta \downarrow \\
E^{p+1}(A) \times F^q(B) & \longrightarrow & H^{p+1+1}(C)
\end{array}
$$

and the vertical maps are isomorphisms.

Proof. Clear.

Of course, we have the dual situation when $\mathbf{M}$ is a coerasing functor for E in dimensions $< p_0$. In this case, we say that $\mathbf{M}$ is **special** if for each θ there exist bilinear maps $\mathbf{M}(\theta)$ and $Y(\theta)$ making the following diagram commutative:

$$
\begin{array}{ccccc}
Y_A \times B & \longrightarrow & \mathbf{M}_A \times B & \longrightarrow & A \times B \\
\Big\downarrow{\scriptstyle Y(\theta)} & & \Big\downarrow{\scriptstyle \mathbf{M}(\theta)} & & \Big\downarrow{\scriptstyle \theta} \\
Y_C & \longrightarrow & \mathbf{M}_C & \longrightarrow & C
\end{array}
$$

One then has:

Theorem 1.5 (left). *Let E, F, H be three exact δ-functors on a multilinear category $\mathfrak{A}$. Suppose given a cupping $E \times F \to H$. Let $\mathbf{M}$ be a special coerasing functor for E and H in all dimensions. Then for each bilinear map $\theta : A \times B \to C$ we have a commutative diagram*

$$
\begin{array}{ccc}
E^p(A) \quad \times \quad F^q(B) & \longrightarrow & H^{p+q}(C) \\
\Big\downarrow{\scriptstyle \delta} \qquad\qquad \Big\downarrow{\scriptstyle \mathrm{id}} & & \Big\downarrow{\scriptstyle \mathrm{id}} \\
E^{p+q+1}(Y_A) \times \quad F^q(B) & \longrightarrow & H^{p+q+1}(C)
\end{array}
$$

and the vertical maps are isomorphisms.

Corollary 1.6. *Let E, F, H be as in the preceding theorems, with values $p = 0, 1$ (resp. $0, -1$); $q = 0$; $r = 0, 1$ (resp. $0, -1$). Let M be an erasing functor (resp. coerasing) for E and H. Suppose given two cuppings $E \times F \to H$ which coincide in dimension 0. Then they coincide in dimension 1 (resp. -1).*

We observe that the choice of indices $0, 1, -1$ is arbitrary, and the corollary applies mutatis mutandis to $p, p+1$, or $p, p-1$ for E, q arbitrary for F, and $p+q, p+q+1$ (resp. $p+q, p+q-1$) for H.

Corollary 1.7. *Let H be a cohomological functor on a multilinear category $\mathfrak{A}$. Suppose there exists a special erasing and coerasing functor on H. Suppose there are two cup functor structures on H, coinciding in dimension 0. Then these cuppings coincide in all dimensions.*

Proposition 1.8. *Let G be a finite group and G' a subgroup. Let $\mathbf{H}$ be the cohomological functor on $\mathrm{Mod}(G)$ such that $\mathbf{H}(A) = \mathbf{H}(G', A)$. Suppose given in addition an additional structure of cup functor on $\mathbf{H}$. Then the erasing functor $A \mapsto \mathbf{M}_G(A)$ (and the similar coerasing functor) are special.*

Proof. Let $\theta : A \times B \to C$ be bilinear. Viewing $\mathbf{M}_G$ as coerasing, we have the commutative diagram

$$
\begin{array}{ccccc}
I_G \otimes A \otimes B & \longrightarrow & \mathbf{Z}[G] \otimes A \otimes B & \longrightarrow & \mathbf{Z} \otimes A \otimes B \\
\downarrow & & \downarrow & & \downarrow \\
I_G \otimes C & \longrightarrow & \mathbf{Z}[G] \otimes C & \longrightarrow & \mathbf{Z} \otimes C
\end{array}
$$

where the vertical maps are defined by $\lambda \otimes a \otimes b \mapsto \lambda \otimes ab$. We have a similar diagram to the right for the erasing functor.

From the general theorems, we then obtain:

Theorem 1.9. *Let G be a group and $H = H_G$ the ordinary cohomological functor on $\mathrm{Mod}(G)$. Then for each multilinear map $A_1 \times \cdots \times A_n \to B$ in $\mathrm{Mod}(G)$, if we define*

$$
\varkappa(a_1) \cdots \varkappa(a_n) = \varkappa(a_1 \cdots a_n) \quad \text{for} \quad a_i \in A_i^G,
$$

then we obtain a multilinear functor, and a cupping

$$
H^0 \times \cdots \times H^0 \to H^0.
$$

A cup functor structure on H which is the above in dimension 0 is uniquely determined. The similar assertion holds when G is finite and H is replaced by the special functor $\mathbf{H} = \mathbf{H}_G$.

As mentioned previously, existence will be proved in the next section. For the rest of this section, we let H denote the ordinary or special cohomology functor on $\mathrm{Mod}(G)$, depending on whether G is arbitrary or finite.

Corollary 1.10. *Let G be a group and H the ordinary or special functor if G is finite. Let $A_1, A_2, A_3, A_{12}, A_{123}$ be G-modules. Suppose given multilinear maps in $\mathrm{Mod}(G)$:*

$$
A_1 \times A_2 \to A_{12} \qquad\qquad A_{12} \times A_3 \to A_{123}
$$

whose composite gives rise to a multilinear map

$$A_1 \times A_2 \times A_3 \rightarrow A_{123}.$$

Let $\alpha_i \in H^{p_i}(A_i)$. Then we have associativity,

$$(\alpha_1\alpha_2)\alpha_3 = \alpha_1(\alpha_2\alpha_3),$$

these cup products being taken relative to the multilinear maps as above.

Proof. One first reduces the theorem to the case when $A_{12} = A_1 \otimes A_2$ and $A_{123} = A_1 \otimes A_2 \otimes A_3$. The products on the right and on the left of the equation satisfy the axioms of a cup product, so we can apply the uniqueness theorem.

Remark. More generally, to define a cup functor structure on a cohomological functor H on an abelian category of abelian groups $\mathfrak{A}$, closed under tensor products, it suffices to give a cupping for two factors, i.e. $H \times H \rightarrow H$. Once this is done, let

$$\alpha_1 \in H^{r_1}(A_1), \ldots, \alpha_n \in H^{r_n}(A_n).$$

We may then define

$$\alpha_1 \ldots \alpha_n = (\ldots(\alpha_1\alpha_2)\alpha_3)\ldots\alpha_n),$$

and one sees that this gives a structure of cup functor on H, using the universal property of the tensor product.

Corollary 1.11. *Let G be a group and $H = H_G$ the ordinary cup functor, or the special one if G is finite. Let $\theta : A \times B \rightarrow C$ be bilinear in $\mathrm{Mod}(G)$. Let $a \in A^G$, and let*

$$\theta_a : B \rightarrow C \quad \text{be defined by} \quad \theta_a(b) = ab.$$

Then θ_a is a morphism in $\mathrm{Mod}(G)$. If

$$H^q(\theta_a) = \theta_{a*} : H^q(B) \rightarrow H^q(C)$$

is the induced homomorphism, then

$$\varkappa(a)\beta = \theta_{a*}(\beta) \quad \text{for} \quad \beta \in H^q(B).$$

Proof. The first assertion is clear from the fact that $a \in A^G$ implies $\sigma(ab) = a\sigma b$. If $q = 0$ the second assertion amounts to the definition of the induced mapping. For the other values of q, we apply the uniqueness theorem to the cuppings $H^0 \times H \rightarrow H$ given either by the cup product or by the induced homomorphism, to conclude the proof.

Corollary 1.12. *Suppose G finite and $\theta : A \times B \to C$ bilinear in* $\mathrm{Mod}(G)$. *Then for $a \in A^G$ and $b \in B_{S_G}$ we have*

$$\varkappa(a) \cup \rtimes(b) = \rtimes(ab).$$

Proof. A direct verification shows that the above formula defines a cupping of $\mathbf{H}^0$ and $(\mathbf{H}^0, \mathbf{H}^{-1})$ into $(\mathbf{H}^0, \mathbf{H}^{-1})$. Under the stated hypotheses, we clearly have $ab \in C_{S_G}$, so the formula makes sense, and is valid.

§2. Existence

We shall see that on $\mathrm{Mod}(G)$ the cup product is induced by a product of cochains, and we shall give the explicit formula for cochains in the standard complex. First we give an axiomatic version so the reader sees the general situation in the framework of abelian categories. But as usual, the reader may think of $\mathrm{Mod}(G)$ and abelian groups for the categories mentioned in the next theorem.

Theorem 2.1. *Let $\mathfrak{A}$ be a multilinear category of abelian groups. Let*

$$A \mapsto Y(A)$$

be an exact functor of $\mathfrak{A}$ into the category of complexes in $\mathfrak{A}$. Suppose that for each bilinear map $\theta : A \times B \to C$ in $\mathfrak{A}$ we are given a bilinear map

$$Y^r(A) \times Y^s(B) \to Y^{r+s}(C)$$

which is functorial in A, B, C (covariant) and such that if $f \in Y^r(A)$ and $g \in Y^s(B)$, then

$$\delta(fg) = (\delta f)g + (-1)^r f(\delta g).$$

Let H be the cohomological functor associated to the functor Y. Then there exists on H a structure of cup functor, induced by the above bilinear map. This structure satisfies the property of the three exact sequences, as it is described below.

The **property of the three exact sequences** is the following. Consider three exact sequences in $\mathfrak{A}$:

$$0 \longrightarrow A' \longrightarrow A \longrightarrow A'' \longrightarrow 0$$

$$0 \longrightarrow B' \longrightarrow B \longrightarrow B'' \longrightarrow 0$$

$$0 \longrightarrow C' \longrightarrow C \longrightarrow C'' \longrightarrow 0$$

Suppose given a bilinear map $A \times B \to C$ in $\mathfrak{A}$ such that

$$A'B = 0, \qquad AB' \subset C', \qquad A'B \subset C', \qquad A''B'' \subset C''.$$

Then for $\alpha'' \in H^r(A'')$ and $\beta'' \in H^s(B'')$ we have

$$\delta(\alpha'' \cup \beta'') = (\delta\alpha'') \cup \beta'' + (-1)^r \alpha'' \cup (\delta\beta'').$$

It is a pain to write down in complete detail what amounts to a routine proof. One has to combine the additive construction with multiplicative considerations on the product of two complexes, cf. for instance Exercise 29 of *Algebra*, Chapter XX. More generally, we observe the following general fact. Let $\mathfrak{A}$ be a multilinear abelian category and let $\mathrm{Com}(\mathfrak{A})$ be the abelian category of complexes in $\mathfrak{A}$. Then we can make $\mathrm{Com}(\mathfrak{A})$ into a multilinear category as follows. Let K, L, M be three complexzes in $\mathfrak{A}$. We **define** a **bilinear map**

$$\theta : K \times L \to M$$

to be a family of bilinear maps

$$\theta_{r,s} : K^r \times L^s \to M^{r+s}$$

satisfying the condition

$$\delta_M(\theta_{r,s}(x,y)) = \theta_{r+1,s}(\delta_K x, y) + (-1)^r \theta_{r,s+1}(x, \delta_L y)$$

for $x \in K^r$ and $y \in L^s$. We have indexed the coboundaries $\delta_M, \delta_L, \delta_K$ according to the complexes to which they belong. If we omit all indices to simplify the notation, then the above condition reads

$$\delta(x \cdot y) = \delta x \cdot y + (-1)^r x \cdot \delta y.$$

Exercise 29 loc. cit. reproduces this formula for the universal bilinear map given by the tensor product. Then the cup product is induced by a product on representing cochains in the cochain complex. We leave the details to the reader. As we shall see, in practice, one can give explicit concrete formulas which allow direct verification.

We shall now make the theorem explicit for the standard complex and $\mathrm{Mod}(G)$.

Lemma 2.2. *Let G be a group and $Y(G, A)$ the homogeneous standard complex for $A \in \mathrm{Mod}(G)$. Let $A \times B \to C$ be bilinear in $\mathrm{Mod}(G)$. For $f \in Y^r(G, A)$ and $g \in Y^s(G, A)$ define the product fg by the formula*

$$(fg)(\sigma_0, \ldots, \sigma_{r+1}) = f(\sigma_0, \ldots, \sigma_r)g(\sigma_r, \ldots, \sigma_{r+s}).$$

Then this product satisfies the relation

$$\delta(fg) = (\delta f)g + (-1)^r f(\delta g).$$

Proof. Straightforward.

In terms of non-homogeneous cochains f, g in the non-homogeneous standard complex, the formula for the product is given by

$$(fg)(\sigma_1, \ldots, \sigma_{r+s}) = f(\sigma_1, \ldots, \sigma_r)(\sigma_1, \ldots, \sigma_r g(\sigma_{r+1}, \ldots, \sigma_{r+s})).$$

Theorem 2.3. *Let G be a group and let H_G be the ordinary cohomological functor on $\mathrm{Mod}(G)$. Then the product defined in Lemma 2.2 induces on H_G a structure of cup functor which satisfies the property of the three exact sequences. Furthermore in dimension 0, we have*

$$\varkappa(ab) = \varkappa(a)\varkappa(b) \qquad \text{for } a \in A^G \text{ and } b \in B^G.$$

Proof. This is simply a special case of Theorem 2.1, taking the lemma into account, giving the explicit expression for the cupping in terms of cochains in the standard complex. The property of the three exact sequences is immediate from the definition of the coboundary. Indeed, if we are given cochains f'' and g'' representing α'' and β'', their coboundaries are defined by taking cochains f, g in A, B respectively mapping on f'', g'', so that fg maps on $f''g''$. The formula of the lemma implies the formula in the property of the three exact sequences.

We have the analogous result for finite groups and the special functor.

Lemma 2.4. *Let G be a finite group. Let X be a complete, $\mathbf{Z}[G]$-free, acyclic resolution of $\mathbf{Z}$, with augmentation ε. Let d be the boundary operation in X and let*

$$d' = d \otimes \mathrm{id}_X \qquad and \qquad d'' = \mathrm{id}_X \otimes d$$

be those induced in $X \otimes X$. Then there is a family of G-morphisms

$$h_{r,s} : X_{r+s} \longrightarrow X_r \otimes X_s \qquad for \quad -\infty < r, s < \infty,$$

satisfying the following conditions:

(i) $h_{r,s} d = d' h_{r+1,s} + d'' h_{r,s+1}$

(ii) $(\varepsilon \otimes \varepsilon) h_{0,0} = \varepsilon$.

Proof. Readers will find a proof in Cartan-Eilenberg [CaE].

Theorem 2.5. *Let G be a finite group and $\mathbf{H}_G$ the special cohomology functor. Then there exists a unique structure of cup functor on $\mathbf{H}_G$ such that if $\theta : A \times B \to C$ is bilinear in $\mathrm{Mod}(G)$ then*

$$\varkappa(a)\varkappa(b) = \varkappa(ab) \quad for \quad a \in A^G, b \in B^G.$$

Furthermore, this cupping satisfies the three exact sequences property.

Proof. Let $A \in \mathrm{Mod}(G)$, and let X be the standard complex. Let $Y(A)$ be the cochain complex $\mathrm{Hom}_G(X, A)$. For

$$f \in Y^r(A) = \mathrm{Hom}_G(X^r, A) \qquad and \qquad g \in Y^s(A),$$

we can define the product $fg \in Y^{r+s}(A)$ by the composition of canonical maps

$$X_{r+s} \xrightarrow{h_{r,s}} X_r \otimes X_s \xrightarrow{f \otimes g} A \otimes B \xrightarrow{\theta'} C,$$

where θ' is the morphism in $\mathrm{Mod}(G)$ induced by θ, that is

$$fg = \theta'(f \otimes g) h_{r,s}.$$

One then verifies without difficulty the formula

$$\delta(fg) = (\delta f)g + (-1)^r f(\delta g),$$

and the rest of the proof is as in Theorem 2.3.

§3. Relations with subgroups

We shall tabulate a list of commutativity relations for the cup product with a group and its subgroups.

First we note that every multilinear map θ in $\mathrm{Mod}(G)$ induces in a natural way a multilinear map θ' in $\mathrm{Mod}(G')$ for every subgroup G' of G. Set theoretically, it is just θ.

Theorem 3.1. *Let G be a group and $\theta : A \times B \to C$ bilinear in* $\mathrm{Mod}(G)$. *Let G' be a subgroup of G. Let H denote the ordinary cup functor, or the special cup functor if G is finite, except when we deal with inflation in which case H denotes only the ordinary functor. Let the restriction res be from G to G'. Then:*

(1) $\mathrm{res}(\alpha\beta) = (\mathrm{res}\ \alpha)(\mathrm{res}\ \beta)$ *for $\alpha \in H^r(G, A)$ and $\beta \in H^s(G, B)$.*

(2) $\mathrm{tr}((\mathrm{res}\ \alpha)\beta') = \alpha(\mathrm{tr}\ \beta')$ *for $\alpha \in H^r(G, A)$ and $\beta' \in H^s(G', B)$,*

the transfer being taken from G' to G. Similarly,

$$\mathrm{tr}(\alpha'(\mathrm{res}\ \beta)) = (\mathrm{tr}\ \alpha')\beta \ \text{for}\ \alpha' \in H^r(G', A) \ \text{and}\ \beta \in H^s(G, B).$$

(3) *Let G' be normal in G. Then θ induces a bilinear map*

$$A^{G'} \times B^{G'} \to C^{G'},$$

and for $\alpha \in H^r(G/G', A^{G'}), \beta \in H^s(G/G', B^{G'})$ we have

$$\inf(\alpha\beta) = (\inf\ \alpha)(\inf\ \beta).$$

Proof. The formulas are immediate in dimension 0, i.e. for $r = s = 0$. In each case, the expressions on the left and on the right of the stated equality define separately a cupping of a cohomological functor into another, coinciding in dimension 0, and satisfying the conditions of the uniqueness theorem. The equalities are therefore valid in all dimensions. For example, in (1) we have two cuppings of $H_G \times H_G \to H_G$, given by

$$(\alpha, \beta) \mapsto \mathrm{res}(\alpha\beta) \quad \text{and} \quad (\alpha, \beta) \mapsto (\mathrm{res}\ \alpha)(\mathrm{res}\ \beta).$$

In (3), we find first a cupping

$$H_{G/G'} \times H_{G/G'} \to H_G$$

to which we apply the uniqueness theorem on the right. We let the reader write out the details. For the inflation, we may write explicitly one of these cuppings:

$$H(G/G', A^{G'}) \times H(G/G', B^{G'}) \xrightarrow{\text{cup}} H(G/G', C^{G'}) \to H(G, C^G).$$

§4. The triplets theorem

We shall formulate for cup products the analogue of the triplets theorem. We shall reduce the proof to the preceding situation.

Theorem 4.1. *Let G be a finite group and $\theta : A \times B \to C$ bilinear in $\text{Mod}(G)$. Fix $\alpha \in \mathbf{H}^p(G, A)$ for some index p. For each subgroup G' of G let $\alpha' = \text{res}^G_{G'}(\alpha)$ be the restriction in $\mathbf{H}^p(G', A)$. For each integer s denote by*

$$\alpha'_s : \mathbf{H}^s(G', B) \to \mathbf{H}^s(G', C)$$

the homomorphism $\beta' \mapsto \alpha'\beta'$. Suppose there exists an index r such that α'_{r+1} is surjective, α'_r is an isomorphism, and α'_{r+1} is injective, for all subgroups G' of G. Then α'_s is an isomorphism for all s.

Proof. Suppose first that $r = 0$. We then know by Corollary 1.11 that α'_s is the homomorphism $(\theta_a)_*$ induced by an element $a \in A^G$, where $\theta_a : B \to C$ is defined by $b \mapsto \theta(a, b) = ab$. The theorem is therefore true if $p = 0$ by the ordinary triplets theorem. We note that the induced homomorphism is compatible with the restriction from G to G'.

We then prove the theorem in general by ascending and descending induction on p. For example, let us give the details in the case of descending induction to the left. We have $E = F = H$. There exists $\xi \in \mathbf{H}^r(G, X_A)$ such that $\alpha = \delta\xi$, where X_A is the cokernel in the dimension shifting exact sequence as in Theorem 1.14 of Chapter II. The restriction being a morphism of functors, we have $\alpha'_s = \delta\xi'_s$ for all s. It is clear that α'_s is an isomorphism (resp. is injective, resp. surjective) if and only if ξ'_s is an isomorphism (resp. is injective, resp. surjective). Thus we have an inductive procedure to prove our assertion.

§5. The cohomology ring and duality

Let A be a ring and suppose that the group G acts on the additive group of A, i.e. that this additive group is in $\mathrm{Mod}(G)$. We say that A is a G-**ring** if in addition we have

$$\sigma(ab) = (\sigma a)(\sigma b) \quad \text{for all} \quad \sigma \in G, a, b \in A.$$

Suppose A is a G-ring. Then multiplication of n elements of A is a multilinear map in the multilinear category $\mathrm{Mod}(G)$.

Let us denote by $H(A)$ the direct sum

$$H(A) = \bigoplus_{-\infty}^{\infty} H^p(A),$$

where H is the ordinary functor on $\mathrm{Mod}(G)$, or special functor in case G is finite. Then $H(A)$ is a graded ring, multiplication being first defined for homogeneous elements $\alpha \in H^p(A)$ and $\beta \in H^q(A)$ by the cup product, and then on direct sums by linearity, that is

$$\left(\sum \alpha^{(p)}\right)\left(\sum \beta^{(q)}\right) = \sum_r \left(\sum_{p+q=r} \alpha^{(p)}\beta^{(q)}\right).$$

We then say that $H(A)$ is the **cohomology ring** of A.

One verifies at once that if A is a commutative ring, then $H(A)$ is anti-commutative, that is if $\alpha \in H^p(A)$ and $\beta \in H^q(A)$ then

$$\alpha\beta = (-1)^{pq}\beta\alpha.$$

Since by definition a ring has a unit element, we have $1 \in A^G$ and $\varkappa(1)$ is the unit element of $H(A)$. Indeed, for $\beta \in H^q(A)$ we have

$$\varkappa(1)\beta = \theta_{1*}\beta = \beta,$$

because $\theta_1 : a \mapsto 1a = a$ is the identity.

Let A be a G-ring and $B \in \text{Mod}(G)$. Suppose that B is a left A-module, compatible with the action of G, that is the map

$$A \times B \to B$$

defined by the action of A on B is bilinear in the multilinear category $\text{Mod}(G)$. We then obtain a product

$$H^p(A) \times H^q(B) \to H^{p+q}(B)$$

which we can extend by linearity so as to make the direct sum $H(B) = \bigoplus H^q(B)$ into a graded $H(A)$-module. The unit element of $H(A)$ acts as the identity on $H(B)$ according to the previous remarks.

Let $B, C \in \text{Mod}(G)$. There is a natural map

$$\text{Hom}(B, C) \times B \to C$$

defined by $(f, b) \mapsto f(b)$. This map is bilinear in the multilinear category $\text{Mod}(G)$, because we have

$$([\sigma]f)(\sigma b) = \sigma f \sigma^{-1} \sigma b = \sigma(fb)$$

for $f \in \text{Hom}(B, C)$ and $b \in B$. Thus we obtain a product

$$(\varphi, \beta) \mapsto \varphi\beta, \quad \text{for} \quad \varphi \in H^p(\text{Hom}(B, C)) \quad \text{and} \quad \beta \in H^q(B).$$

Theorem 5.1. *Let*

$$0 \to B' \to B \to B'' \to 0$$

be a short exact sequence in $\text{Mod}(G)$, *let* $C \in \text{Mod}(G)$, *and suppose the* Hom *sequence*

$$0 \to \text{Hom}(B'', C) \to \text{Hom}(B, C) \to \text{Hom}(B', C') \to 0$$

is exact. Then for $\beta'' \in H^{q-1}(B'')$ *and* $\varphi' \in H^p(\text{Hom}(B', C))$, *we have*

$$(\delta\varphi')\beta'' + (-1)^p \varphi'(\delta\beta'') = 0.$$

Proof. We consider the three exact sequences:

$$0 \to \operatorname{Hom}(B'', C) \to \operatorname{Hom}(B, C) \to \operatorname{Hom}(B', C) \to 0$$

$$0 \to B' \to B \to B'' \to 0.$$

$$0 \to C \to C \to 0 \to 0,$$

and a bilinear map from $\operatorname{Mod}(G)$, in the middle, inducing mappings as in the existence theorem for the cup product. Since $\varphi'\beta'' = 0$, we find the present result as a special case.

We may rewrite the result of Theorem 5.1 in the form of a diagram

$$
\begin{array}{ccccc}
H^r(\operatorname{Hom}(B', C)) & \longrightarrow & H^s(B') & \longrightarrow & H^{r+s}(C) \\
\delta \downarrow & & \delta \downarrow & & \operatorname{id} \downarrow \\
H^{r+1}(\operatorname{Hom}(B'', C)) & \longrightarrow & H^{s-1}(B'') & \longrightarrow & H^{r+s}(C)
\end{array}
$$

which has character $(-1)^{r+1}$.

In dimension 0, we find:

Proposition 5.2. *Let $f \in \operatorname{Hom}_G(B, C) = (\operatorname{Hom}(B, C))^G$, and $\beta \in H^r(B)$. Then*

$$\varkappa(f)\beta = f_*\beta.$$

If G is finite and $\mathbf{H}$ is the special functor, then

$$\varkappa(f) \cup \varkappa(b) = \varkappa(f(b)) \quad \text{for} \quad b \in B_{\mathbf{S}_B}.$$

Proof. In dimension 0, this is an old result. The assertion concerning dimensions -1 and 0 is a special case of the uniqueness theorem, Corollary 1.12.

There is a homomorphism

$$h_{r,s} : H^r(\operatorname{Hom}(B, C)) \to \operatorname{Hom}(H^r(B), H^{r+s}(C)),$$

obtained from the bilinear map

$$H^r(\operatorname{Hom}(B, C)) \times H^s(B) \to H^{r+s}(C).$$

In important cases, we shall see that $h_{r,s}$ is an isomorphism. We shall now give a criterion for this in the case of the special functor.

Theorem 5.3. *Let G be finite, and $\mathbf{H} = \mathbf{H}_G$ the special functor. Suppose that for C fixed and B variable in $\mathrm{Mod}(G)$, and two fixed integers p_0, q_0 the map h_{p_0,q_0} is an isomorphism. Then $h_{p,q}$ is an isomorphism for all p, q such that $p + q = p_0 + q_0$.*

Proof. We are going to use dimension shifting. We consider the exact sequence

$$0 \to I \otimes B \to \mathbf{Z}[G] \otimes B \to \mathbf{Z} \otimes B = B \to 0,$$

which we hom into C. Since the sequence splits, we obtain an exact sequence

$$0 \to \mathrm{Hom}(B, C) \to \mathrm{Hom}(\mathbf{Z}[G] \otimes B, C) \to \mathrm{Hom}(I \otimes B, C) \to 0.$$

Applying the diagram following theorem 5.2, we find:

$$
\begin{array}{ccc}
\mathbf{H}^p(\mathrm{Hom}(I \otimes B, C)) & \xrightarrow{\ h_{p,q}\ } & \mathrm{Hom}(\mathbf{H}^q(I \otimes B), \mathbf{H}^{p+q}(C)) \\
\delta \downarrow & & \downarrow (\delta, 1) \\
\mathbf{H}^{p+1}(\mathrm{Hom}(B, C)) & \xrightarrow[h_{p+1, q+1}]{} & \mathrm{Hom}(\mathbf{H}^{q-1}(B), \mathbf{H}^{p+q}(C)).
\end{array}
$$

The vertical coboundaries are isomorphisms because the middle object in the exact sequence is G-regular, and so annuls the cohomology. This concludes the proof going from p to $p+1$. Going the other way, we use the other exact sequence

$$0 \to B \to \mathbf{Z}[G] \otimes B \to J \otimes B \to 0,$$

and we let the reader finish this side of the proof.

As an application, we shall prove a duality theorem. Let B be an abelian group. As before, we define its dual group by $\hat{B} = \mathrm{Hom}(B, \mathbf{Q}/\mathbf{Z})$. It is the group of characters of finite order, which we consider as a discrete group. Its elements will be called simply **characters**. Let $B \in \mathrm{Mod}(G)$. We consider B as an abelian group to get $\hat{B}$.

We have an isomorphism

$$\maltese^{-1} : \mathbf{H}^{-1}(\mathbf{Q}/\mathbf{Z}) \to (\mathbf{Q}/\mathbf{Z})_n$$

between $\mathbf{H}^{-1}(\mathbf{Q}/\mathbf{Z})$ and the elements of order $n = (G : e)$ in $\mathbf{Q}/\mathbf{Z}$.

In addition, we have a bilinear map in $\mathrm{Mod}(G)$:

$$\hat{B} \times B \to \mathbf{Q}/\mathbf{Z},$$

and consequently a corresponding bilinear map of abelian groups

$$\mathbf{H}^{-q}(\hat{B}) \times \mathbf{H}^{q-1}(B) \to \mathbf{H}^{-1}(\mathbf{Q}/\mathbf{Z}).$$

Theorem 5.4. Duality Theorem. *The homomorphism*

$$\mathbf{H}^{-q}(\hat{B}) \to \mathbf{H}^{q-1}(B)^{\wedge}$$

which to each $\varphi \in \mathbf{H}^{-q}(B)$ associates the character $\beta \mapsto \text{ж}^{-1}(\varphi\beta)$, is an isomorphism, so we have

$$\mathbf{H}^{-q}(\hat{B}) = \mathbf{H}^{q-1}(B)^{\wedge}.$$

Proof. From the definitions, we see that the theorem amounts to proving that $h_{-q,q-1}$ is an isomorphism. According to the preceding theorem, it suffices to show that $h_{0,-1}$ is an isomorphism. Since $h_{0,q}$ is an induced homomorphism, we can make $h_{0,-1}$ explicit in the present case, as follows. We have a homomoprhism

$$\hat{B}^G / \mathbf{S}_G \hat{B} \to (B_{\mathbf{S}_G}/IB)^{\wedge},$$

obtained by associating to an element $f \in \hat{B}$ the character $b \mapsto f(b)$ for $b \in B_{\mathbf{S}_G}$. We have to prove that this map is an isomorphism.

For the surjectivity, let $f_0 : B_{\mathbf{S}} \to (\mathbf{Q}/\mathbf{Z})_n$ be a homomorphism vanishing on IB. We can extend f_0 to a homomorphism f of B into $\mathbf{Q}/\mathbf{Z}$ because $\mathbf{Q}/\mathbf{Z}$ is injective. Furthermore f is in $\hat{B}^G$ because

$$f(\sigma b) - \sigma f(b) = f(\sigma b) - f(b) = f(\sigma b - b) = 0$$

by hypothesis. This proves surjectivity.

For the injectivity, let $f \in \hat{B}^G$ and suppose $f(B_{\mathbf{S}}) = 0$. Since $B/B_{\mathbf{S}}$ is isomorphic to $\mathbf{S}B$, there exists $g \in (\mathbf{S}B)^{\wedge}$ such that

$$f(b) = g(Sb) \quad \text{for} \quad b \in B.$$

We extend g to a homomorphism of B into $\mathbf{Q}/\mathbf{Z}$, denoted by the same letter. Then $f = \mathbf{S}g$, because

$$(\mathbf{S}g)(b) = \sum \sigma g \sigma^{-1} b = \sum g \sigma^{-1} b = g\left(\sum \sigma^{-1} b\right) = g\mathbf{S}b = f(b).$$

This concludes the proof of hte duality theorem.

Consider the special case when $B = \mathbf{Z}$. Then

$$\hat{B} = \hat{Z} = \mathrm{Hom}(\mathbf{Z}, \mathbf{Q}/\mathbf{Z})$$

and we find:

Corollary 5.5. $\mathbf{H}^{-q}(\mathbf{Q}/\mathbf{Z}) \approx \mathbf{H}^{q-1}(\mathbf{Z})^\wedge$.

Applying the coboundary homomorphism arising from the exact sequence

$$0 \to \mathbf{Z} \to \mathbf{Q} \to \mathbf{Q}/\mathbf{Z} \to 0,$$

we obtain:

Corollary 5.6. *The following diagram is commutative:*

$$
\begin{array}{ccc}
\mathbf{H}^{-p-1}(\mathbf{Q}/\mathbf{Z}) \times \mathbf{H}^p(\mathbf{Z}) & \longrightarrow & \mathbf{H}^{-1}(\mathbf{Q}/\mathbf{Z}) \\
\downarrow{\delta} \qquad\qquad \downarrow{\mathrm{id}} & & \downarrow{\delta} \\
\mathbf{H}^{-p}(\mathbf{Z}) \quad \times \mathbf{H}^p(\mathbf{Z}) & \longrightarrow & \mathbf{H}^0(\mathbf{Z})
\end{array}
$$

The vertical maps are isomorphisms, and thus

$$\mathbf{H}^{-p}(\mathbf{Z}) \approx \mathbf{H}^p(\mathbf{Z})^\wedge.$$

Proof. Since $\mathbf{Q}$ is uniquely divisible by n, its cohomology groups are trivial, and hence the coboundaries are isomorphisms, so the corollary is clear.

Corollary 5.7. *Let $M \in \mathrm{Mod}(G)$ be $\mathbf{Z}$-free. Then one has a commutative diagram:*

$$
\begin{array}{ccc}
\mathbf{H}^{p-1}(\mathrm{Hom}(M, Q/Z)) \times \mathbf{H}^{-p}(M) & \longrightarrow & \mathbf{H}^{-1}(\mathbf{Q}/\mathbf{Z}) \\
\downarrow{\delta} \qquad\qquad \downarrow{\mathrm{id}} & & \downarrow{\delta} \\
\mathbf{H}^p(\mathrm{Hom}(M, \mathbf{Z})) \quad \times \mathbf{H}^{-p}(M) & \longrightarrow & \mathbf{H}^0(\mathbf{Z})
\end{array}
$$

where the vertical maps are isomorphisms. Thus we obtain a canonical isomorphism

$$\mathbf{H}^p(\mathrm{Hom}(M, \mathbf{Z})) \approx \mathbf{H}^{-p}(M)^{\wedge}.$$

Proof. This is an immediate consequence of the fact that $\mathbf{Q}$ is G-regular (the identity is the trace of $1/n$), and so $\mathrm{Hom}(M, A)$ is also G-regular, and so annuls cohomology. The sequence

$$0 \longrightarrow \mathrm{Hom}(M, \mathbf{Z}) \longrightarrow \mathrm{Hom}(M, \mathbf{Q}) \longrightarrow \mathrm{Hom}(M, \mathbf{Q}/\mathbf{Z}) \longrightarrow 0$$

is exact. We can then apply the definition of the cup functor to conclude the proof.

It will be convenient to use the following terminology. Suppose given a bilinear map of finite abelian groups

$$F' \times F \to \mathbf{Q}/\mathbf{Z}.$$

This induces a homomorphism $F' \to F^{\wedge}$. If $F' \to F^{\wedge}$ is an isomorphism, then we say that F' and F are in **perfect duality** under the bilinear map. Each of Theorem 5.4, Corollary 5.5 and Corollary 5.6 establish a perfect duality of cohomology groups in their conclusions, when B is, say, finitely generated.

§6. Periodicity

In Chapter I, we saw that the cohomology of a finite cyclic group is periodic. We shall now give a general criterion for periodicity for an arbitrary finite group G. *This section will not be used in what follows.*

Let $r \in \mathbf{Z}$ be fixed. An element $\zeta \in \mathbf{H}^r(G, \mathbf{Z})$ will be said to be a **maximal generator** if ζ generates $\mathbf{H}^r(G, \mathbf{Z})$ and if ζ is of finite order $(G : e)$.

Theorem 6.1. *Let G be a finite group and $\zeta \in \mathbf{H}^r(G, \mathbf{Z})$. The following properties are equivalent.*

MAX 1. ζ *is a maximal generator.*

MAX 2. ζ *is of order $(G : e)$.*

MAX 3. *There exists an element $\zeta^{-1} \in \mathbf{H}^{-r}(G, \mathbf{Z})$*

such that $\zeta^{-1}\zeta = 1$.

MAX 4. For all $A \in \mathrm{Mod}(G)$, the map

$$\alpha \mapsto \zeta\alpha \quad \text{of} \quad \mathbf{H}^i(G, A) \to \mathbf{H}^{i+r}(G, A)$$

is an isomorphism for all i.

Proof. That **MAX 1** implies **MAX 2** is trivial.

Assume **MAX 2.** Suppose ζ has order $(G : e)$. Since $\mathbf{H}^{-r}(\mathbf{Z})$ is dual to $\mathbf{H}^r(\mathbf{Z})$ by Corollary 5.6, the existence of ζ^{-1} follows from the definition of the dual group, so **MAX 3** is satisfied.

Assume **MAX 3.** The maps

$$\alpha \mapsto \zeta\alpha \text{ for } \alpha \in H^i(A) \text{ and } \beta \mapsto \zeta^{-1}\beta \text{ for } \beta \in \mathbf{H}^{i+r}(A)$$

are inverse to each other, up to a power $(-1)^r$, and are therefore isomorphisms, thus proving **MAX 4.**

Assume **MAX 4.** We take $A = \mathbf{Z}$ and $i = 0$ in the preceding assertion, and we use the fact that $\mathbf{H}^0(\mathbf{Z})$ is cyclic of order $(G : e)$. Then **MAX 1** follows at once.

The uniqueness of ζ^{-1} in Theorem 6.1, satisfying condition **MAX 3**, is clear, taking into account that $\mathbf{H}^{-r}(\mathbf{Z})$ is the dual group of $\mathbf{H}^r(\mathbf{Z})$.

Proposition 6.2. *Let $\zeta \in \mathbf{H}^r(G, \mathbf{Z})$ be a maximal generator. Then so is ζ^{-1}. If ζ_1 is a maximal generator of $\mathbf{H}^s(G, \mathbf{Z})$ for some s, then $\zeta\zeta_1$ is a maximal generator.*

Proof. The first assertion follows from **MAX 3**; the second from **MAX 4.**

An integer m will be said to be a **cohomology period** of G if $\mathbf{H}^m(G, \mathbf{Z})$ contains a maximal generator, or in other words, $\mathbf{H}^m(G, \mathbf{Z})$ is cyclic of order $(G : e)$. The anticommutativity of the cup product shows that a period is even.

Proposition 6.3. *Suppose m is a cohomology period of G. Let U be a subgroup of G and let $\zeta \in \mathbf{H}^m(G, \mathbf{Z})$ be of order $(G : e)$.*

Then $\operatorname{res}_U^G(\zeta)$ *has order* $(U : e)$ *and* m *is a cohomology period of* U.

Proof. Since
$$\operatorname{tr}_G^U \operatorname{res}_U^G(\zeta) = (G : U)\zeta,$$
it follows that the order of the restriction of ζ to U is at least $(U : e)$. Since it is at most equal to $(U : e)$, it is a period.

Proposition 6.4. *Let* G_p *be a* p-*Sylow subgroup of* G *and let* $\zeta \in \mathbf{H}^r(G_p, \mathbf{Z})$ *be a maximal generator. Let* n *be a positive integer such that*
$$k^n \equiv 1 \mod (G_p : e)$$
for all integers k *prime to* p. *Then* $\zeta^n \in \mathbf{H}^{nr}(G_p, \mathbf{Z})$ *is stable, i.e.* $\sigma_*(\zeta^n) = \zeta^n$ *for all* $\sigma \in G$, *and*
$$\operatorname{tr}_G^{G_p}(\zeta^n)$$
has order $(G_p : e)$.

Proof. Since σ_* is an isomorphism, and since the restriction of a maximal generator is a maximal generator, one concludes that the elements
$$\beta = \operatorname{res}_{G_p}^{G_p \cap G_p[\sigma]}(\zeta^n) \quad \text{and} \quad \operatorname{res}_{G_p \cap G_p[\sigma]}^{G_p[\sigma]} \circ \sigma_*(\zeta^n) = \lambda$$
are both maximal generators in $\mathbf{H}^r(G_p \cap G_p[\sigma], \mathbf{Z})$. Hence there exists an integer k prime to p such that one is equal to k times the other, i.e. $k\beta = \lambda$. Taking the n-th power we get
$$k^n \beta^n = \lambda^n.$$
From the definition of n, with the fact that $(G_p : e)$ kills β and λ, and with the commutativity of the cup product and the indicated operations, we find that ζ^n is stable. This being the case, we know from Proposition 1.10 of Chapter II that
$$\operatorname{res}_{G_p}^G \circ \operatorname{tr}_G^{G_p}(\zeta^n) = (G : G_p)\zeta^n.$$
Since $(G : G_p)$ is prime to p, it follows that the transfer followed by the restriction is injective on the subgroup generated by ζ^n. Thus the transfer is injective on this subgroup. From this one sees that the period of this transfer is the same as that of ζ^n, whence the same as that of ζ, thus concluding the proof.

Corollary 6.5. *Let G be a finite group. Then G admits a cohomological period > 0 if and only if each Sylow subgroup G_p has a period > 0.*

Proof. If G has a period > 0, the proposition shows that G_p also has one. Conversely, suppose that $\zeta'_p \in \mathbf{H}^r(G_p, \mathbf{Z})$ is a maximal generator. Let

$$\zeta_p = \mathrm{tr}_G^{G_p}(\zeta'_p).$$

Then the order of ζ_p is the same as that of ζ'_p by Proposition 6.4. Let

$$\zeta = \sum \zeta_p.$$

Then ζ is an element of order $(G : e)$, so G has a cohomological period > 0, as was to be shown.

The preceding corollary reduces the study of periodicity to p-groups. One can show easily:

Proposition 6.6. *Let $G = G_p$ be a p-group. Then G admits a cohomological period > 0 if and only if G is cyclic or G is a generalized quaternion group.*

We omit the proof.

§7. The theorems of Tate-Nakayama

We shall now go back to the theorem concerning the splitting module for a class module as in Chapter III, §3. We recall that if $A' \in \mathrm{Mod}(G)$ is cohomologically trivial and M is a G-module without torsion, then $A' \otimes M$ is cohomologically trivial.

Theorem 7.1. (Tate). *Let G be a finite group, $M \in \mathrm{Mod}(G)$ without torsion, $A \in \mathrm{Mod}(G)$ a class module, and $\alpha \in \mathbf{H}^2(G, A)$ fundamental. Let*

$$\alpha_r : \mathbf{H}^r(G, M) \to \mathbf{H}^{r+2}(G, A \otimes M)$$

be the cup product relative to the bilinear map $A \times M \to A \otimes M$, i.e. such that

$$\alpha_r(\lambda) = \alpha \cup \lambda \quad for \quad \lambda \in \mathbf{H}^r(G, M).$$

Then α_r is an isomorphism for all $r \in \mathbf{Z}$.

Proof. As in the main theorem on cohomological triviality (Theorem 3.1 of Chapter III, we have exact sequences

$$0 \longrightarrow A \longrightarrow A' \longrightarrow I \longrightarrow 0$$

$$0 \longrightarrow A \otimes M \longrightarrow A' \otimes M \longrightarrow I \otimes M \longrightarrow 0.$$

The exactness of the second sequence is due to the fact that the first one splits.

In addition, $A' \otimes M$ is cohomologically trivial. Let us put $\beta = \delta\zeta$. Then

$$\alpha_r(\lambda) = \alpha\lambda = (\delta\beta)\lambda = \delta(\beta\lambda).$$

If we now use the exact sequences

$$0 \longrightarrow I \longrightarrow \mathbf{Z}[G] \longrightarrow \mathbf{Z} \longrightarrow 0$$

$$0 \longrightarrow I \otimes M \longrightarrow \mathbf{Z}[G] \otimes M \longrightarrow \mathbf{Z} \otimes M \longrightarrow 0,$$

we find

$$\alpha_r(\lambda) = \delta\delta(\zeta\lambda).$$

The coboundaries δ are isomorphisms, in one case because $\mathbf{Z}[G] \otimes M$ is G-regular, in the other case by the main theorem on cohomological triviality. To show that α_r is an isomorphism, it will suffice to show that $\zeta_r : \lambda \mapsto \zeta\lambda$ is an isomorphism. But this is clear because it is the identity, as one sees by making explicit the canonical isomorphism $\mathbf{Z} \otimes M \to M$. This concludes the proof of Theorem 7.1.

We can rewrite the commutative diagram arising from the theorem in the following manner.

$$
\begin{array}{ccccc}
\mathbf{H}^0(Z) & \times & \mathbf{H}^r(M) & \longrightarrow & \mathbf{H}^r(\mathbf{Z} \otimes M) = \mathbf{H}^r(M) \\
\delta \downarrow & & \downarrow \text{id} & & \delta \downarrow \\
\mathbf{H}^1(I) & \times & \mathbf{H}^r(M) & \longrightarrow & \mathbf{H}^{r+1}(I \otimes M) \\
\delta \downarrow & & \downarrow \text{id} & & \delta \downarrow \\
\mathbf{H}^2(A) & \times & \mathbf{H}^r(M) & \longrightarrow & \mathbf{H}^{r+2}(A \otimes M).
\end{array}
$$

The vertical maps δ are isomorphisms, and the cup product on top corresponds to the bilinear map $\mathbf{Z} \times M \to \mathbf{Z} \otimes M = M$, so the isomorphism induced by ζ_r is the identity.

If we take $M = \mathbf{Z}$ and $r = -2$, we find

$$\mathbf{H}^2(A) \times \mathbf{H}^{-2}(\mathbf{Z}) \to \mathbf{H}^0(A).$$

We know that $\mathbf{H}^{-2}(\mathbf{Z}) = G/G^c$ and so we find an isomorphism

$$G/G^c \approx \mathbf{H}^0(A) = A^G/\mathbf{S}_G A.$$

We shall make this isomorphism more explicit below.

We also obtain an analogous theorem by taking Hom instead of the tensor product, and by using the duality theorem.

Theorem 7.2. *Let G be a finite group, $M \in \mathrm{Mod}(G)$ and $\mathbf{Z}$-free, $A \in \mathrm{Mod}(G)$ a class module. Then for all $r \in \mathbf{Z}$, the bilinear map of the cup product*

$$\mathbf{H}^r(G, \mathrm{Hom}(M, A)) \times \mathbf{H}^{2-r}(G, M) \to \mathbf{H}^2(G, A)$$

induces an isomorphism

$$\mathbf{H}^r(G, \mathrm{Hom}(M, A)) \approx \mathbf{H}^{2-r}(G, M)^\wedge.$$

Proof. We shift dimensions on A twice. Since A' and $\mathbf{Z}[G]$ are $\mathbf{Z}$-free, it follows that the sequences

$$0 \longrightarrow \mathrm{Hom}(M,\mathbf{Z}) \longrightarrow \mathrm{Hom}(M,A') \longrightarrow \mathrm{Hom}(M,I) \longrightarrow 0$$

$$0 \longrightarrow \mathrm{Hom}(M,I) \longrightarrow \mathrm{Hom}(M,\mathbf{Z}[G]) \longrightarrow \mathrm{Hom}(M,\mathbf{Z}) \longrightarrow 0$$

are exact. By the definition of the cup product one finds commutative diagrams as follows, where the vertical maps are isomorphisms.

$$
\begin{array}{ccccc}
\mathbf{H}^{r-2}(\mathrm{Hom}(M,\mathbf{Z})) & \times & \mathbf{H}^{2-r}(M) & \longrightarrow & \mathbf{H}^0(\mathbf{Z}) \\
\downarrow{\scriptstyle\delta} & & \downarrow{\scriptstyle\mathrm{id}} & & \downarrow{\scriptstyle\delta} \\
\mathbf{H}^{r-1}(\mathrm{Hom}(M,I)) & \times & \mathbf{H}^{2-r}(M) & \longrightarrow & \mathbf{H}^1(I) \\
\downarrow{\scriptstyle\delta} & & \downarrow{\scriptstyle\mathrm{id}} & & \downarrow{\scriptstyle\delta} \\
\mathbf{H}^r(\mathrm{Hom}(M,A)) & \times & \mathbf{H}^{2-r}(M) & \longrightarrow & \mathbf{H}^2(A).
\end{array}
$$

The bilinear map on top is that of Corollary 5.7, and the theorem follows.

Selecting $M = \mathbf{Z}$ we get for $r = 0$:

$$\mathbf{H}^0(A) \times \mathbf{H}^2(\mathbf{Z}) \to \mathbf{H}^2(A),$$

this being compatible with the bilinear map

$$A \otimes \mathbf{Z} \to A \quad \text{such that} \quad (a, n) \mapsto na.$$

We know that $\delta : \mathbf{H}^1(\mathbf{Q}/\mathbf{Z}) \to \mathbf{H}^2(\mathbf{Z})$ is an isomorphism, so we find the pairing

$$\mathbf{H}^0(G, A) \times \mathbf{H}^1(\mathbf{Q}/\mathbf{Z}) = \hat{G} \to \mathbf{H}^2(G, A)$$

which is a perfect duality since G is finite. One can give an explicit determination of this duality in terms of standard cocycles as follows.

Theorem 7.3. *Let A be a class module in $\mathrm{Mod}(G)$. Then the perfect duality between $H^0(A)$ and $\hat{G}$ is induced by the following pairing. For $a \in A^G$ and a character $\chi : G \to \mathbf{Q}/\mathbf{Z}$, we get a 2-cocycle*

$$a_{\sigma,\tau} = [\chi'(\sigma) + \chi'(\tau) - \chi'(\sigma\tau)]a,$$

where χ' is a lifting of χ in $\mathbf{Q}$. The expression in brackets is a 2-cocycle of G in $\mathbf{Z}$. The cocycle $(a_{\sigma,\tau})$ represents the class $\varkappa(a) \cup \delta\chi$.

Thus the perfect duality arises from a bilinear map

$$A^G \times \hat{G} \to \mathbf{H}^2(G, A)$$

which we may write

$$(a, \chi) \mapsto a \cup \delta\chi,$$

whose kernel on the left is $\mathbf{S}_G A$, and the kernel on the right is 0.

§8. Explicit Nakayama maps

Throughout this section we let G be a finite group.

In Chapter I, §4 we had an isomorphism

$$\mathbf{H}^{-1}(G, \mathbf{Z}) \xrightarrow{\approx} G/G^c$$

by means of a sequence of isomorphisms

$$\mathbf{H}^{-2}(\mathbf{Z}) \approx \mathbf{H}^{-1}(I) \approx I/I^2 \approx G/G^c.$$

If $\tau \in G$, we denote by ζ_τ the element of $\mathbf{H}^{-2}(\mathbf{Z})$ corresponding to the coset τG^c in G/G^c. So by definition

$$\zeta_\tau = \delta^{-1}(\varkappa(\tau - e)),$$

where δ is the coboundary associated to the exact sequence

$$0 \to I_G \to \mathbf{Z}[G] \to \mathbf{Z} \to 0.$$

On the other hand, we now have a cup product

$$\mathbf{H}^r(A) \times \mathbf{H}^{-2}(\mathbf{Z}) \to \mathbf{H}^{r-2}(A)$$

for $A \in \mathrm{Mod}(G)$, associated to the natural bilinear map $\mathbf{Z} \times A \to A$. We are going to make this cup product explicit for $r \geq 1$, in terms of cochains from the standard complex, and the description of $\mathbf{H}^{-2}(\mathbf{Z})$ given above.

To start, we give a special case of the cup product under dimension shifting. We consider as usual the exact sequence

$$0 \to I \to \mathbf{Z}[G] \to \mathbf{Z} \to 0$$

and its dual

$$0 \to \mathrm{Hom}(\mathbf{Z}, A) \to \mathrm{Hom}(\mathbf{Z}[G], A) \to \mathrm{Hom}(I, A) \to 0.$$

We have a bilinear map in $\mathrm{Mod}(G)$

$$\mathrm{Hom}(\mathbf{Z}[G], A) \times \mathbf{Z}[G] \to A \quad \text{given by} \quad (f, \lambda) \mapsto f(\lambda).$$

There results a pairing of these two exact sequences into the exact sequence

$$0 \to A \to A \to 0 \to 0,$$

to which one can apply the commutative diagram following Theorem 5.1 to get:

Proposition 8.1. *The following diagram has character -1:*

$$
\begin{array}{ccccc}
H^0(\mathrm{Hom}(I,A)) & \times & H^{-1}(I) & \longrightarrow & H^{-1}(A) \\
\downarrow{\scriptstyle\delta} & & \downarrow{\scriptstyle\delta} & & \downarrow{\scriptstyle\mathrm{id}} \\
H^1(\mathrm{Hom}(\mathbf{Z},A)) & \times & H^{-2}(\mathbf{Z}) & \longrightarrow & H^{-1}(A)
\end{array}
$$

On the other hand, we know that in dimension 0, the cup product is given by the induced morphisms. By Corollary 1.12, we see that the cup product in the top line is given by the maps

$$\varkappa(f) \cup \varkappa(\sigma - e) = \varkappa(f(\sigma - e)) \quad \text{for} \quad f \in \mathrm{Hom}_G(I,A).$$

We now pass to the general case.

Theorem 8.2. *Let* $a = a(\sigma_1, \ldots, \sigma_r)$ *be a standard cochain in* $C^r(G,A)$ *for* $r \geq 1$. *For each* $\tau \in G$, *define a map*

$$a \mapsto a * \tau \quad \text{of} \quad C^r(G,A) \to C^{r-2}(G,A)$$

by the formulas:

$$
\begin{array}{ll}
(a * \tau)(\cdot) = a(\tau) & \text{if} \quad r = 1 \\
(a * \tau)(\cdot) = \displaystyle\sum_{\rho \in G} a(\rho, \tau) & \text{if} \quad r = 2 \\
(a * \tau)(\sigma_1, \ldots, \sigma_{r-2}) = \displaystyle\sum_{\rho \in G} a(\sigma_1, \ldots, \sigma_{r-2}, \rho, \tau) & \text{if} \quad r > 2.
\end{array}
$$

Then for $r \geq 1$ *we have the relation*

$$(\delta a) * \tau = \delta(a * \tau).$$

If a *is a cocycle representing an element* α *of* $H^r(G,A)$, *then* $(a * \tau)$ *represents* $\alpha \cup \zeta_r \in H^{r-2}(G,A)$.

Proof. Let us first give the proof for $r = 1$ and 2. We let $a = a(\sigma)$ be a 1-cochain. We find

$$
\begin{aligned}
((\delta a) * \tau)(\cdot) &= \sum_{\rho}(\delta a)(\rho, \tau) \\
&= \sum_{\rho}(\rho a(\tau) - a(\rho\tau) + a(\rho)) \\
&= \sum_{\rho}\rho a(\tau) \\
&= S_G(a(\tau))) = S_G((a * \tau)(\cdot)) \\
&= (\delta(a * \tau))(\cdot),
\end{aligned}
$$

which proves the commutativity for $r = 1$.

Next let $r = 2$ and let $a(\sigma, \tau)$ be a 2-cochain. Then:

$$
\begin{aligned}
((\delta a) * \tau)(\sigma) &= \sum_{\rho}(\delta a)(\sigma, \rho, \tau) \\
&= \sum_{\rho}(\sigma a(\rho, \tau) - a(\sigma\rho, \tau) + a(\sigma, \rho\tau) - a(\sigma, \rho)) \\
&= \sum_{\rho}\sigma a(\rho, \tau) - a(\rho, \tau) \\
&= (\sigma - e)\sum_{\rho} a(\rho, \tau) \\
&= (\sigma - e)((a * \tau)(\cdot)) \\
&= (\delta(a * \tau))(\sigma),
\end{aligned}
$$

which proves the commutativity relation for $r = 2$. For $r > 2$, the proof is entirely similar and is left to the reader.

From the commutativity relation, one obtains an induced homomorphism on the cohomology groups, namely

$$
\varphi_\tau : \mathbf{H}^r(A) \to \mathbf{H}^{r-2}(A) \quad \text{for} \quad r \geqq 2.
$$

For $r = 2$, we have to note that if the cochain $a(\sigma)$ is a coboundary, that is

$$
a(\sigma) = (\sigma - e)b \quad \text{for some} \quad b \in A,
$$

then $(a * \tau)(\cdot) = (\tau - e)b$ is in $I_G A$. Thus φ_τ is a morphism of functors. It is also a δ-morphism, i.e. φ_τ commutes with the coboundary associated with a short exact sequence. Since

$$\alpha \mapsto \alpha \cup \zeta_\tau$$

is also a δ-morphism of $\mathbf{H}^r$ to $\mathbf{H}^{r-2}$, to show that they are equal, it suffices to show that they coincide for $r = 1$, because of the uniqueness theorem.

Explicitly, we have to show that if $\alpha \in \mathbf{H}^1(A)$ is represented by the cocycle $a(\sigma)$, then $\alpha \cup \zeta_\tau$ is represented by $(a * \tau)(\cdot)$, that is

$$\alpha \cup \zeta_\tau = \varkappa(a(\tau)).$$

This is now clear, because of the diagram:

$$
\begin{array}{ccccc}
\varkappa(f) & \times & (\tau - e) & \longmapsto & (f(\tau - e)) \\
\downarrow{\scriptstyle\delta} & & \downarrow{\scriptstyle\delta^{-1}} & & \downarrow{\scriptstyle\text{id}} \\
[\delta\varkappa(f) = -\alpha] & \times & \delta^{-1}(\tau - e) & \longmapsto & \alpha \cup \zeta_\tau.
\end{array}
$$

The coboundary on the left comes from Proposition 8.1. This concludes the proof.

Corollary 8.3. *If $\alpha \in \mathbf{H}^1(A)$ is represented by a standard cocycle $a(\sigma)$, then for each $\tau \in G$ we have $a(\tau) \in A_{\mathbf{S}_G}$ and*

$$\alpha \cup \zeta_\tau = \varkappa(a(\tau)) \in \mathbf{H}^{-1}(A).$$

If $\alpha \in \mathbf{H}^2(A)$ is represented by a standard cocycle $a(\sigma, \tau)$, then for each $\tau \in G$ we have $\sum_\rho a(\rho, \tau) \in A^G$, and

$$\alpha \cup \zeta_\tau = \varkappa\left(\sum_\rho a(\rho, \tau)\right) \in \mathbf{H}^0(A).$$

Corollary 8.4. *The duality between $\mathbf{H}^1(G, \mathbf{Q}/\mathbf{Z})$ and $\mathbf{H}^{-2}(G, \mathbf{Z})$ in the duality theorem is consistent with the identification of $\mathbf{H}^1(G, \mathbf{Q}/\mathbf{Z})$ with $\hat{G}$ and of $\mathbf{H}^{-2}(G, \mathbf{Z})$ with G/G^c.*

The above corollary pursues the considerations of Theorem 1.17, Chapter II, in the context of the cup product. We also obtain further commutativity relations in the next theorem.

Proposition 8.5. *Let U be a subgroup of G.*

(i) *For $\tau \in U, A \in \mathrm{Mod}(G), \alpha \in \mathbf{H}^r(G, A)$ we have*

$$\mathrm{tr}_G^U(\zeta_\tau \cup \mathrm{res}_U^G(\alpha)) = \zeta_\tau \cup \alpha.$$

(ii) *If U is normal, $m = (U : e)$, and $\alpha \in \mathbf{H}^r(G/U, A^U)$ with $r \geqq 2$, then*

$$m \cdot \mathrm{inf}_G^{G/U}(\zeta_{\bar\tau} \cup \alpha) = \zeta_\tau \cup \mathrm{inf}_G^{G/U}(\alpha).$$

If $r = 2$, $m \cdot \mathrm{inf}_G^{G/U}$ is induced by the maps

$$B^G \to mB^G \quad and \quad m\mathbf{S}_U B \to \mathbf{S}_G B$$

for $B \in \mathrm{Mod}(G/U)$.

Proof. As to the first formula, since the transfer corresponds to the map induced by inclusion, we can apply directly the cup product formula from Theorem 3.1, that is

$$\mathrm{tr}(\alpha \cup \mathrm{res}(\beta)) = \mathrm{tr}(\alpha) \cup \beta.$$

One can also use the Nakayama maps, as follows. For τ fixed, we have two maps

$$\alpha \mapsto \zeta_\tau \cup \alpha \quad and \quad \alpha \mapsto \mathrm{tr}_G^U(\zeta_\tau \cup \mathrm{res}_U^G(\alpha))$$

which are immediately verified to be δ-morphisms of the cohomological functor $\mathbf{H}_G$ into $\mathbf{H}_G$ with a shift of 2 dimensions. To show that they are equal, it will suffice to do so in dimension 2. We apply Nakayama's formula. We use a coset decomposition $G = \bigcup \bar{c}U$ as usual. If f is a cocycle representing α, the first map corresponds to

$$f \mapsto \sum_{\rho \in G} f(\rho, \tau).$$

The second one is

$$\mathbf{S}_U^G \left(\sum_{\rho \in U} f(\rho, \tau) \right) = \sum_{c, \rho \in U} \bar{c} f(\rho, \tau).$$

Using the cocycle relation

$$f(\bar{c}, \rho) + f(\bar{c}\rho, \tau) - f(\bar{c}, \rho\tau) = \bar{c}f(\rho, \tau),$$

the desired equality falls out.

This method with the Nakayama map can also be used to prove the second part of the proposition, with the lifting morphism $\mathrm{lif}_G^{G/U}$ replacing the inflation $\inf_G^{G/U}$. One sees that $m \cdot \mathrm{lif}_G^{G/U}$ is a δ-morphism of $H_{G/U}$ to H_G on the category $\mathrm{Mod}(G/U)$, and it will suffice to show that the δ-morphism.

$$\alpha \mapsto \zeta_\tau \cup \mathrm{lif}_G^{G/U}(\alpha) \quad \text{and} \quad \alpha \mapsto m \cdot \mathrm{lif}_G^{G/U}(\zeta_{\bar{\tau}} \cup \alpha)$$

coincide in dimension 2. This follows by using the Nakayama maps as in the first case.

If G is cyclic, then $\mathbf{H}^{-2}(G, \mathbf{Z})$ has a maximal generator, of order $(G : e)$, and -2 is a cohomological period. If σ is a generator of G, then for all $r \in \mathbf{Z}$, the map

$$\mathbf{H}^r(G, A) \to \mathbf{H}^{r-2}(G, A) \quad \text{given by} \quad \alpha \mapsto \zeta_\sigma \cup \alpha$$

is an isomorphism. Hence to compute the restriction, inflation, transfer, conjugation, we can use the commutativity formulas and the explicit formulas of Chapter II, §2.

Corollary 8.6. *Let G be cyclic and suppose $(U : e)$ divides the order of G/U. Then the inflation*

$$\inf_G^{G/U} : H^s(G/U, A^U) \to H^s(G, A)$$

is 0 for $s \geqq 3$.

Proof. Write $s = 2r$ or $s = 2r + 1$ with $r \geqq 1$. Let σ be a generator of G. By Proposition 8.5, we find

$$\zeta_\sigma \cup \inf(\alpha) = \inf(\zeta_{\bar{\sigma}} \cup \alpha).$$

But $\zeta_{\bar{\sigma}} \cup \alpha$ has dimension $s - 2$. By induction, its inflation is killed by $(U : e)^{r-1}$, from which the corollary follows.

The last theorem of this section summarizes some commutativities in the context of the cup product, extending the table from Chapter I, §4.

Theorem 8.7. *Let G be finite of order n. Let $A \in \mathrm{Mod}(G)$ and $\alpha \in \mathbf{H}^2(A) = \mathbf{H}^2(G, A)$. The following diagram is commutative.*

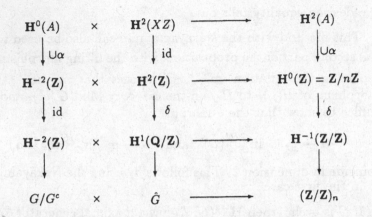

The vertical maps are isomorphisms in the two lower levels. If A is a class module and α a fundamental element, i.e. a generator of $\mathbf{H}^2(A)$, then the cups with α on the first level are also isomorphisms.

Proof. The commutative on top comes from the fact that all elements have even dimension, and that one has commutativity of the cup product for even dimension. The lower commutativities are an old story. If A is a class module, we know that cupping with α gives an isomorphism, this being Tate's Theorem 7.1.

Remark. Theorem 8.7 gives, in an abstract context, the reciprocity isomorphism of class field theory. If G is abelian, then $G^c = e$ and $\mathbf{H}^0(A) = A^G/\mathbf{S}_G A$ is both isomorphic to G and dual to G. On one hand, it is isomorphic to G by cupping with α, and identifying $\mathbf{H}^{-2}(\mathbf{Z})$ with G. On the other hand, if χ is a character of G, i.e. a cocycle of dimension 1 in $\mathbf{Q}/\mathbf{Z}$, then the cupping

$$\varkappa(a) \times \delta\chi \mapsto \varkappa(a) \cup \delta\chi$$

gives the duality between $A^G/\mathbf{S}_G A$ and $\mathbf{H}^1(\mathbf{Q}/\mathbf{Z})$, the values being taken in $\mathbf{H}^2(A)$. The diagram expressed the fact that the identification of $\mathbf{H}^0(G, A)$ with G made in these two ways is consistent.

CHAPTER V
Augmented Products

§1. Definitions

In Tate's work a new cohomological operation was defined, satisfying properties similar to those of the cup product, but especially adjusted to the applications to class field theory and to the duality of cohomology on connection with abelian varieties. As usual here, we give the general setting which requires no knowledge beyond the basic elementary theory we are carrying out.

Let $\mathfrak{A}$ be an abelian bilinear category, and let H, E, F be three δ-functors on $\mathfrak{A}$ with values in the same abelian category $\mathfrak{B}$. For each integers r, s such that H^r, E^s are defined, we suppose that F^{r+s+1} is also defined. By a **Tate product**, we mean the data of two exact sequences

$$0 \to A' \xrightarrow{i} A \xrightarrow{j} A'' \to 0$$

$$0 \to B' \xrightarrow{i} B \xrightarrow{j} B'' \to 0$$

and two bilinear maps

$$A' \times B \to C \quad \text{and} \quad A \times B' \to C$$

coinciding on $A' \times B'$. Such data, denoted by (A, B, C), form a category in the obvious sense. An **augmented cupping**

$$H \times E \to F$$

110

associates to each Tate product a bilinear map

$$U_{\text{aug}} : H^r(A'') \times E^S(B'') \to F^{r+s+1}(C)$$

satisfying the following conditions.

ACup 1. The association is functorial, in other words, if $u : (A, B, C) \to (\bar{A}, \bar{B}, \bar{C})$ is a morphism of a Tate product to another, then the diagram

$$
\begin{array}{ccc}
H^r(A'') \times E^s(B'') \longrightarrow & F^{r+s+1} \\
\Big\downarrow H(u) \quad \Big\downarrow E(u) & \Big\downarrow F(u) \\
H^r(\bar{A}'') \times E^s(\bar{B}'') \longrightarrow & F^{r+s+1}(\bar{C})
\end{array}
$$

is commutative.

ACup 2. The augmented cupping satisfies the property of dimension shifting namely: Suppose given an exact and commutative diagram:

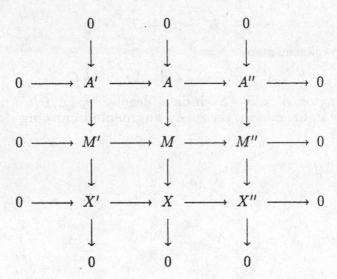

and two exact sequences

$$0 \to B' \to B \to B'' \to 0$$
$$0 \to C' \to M_C \to X_C \to 0,$$

as well as bilinear maps

$$A' \times B \to C \qquad\qquad B' \times B \to M_C$$
$$A \times B' \to C \qquad\qquad M \times B' \to M_C$$
$$X' \times B \to X_C$$
$$X \times B' \to X_C$$

which are compatible in the obvious sense, left to the reader, and coincide on $A' \times B'$ resp. $M' \times B'$, resp. $X' \times B'$, then

$$
\begin{array}{ccc}
H^r(X'') \quad \times \quad E^s(B'') & \longrightarrow & F^{r+s+1}(X_C) \\
\Big\downarrow \delta \qquad\qquad \Big\downarrow \mathrm{id} & & \Big\downarrow \delta \\
H^{r+1}(A'') \quad \times \quad E^s(B'') & \longrightarrow & F^{r+s+2}(C)
\end{array}
$$

is commutative. Similarly, if we shift dimensions on E, then the similar diagram will have character -1.

If H is erasable by an erasing functor M which is exact, and whose cofunctor X is also exact, then we get a uniqueness theorem as in the previous situations.

Thus the agreed cupping behaves like the cup product, but in a little more complicated way. All the relations concerning restriction, transfer, etc. can be formulated for the augmented cupping, and are valid with similar proofs, based as before on the uniqueness theorem. For example, we have:

Proposition 1.1. *Let G be a group. Suppose given on H_G an augmented product. Let U be a subgroup of finite index. Given a Tate product (A, B, C), let*

$$\alpha'' \in H^r(G, A'')$$

and

$$\beta'' \in H^s(U, B'').$$

Then

Then

$$\mathrm{tr}_G^U(\mathrm{res}_U^G(\alpha'') \cup_{\mathrm{aug}} \beta'') = \alpha'' \cup_{\mathrm{aug}} \mathrm{tr}_G^U(\beta'').$$

Proof. Both sides of the above equality define an augmented cupping $H_G \times H_U \to H_G$, these cohomological functors being taken on the multilinear category $\mathrm{Mod}(G)$. They coincide in dimension $(0, 0)$ and 1, as one determines by an explicit computation, so the general uniqueness theorem applies.

Proposition 1.2. *For $\alpha'' \in H^r(U, A)$ and $\beta'' \in H^s(U, B'')$ and $\sigma \in G$, we have*

$$\sigma_*(\alpha'' \cup_{\mathrm{aug}} \beta'') = \sigma_*\alpha'' \cup_{\mathrm{aug}} \sigma_*\beta''.$$

Similarly, if U is normal in G and $\alpha'' \in H^r(G/U, A''^U)$, $\beta'' \in H^s(G/U, B''^U)$, we have for the inflation

$$\inf(\alpha'' \cup_{\mathrm{aug}} \beta'') = \inf(\alpha'') \cup_{\mathrm{aug}} \inf(\alpha'').$$

Of course, the above statements hold for the special functor **H**. when G is finite, except when we deal with inflation.

We make the augmented product more explicit in dimensions $(-1, 0)$ and 0, as well as $(0, 0)$ and 1, for the special functor $\mathbf{H}_G$.

Dimensions $(-1, 0)$ and 0. We are given two exact sequences

$$0 \to A' \xrightarrow{i} A \xrightarrow{j} A'' \to 0$$

$$0 \to B' \xrightarrow{i} B \xrightarrow{j} B'' \to 0$$

as well as a Tate product, that is bilinear maps in $\mathrm{Mod}(G)$:

$$A' \times B \to C \quad \text{and} \quad A \times B' \to C$$

coinciding on $A' \times B'$. We then define the augmented product by

$$\varkappa(a'') \cup_{\mathrm{aug}} \varkappa(b'') = \varkappa(a'b - ab'),$$

where a', b' are determined as follows. We choose $a \in A$ such that $ja = a''$ and $b \in B$ such that $jb = b''$. Then a', b' are uniquely determined by the conditions

$$ia' = \mathbf{S}_G(a) \quad \text{and} \quad ib' = \mathbf{S}_G(b).$$

Dimensions $(0, 0)$ and 1. We define

$$\varkappa(a'') \cup_{\mathrm{aug}} \varkappa(b'') = \text{cohomology class of the cocycle} \quad a'_\sigma b + ab'_\sigma,$$

where the cocycle a'_σ is determined by the formulas

$$ja = a'' \quad \text{and} \quad ia''_\sigma = \sigma a - a,$$

and similarly for b'.

§2. Existence

The existence is given in a way similar to that of the cup product. We shall be very brief. First an abstract statement:

Theorem 2.1. *Let $\mathfrak{A}$ be a multilinear abelian category, and suppose given an exact bilinear functor $A \mapsto Y(A)$ from $\mathfrak{A}$ into the bilinear category of complexes in an abelian category $\mathfrak{B}$. Then the corresponding cohomological functor H on $\mathfrak{A}$ has a structure of augmented cup functor, in the manner described below.*

Recall from Chapter IV, §2 that we already described how the category of complexes forms a bilinear category. For the application to the augmented cup functor, suppose that (A, B, C) is a Tate product. We want to define a bilinear map

$$H^r(A'') \times H^s(B'') \to H^{r+s+1}(C).$$

We do so by a bilinear map defined on the cochains as follows. Let α'' and β'' be cohomology classes in $H^r(A'')$ and $H^s(B'')$ respectively, and let f'', g'' be representative cochains in $Y^r(A), Y^s(B)$ respectively, so that $jf = f''$ and $jg = g''$. We view i as an inclusion, and we let

$$h = \delta f . g + (-1)^r f . \delta g,$$

the products on the right being the Tate product. Then we define $\alpha'' \cup_{\mathrm{aug}} \beta''$ to be the cohomology class of h.

One verifies tediously that this class is independent of the choices made in its construction, and one also proves the dimension shifting property, which is actually a pain, which we do not carry out.

§3. Some properties

Theorem 3.1. *Let the notation be as in Theorem 2.1 with a Tate product (A, B, C). Then the squares in the following diagram from left to right are commutative, resp. of character $(-1)^r$, resp. commutative.*

$$
\begin{array}{cccc}
H^r(A') & \to & H^r(A) & \to & H^r(A'') & \xrightarrow{\delta} & H^{r+1}(A') \\
\times & & \times & & \times & & \times \\
H^s(B) & \leftarrow & H^{s+1}(B') & \xleftarrow{\delta} & H^s(B'') & \leftarrow & H^s(B) \\
\cup\downarrow & & \cup\downarrow & & \downarrow\cup_{\mathrm{aug}} & & \downarrow\cup \\
H^{r+s+1}(C) & \to & H^{r+s+1}(C) & \to & H^{r+s+1}(C) & \to & H^{r+s+1}(C)
\end{array}
$$

The morphisms on the bottom line are all the identity.

Proof. The result follows immediately from the definition of the cup and augmented cup in terms of cochain representatives, both for the ordinary cup and the augmented cup.

The next property arose in Tate's application of cohomology theory to abelian varieties. See Chapter X.

Theorem 3.2. *Let the multilinear categories be those of abelian groups. Let m be an integer ≥ 1, and suppose that the following sequences are exact:*

$$0 \to A''_m \to A'' \xrightarrow{m} A'' \to 0$$

$$0 \to B''_m \to B'' \xrightarrow{m} B'' \to 0.$$

Given a Tate product (A, B, C), one can define a bilinear map

$$A''_m \times B''_m \to C$$

as follows. Let $a'' \in A''_m$ and $b'' \in B''_m$. Choose $a \in A$ and $b \in B$ such that $ja = a''$ and $jb = b''$. We define

$$\langle a'', b'' \rangle = ma.b - a.mb.$$

Then the map $(a'', b'') \mapsto \langle a'', b'' \rangle$ is bilinear.

Proof. Immediate from the definitions and the hypothesis on a Tate product.

Theorem 3.3. *Let (A, B, C) be a Tate product in a multilinear abelian category of abelian groups. Notation as in Theorem 3.2, we have a diagram of character $(-1)^{r-1}$:*

$$
\begin{array}{ccc}
H^r(A''_m) & \longrightarrow & H^r(A'') \\
\times & & \times \\
H^{s+1}(B''_m) & \xleftarrow{\ \delta\ } & H^s(B'') \\
\cup \downarrow & & \downarrow \cup_{\text{aug}} \\
H^{r+s+1}(C) & \xrightarrow[\text{id}]{} & H^{r+s+1}(C)
\end{array}
$$

Note that the coboundary map in the middle is the one associated with the exact sequence involving B''_m and B'' in Theorem 3.2.
The cup product on the left is the one obtained from the bilinear
map as in Theorem 3.2.

CHAPTER VI
Spectral Sequences

We recall some definitions, but we assume that the reader knows the material of *Algebra*, Chapter XX, §9 on spectral sequences, their basic constructions and more elementary properties.

§1. Definitions

Let $\mathfrak{A}$ be an abelian category and A an object in $\mathfrak{A}$. A **filtration** of A consists in a sequence

$$F = F^0 \supset F^1 \supset F^2 \supset \ldots \supset F^n \supset F^{n+1} = 0.$$

If F is given with a differential (i.e. endomorphism) d such that $d^2 = 0$, we also assume that $dF^p \subset F^p$ for all $p = 0, \ldots, n$, and one then calls F a **filtered differential object.** We define the graded object

$$\mathrm{Gr}(F) = \bigoplus_{p \geqq 0} \mathrm{Gr}^p(F) \quad \text{where} \quad \mathrm{Gr}^p(F) = F^p/F^{p+1}.$$

We may view $\mathrm{Gr}(F)$ as a complex, with a differential of degree 0 induced by d itself, and we have the homology $H(\mathrm{Gr}^p F)$.

Filtered objects form an additive category, which is not necessarily abelian. The family $\mathrm{Gr}(A)$ defines a covariant functor on the category of filtered objects.

A **spectral sequence** in A is a family $E = (E_r^{p,q}, E^n)$ consisting of:

(1) Objects $E_r^{p,q}$ defined for integers p, q, r with $r \geqq 2$.

(2) Morphisms $d_r^{p,q} : E_r^{p,q} \to E_r^{p+r,q-r+1}$ such that

$$d_r^{p+r,q-r+1} \circ d_r^{p,q} = 0.$$

(3) Isomorphisms

$$\alpha_r^{p,q} : \mathrm{Ker}(d_r^{p,q})/\mathrm{Im}(d_r^{p-r,q+r-1}) \to E_{r+1}^{p,q}.$$

(4) Filtered objects E^n in A defined for each integer n.

We suppose that for each pair (p,q) we have $d_r^{p,q} = 0$ and $d_r^{p-r,q+r-1} = 0$ for r sufficiently large. It follows that $E_r^{p,q}$ is independent of r for r sufficiently large, and one denotes this object by $E_\infty^{p,q}$. We assume in addition that for n fixed, $F^p(E^n) = E^n$ for p sufficiently small, and is equal to 0 for p sufficiently large.

Finally, we suppose given:

(5) Isomorphisms $\beta^{p,q} : E_\infty^{p,q} \to \mathrm{Gr}^p(E^{p+q})$.

The family $\{E^n\}$, with filtration, is called the **abutment** of the spectral sequence E, and we also say that E **abuts to** $\{E^n\}$ or converges to $\{E^n\}$.

By general principles concerning structures defined by arrows, we know that spectral sequences in $\mathfrak{A}$ form a category. Thus a morphism $u : E \to E'$ of a spectral sequence into another consists in a system of morphisms.

$$u_r^{p,q} : E_r^{p,q} \to E_r^{p,q} \quad \text{and} \quad u^n : E^n \to E''^n.$$

compatible with the filtrations, and commuting with the morphisms $d_r^{p,q}, \alpha_r^{p,q}$ and $\beta_r^{p,q}$. Spectral sequences in $\mathfrak{A}$ then form an additive category, but not an abelian category.

A **spectral functor** is an additive functor on an abelian category, with values in a category of spectral sequences.

We refer to *Algebra*, Chapter XX, §9 for constructions of spectral sequences by means of double complexes.

A spectral sequence is called **positive** if $E_r^{p,q} = 0$ for $p < 0$ and $q < 0$. This being the case, we get:

$$E_r^{p,q} \approx E_\infty^{p,q} \qquad \text{for } r > \sup(p, q+1)$$
$$E^n = 0 \qquad \text{for } n < 0$$
$$F^m(E^n) = 0 \qquad \text{if } m > n$$
$$F^m(E^n) = E^n \qquad \text{if } m \leqq 0.$$

In what follows, we assume that all spectral sequences are positive.

We have inclusions

$$E^n = F^0(E^n) \supset F^1(E^n) \supset \dots \supset F^n(E^n) \supset F^{n+1}(E^n) = 0.$$

The isomorphisms

$$\beta^{0,n} : E_\infty^{0,n} \to \operatorname{Gr}^0(E^n) = F^0(E^n)/F^1(E^n) = E^n/F^1(E^n)$$
$$\beta^{n,0} : E_\infty^{n,0} \to \operatorname{Gr}^n(E^n) = F^n(E^n)$$

will be called the **edge**, or **extreme**, isomorphisms of the spectral sequence.

Theorem 9.6 of *Algebra*, Chapter XX, shows how to obtain a spectral sequence from a composite of functors under certain conditions, the **Grothendieck spectral sequence**. We do not repeat this result here, but we shall use it in the next section.

§2. The Hochschild-Serre spectral sequence

We now apply spectral sequence to the cohomology of groups. Let G be a group and H_G the cohomological functor on $\operatorname{Mod}(G)$. Let N be a normal subgroup of G. Then we have two functors:

$$A \mapsto A^N \text{ of } \operatorname{Mod}(G) \text{ into } \operatorname{Mod}(G/N)$$

$$B \mapsto B^{G/N} \text{ of } \operatorname{Mod}(G/N) \text{ into Grab (abelian groups).}$$

Composing these functors yields $A \mapsto A^G$. Therefore, we obtain the Grothendieck spectral sequence associated to a composite of functors, such that for $A \in \operatorname{Mod}(G)$:

$$E_2^{p,q}(A) = H^p(G/N, H^q(N, A)),$$

with G/N acting on $H^q(N, A)$ by conjugation as we have seen in Chapter II, §1. Furthermore, this spectral functor converges to

$$E^n(A) = H^n(G, A).$$

One now has to make explicit the edge homomorphisms. First we have an isomorphism

$$\beta^{0,n} : E_\infty^{0,n}(A) \to H^n(G, A)/F^1(H^n(G, A)),$$

where F^1 denotes the first term of the filtration. Furthermore

$$E_2^{0,n}(A) = H^n(N, A)^{G/N}$$

and $E_\infty^{0,n}(A)$ is a subgroup of $E_2^{0,n}(A)$, taking into account that the spectral sequence is positive. Hence the inverse of $\beta^{0,n}$ yields a monomorphism of $H^n(G, A)/F^1(H^n(G, A))$ into $H^n(N, A)$, and induces a homomorphism

$$H^n(G, A) \to H^n(N, A).$$

Proposition 2.1. *This homomorphism induced by the inverse of $\beta^{0,n}$ is the restriction.*

Proof. This is a routine tedious verification of the edge homomorphism in dimension 0, left to the reader.

In addition, we have an isomorphism

$$\beta^{n,0} : E_\infty^{n,0}(A) \to F^n(H^n(G, A)),$$

whose image is a subgroup of $H^n(G, A)$. Dually to what we had previously, $E_\infty^{n,0}(A)$ is a factor group of $E_2^{n,0}(A) = H^n(G/N, A^N)$. Composing the canonical homomorphism coming from the $d_r^{n,0}$ and $\beta^{n,0}$, we find a homomorphism

$$H^n(G/N, A^N) \to H^n(G, A).$$

Proposition 2.2. *This homomorphism is the inflation.*

Proof. Again omitted.

Besides the above edge homomorphisms, we can also make the spectral sequence more explicit, both in the lowest dimension and under other circumstances, as follows.

Theorem 2.3. *Let G be a group and N a normal subgroup. Then for $A \in Mod(G)$ we have an exact sequence:*

$$0 \to H^1(G/N, A^N) \xrightarrow{\text{inf}} H^1(G, A) \xrightarrow{\text{res}} H^1(N, A)^{G/N} \xrightarrow{d_2}$$

$$\xrightarrow{d_2} H^2(G/N, A^N) \xrightarrow{\text{inf}} H^2(G, A).$$

The homomorphism d_2 in the above sequence is called the **transgression**, and is denoted by tg, so

$$\text{tg} : H^1(N, A)^{G/N} \to H^2(G/N, A^N).$$

This map tg can be defined in higher dimensions under the following hypothesis.

Theorem 2.4. *If $H^r(N, A) = 0$ for $1 \leqq r < s$, then we have an exact sequence:*

$$0 \to H^s(G/N, A^N) \xrightarrow{\text{inf}} H^s(G, A) \xrightarrow{\text{res}} H^s(N, A)^{G/N} \xrightarrow{\text{tg}}$$

$$\xrightarrow{\text{tg}} H^{s+1}(G/N, A^N) \xrightarrow{\text{inf}} H^{s+1}(G, A).$$

For computations, it is useful to describe tg in dimension 1 in terms of cochains, so we consider tg as in Theorem 2.3, in dimension 1, and we have:

An element $\alpha \in H^2(G/N, A^N)$ can be written as

$$\alpha = \text{tg}(\beta) \text{ with } \beta \in H^1(N, A)^{G/N}$$

if and only if there exists a cochain $f \in C^1(G, A)$ such that:

1. The restriction of f to N is a 1-cocycle representing β.

2. We have $\delta f = $ inflation of a 2-cocycle representing α.

In case many groups $H^r(N, A)$ are trivial, the spectral sequence gives isomorphisms and exact sequences as in the next two theorems.

Theorem 2.5. *Suppose* $H^r(N, A) = 0$ *for* $r > 0$. *Then*

$$\alpha_2^{p,0} : H^p(G/N, A^N) \to H^p(G, A)$$

is an isomorphism for all $p \geqq 0$.

The hypothesis in Theorem 2.5 means that all points of the spectral sequence are 0 except those of the bottom line. Furthermore:

Theorem 2.6. *Suppose that* $H^r(N, A) = 0$ *for* $r > 1$. *Then we have an infinite exact sequence:*

$$0 \to H^1(G/N, H^0(N,A)) \to H^1(G,A) \to H^0(G/N, H^1(N,A)) \to$$

$$\overset{d_2}{\to} H^2(G/N, H^0(N,A)) \to H^2(G,A) \to H^1(G/N, H^1(N,A)) \to$$

$$\overset{d_2}{\to} H^3(G/N, H^0(N,A)) \to H^3(G,A) \to H^2(G/N, H^1(N,A)) \to$$

The hypothesis in Theorem 2.6 means that all the points of the spectral sequence are 0 except those of the two bottom lines.

§3. Spectral sequences and cup products

In this section we state two theorems where cup products occur within spectral sequences. We deal with the multilinear category $\text{Mod}(G)$, a normal subgroup N of G, and the Hochschild-Serre spectral sequence.

Theorem 3.1. *The spectral sequence is a cup functor (in two dimensions) in the following sense. To each bilinear map*

$$A \times B \to C$$

there is a cupping determined functorially

$$E_r^{p,q}(A) \times E_r^{p',q'}(B) \to E_r^{p+p',q+q'}(C)$$

such that for $\alpha \in E_r^{p,q}(A)$ *and* $\beta \in E_r^{p',q'}(B)$ *we have*

$$d_r(\alpha \cdot \beta) = (d_r\alpha) \cdot \beta + (-1)^{p+q} \alpha \cdot (d_r\beta).$$

If we denote by $\cup$ the usual cup product, then for $r = 2$

$$\alpha \cup \beta = (-1)^{q'p} \alpha \cdot \beta.$$

The cupping is induced by the bilinear map

$$H^q(N, A) \times H^{q'}(N, B) \to H^{q+q'}(N, C).$$

Finally, suppose G is a finite group, and $B \in \text{Mod}(G)$. We have an exact sequence with arrows pointing to the left:

$$0 \leftarrow B^G/S_{G/N} B^N \xleftarrow{\text{can}} B^G \xleftarrow{S_{G/N}} B^N/I_{G/N} B^N \xleftarrow{\text{inc}} B^N_{S_{G/N}}/I_{G/N} B^N \leftarrow 0,$$

or in other words

$$0 \leftarrow \mathbf{H}^0(G/N, B^N) \leftarrow H^0(G, B) \leftarrow H_0(G/N, B^N) \leftarrow \mathbf{H}^{-1}(G/N, B^N) \leftarrow 0$$

This exact sequence is dual to the inflation-restriction sequence, in the following sense.

Theorem 3.2. *Let G be a group, U a normal subgroup of finite index, and (A, B, C) a Tate product in $\text{Mod}(G)$. Suppose that (A^U, B^U, C^U) is also a Tate product. Then the following diagram is commutative:*

$$
\begin{array}{ccccc}
H^1(G/U, A''^U) & \xrightarrow{\text{inf}} & H^1(G, A'') & \xrightarrow{\text{res}} & H^1(U, A'') \\
\times & & \times & & \times \\
\end{array}
$$

$$
\begin{array}{ccccc}
0 \longleftarrow & H^0(G/U, B''^U) & \xleftarrow{\text{can}} & H^0(G, B'') & \xleftarrow{S_G^U} & H^0(U, B'') \\
& \Big\downarrow{\cup_{\text{aug}}} & & \Big\downarrow & & \Big\downarrow \\
& H^2(G/U, C^U) & \xrightarrow[\text{inf}]{} & H^2(G, C) & \xrightarrow[\text{tr}]{} & H^2(U, C).
\end{array}
$$

The two horizontal sequences on top are exact.

CHAPTER VII

Groups of Galois Type

(Unpublished article of Tate)

§1. Definitions and elementary properties

We consider here a new category of groups and a cohomological functor, obtained as limits from finite groups.

A topological group G will be said to be of **Galois type** if it is compact, and if the normal open subgroups form a fundamental system of neighborhoods of the identity e. Since such a group is compact, it follows that every open subgroup is of finite index in G, and is therefore closed.

Let S be a closed subgroup of G (no other subgroups will ever be considered). *Then S is the intersection of the open subgroups U containing S.* Indeed, if $\sigma \in G$ and $\sigma \notin S$, we can find an open normal subgroup U of G such that $U\sigma$ does not intersect S, and so $US = SU$ does not contain σ. But US is open and contains S, whence the assertion.

We observe that every closed subgroup of finite index is also open. Warning: There may exist subgroups of finite index which are not open or closed, for instance if we take for G the invertible power series over a finite field with p elements, with the usual topology of formal power series. The factor group G/G^p is a vector space over

F_p, one can choose an intermediate subgroup of index p which is not open.

Examples of groups of Galois type come from Galois groups of infinite extensions in field theory, p-adic integers, etc.

Groups of Galois type form a category, the morphisms being the continuous homomorphisms. This category is stable under the following operations:

1. Taking factor groups by closed normal subgroups.
2. Products.
3. Taking closed subgroups.
4. Inverse limits (which follows from conditions 2 and 3).

Finite groups are of Galois type, and consequently every inverse limit of finite groups is of Galois type. Conversely, every group of Galois type is the inverse limit of its factor groups G/U taken over all open normal subgroups. Thus one often says that a Group of Galois type is **profinite**.

The following result will allow us to choose coset representatives as in the theory of discrete groups, which is needed to make the cohomology of finite groups go over formally to the groups of Galois type.

Proposition 1.1. *Let G be of Galois type, and let S be a closed subgroup of G. Then there exists a continuous section*

$$G/S \to G,$$

i.e. one can choose representatives of left cosets of S in G in a continuous way.

Proof. Consider pairs (T, f) formed by a closed subgroup T and a continuous map $f : G/S \to G/T$ such that for all $x \in G$ the coset $f(xS) = yT$ is contained in xS. We define a partial order by putting $(T, F) \leq (T_1, f_1)$ if $T \subset T_1$ and $f_1(xS) \subset f(xS)$. We claim that these pairs are then inductively ordered. Indeed, let $\{(T_i, f_i)\}$ be a totally ordered subset. Let $T = \bigcap_i T_i$. Then T is a closed subgroup of G. For each $x \in G$, the intersection

$$\bigcap f_i(xS)$$

is a coset of T_i, and is closed in S. Indeed, the finite intersection of such cosets $f_i(xS)$ is not empty because of the hypothesis on the maps f_i. The intersection $\bigcap f_i(xS)$ taken over all indices i is therefore not empty. Let y be an element of this intersection. Then by definition, $yT_i = f_i(xS)$ for all i, and hence $yT \subset f_i(xS)$ for all i. We define $f(xS) = yT$. Then $f(xS) \subset xS$.

The projective limit of the homogeneous spaces G/T_i is then canonically isomorphic to G/T, as one verifies immediately by the compactness of the objects involved. Hence the continuous sections $G/S \to G/T_i$ which are compatible can be lifted to a continuous section $G/S \to G/T$. By Zorn's lemma, we may suppose G/T is maximal, in other words, T is minimal. We have to show that $T = e$.

In other words, with the subgroup S given as at the beginning, if $S \neq e$ it will suffice to find $T \neq S$ and T open in S, closed in G, such that we can find a section $G/S \to S/T$. Let U be a normal open in G, $U \cap S \neq S$, and put $U \cap S = T$. If $G = \bigcup x_i US$ is a coset decomposition, then the map

$$x_i uS \mapsto x_i uT \quad \text{for} \quad u \in U$$

gives the desired section. This concludes the proof of Proposition 1.1.

We shall now extend to closed subgroups of groups of Galois type the notion of index. By a **supernatural number**, we mean a formal product

$$\prod p^{n_p}$$

taken over all primes p, the exponents n_p being integers ≥ 0 or ∞. One multiplies such products by adding the exponents, and they are ordered by divisibility in the obvious manner. The sup and inf of an arbitrary family of such products exist in the obvious way. If S is a closed subgroup of G, then we define the **index** $(G : S)$ to be equal to the supernatural number

$$(G : S) = \underset{V}{\text{l.c.m.}} \ (G : V),$$

the least common multiple l.c.m. being taken over open subgroups V containing S. Then one sees that $(G : S)$ is a natural number if and only if S is open. One also has:

Proposition 1.2. *Let* $T \subset S \subset G$ *be closed subgroups of* G. *Then*

$$(G : S)(S : T) = (G : T).$$

If (S_i) *is a decreasing family of closed subgroups of* G, *then*

$$(G : \bigcap_i S_i) = \text{l.c.m.} \, (G : S_i).$$

Proof. Let us prove the first assertion. Let m, n be integers ≥ 1 such that m divides $(G : S)$ and n divides $(S : T)$. We can find two open subgroups U, V of G such that $U \supset S, V \supset T$, m divides $(G : U)$ and n divides $(S : V \cap S)$. We have

$$(G : S \cap V) = (G : U)(U : U \cap V).$$

But there is an injection $S/(V \cap S) \to U/(U \cap V)$ of homogeneous spaces. By definition, one sees that mn divides $(G : T)$, and it follows that

$$(G : S)(S : T) \quad \text{divides} \quad (G : T).$$

One shows the converse divisibility by observing that if $U \supset T$ is open, then

$$(G : U) = (G : US)(US : U) \quad \text{and} \quad (US : U) = (S : S \cap U),$$

whence $(G : T)$ divides the product. This proves the first assertion of Proposition 1.2. The second assertion is proved by applying the first.

Let p be a fixed prime number. We say that G is a p**-group** if $(G : e)$ is a power of p, which is equivalent to saying that G is the inverse limit of finite p-groups. We say that S is a **Sylow** p **-subgroup** of G if S is a p-group and $(G : S)$ is prime to p.

Proposition 1.3. *Let* G *be a group of Galois type and* p *a prime number. Then* G *has a* p-*Sylow subgroup, and any two such subgroups are conjugate. Every closed* p-*subgroup* S *of* G *is contained in a* p-*Sylow subgroup.*

Proof. Consider the family of closed subgroups T of G containing S and such that $(G : T)$ is prime to p. It is partially ordered by descending inclusion, and it is actually inductively ordered since the intersection of a totally ordered family of such subgroups contains S and has index prime to p by Proposition 1.2. Hence the family contains a minimal element, say T. Then T is a p-group. Otherwise, there would exist an open normal subgroup U of G such that $(T : T \cap U)$ is not a p-power. Taking a Sylow subgroup of the finite group $T/(T \cap U) = TU/U$, for a prime number $\neq p$, once can find an open subgroup V of GH such that $(T : T \cap V)$ is prime to p, and hence $(S : S \cap V)$ is also prime to p. Since S is a p-group, one must have $S = S \cap V$, in other words $V \supset S$, and hence also $T \cap V \supset S$. This contradicts the minimality of T, and shows that T is a p-group of index prime to p, in other words, a p-Sylow subgroup.

Next let S_1, S_2 be two p-Sylow subgroups of G. Let $S_1(U)$ be the image of S under the canonical homomorphism $G \to G/U$ for U open normal in G. Then

$$(G/U : S_1U/U) \quad \text{divides} \quad (G : S_1U),$$

and is therefore prime to p. Hence $S_1(U)$ is a Sylow subgroup of G/U. Hence there exists an element $\sigma \in G$ such that $S_2(U)$ is conjugate to $S_1(U)$ by $\sigma(U)$. Let F_U be the set of such σ. It is a closed subset, and the intersection of a finite number of F_U is not empty, again because the conjugacy theorem is known for finite groups. Let σ be in the intersection of all F_U. Then S_1^σ and S_2 have the same image by all homomorphisms $G \to G/U$ for U open normal in G, whence they are equal, thus proving the theorem.

Next, we consider a new category of modules, to take into account the topology on a group of Galois type G. Let $A \in \text{Mod}(G)$ be an ordinary G-module. Let

$$A_0 = \bigcup A^U,$$

the union being taken over all normal open subgroups U. Then A_0 is a G-submodule of A and $(A_0)_0 = A_0$. We denote by $\text{Galm}(G)$ the category of G-modules A such that $A = A_0$, and call it the category of **Galois modules**. Note that if we give A the discrete topology, then $\text{Galm}(G)$ is the subcategory of G-modules such that G operates continuously, the orbit of each element being finite, and the isotropy group being open. The morphisms in $\text{Galm}(G)$ are the ordinary G-homomorphisms, and we still write $\text{Hom}_G(A, B)$ for $A, B \in \text{Galm}(G)$. Note that $\text{Galm}(G)$ is an abelian category (the kernel and cokernel of a homomorphism of Galois modules are again Galois modules).

Let $A \in \mathrm{Galm}(G)$ and $B \in \mathrm{Mod}(G)$. Then

$$\mathrm{Hom}_G(A, B) = \mathrm{Hom}_G(A, B_0)$$

because the image of A by a G-homomorphism is automatically contained in B_0. From this we get the existence of enough injectives in $\mathrm{Galm}(G)$, as follows:

Proposition 1.4. *Let G be of Galois type. If $B \in \mathrm{Mod}(G)$ is injective in $\mathrm{Mod}(G)$, then B_0 is injective in $\mathrm{Galm}(G)$. If $A \in \mathrm{Galm}(G)$, then there exists an injective $M \in \mathrm{Galm}(G)$ and a monomorphism $u : A \to M$.*

Thus we can define the derived functor of $A \mapsto A^G$ in $\mathrm{Galm}(G)$, and we denote this functor again by H_G, so

$$H^0(G, A) = H_G^0(A) = A^G$$

as before.

Proposition 1.5. *Let G be of Galois type and N a closed normal subgroup of G. Let $A \in \mathrm{Galm}(G)$. If A is injective in $\mathrm{Galm}(G)$, then A^N is injective in $\mathrm{Galm}(G/N)$.*

Proof. If $B \in \mathrm{Galm}(G/N)$, we may consider B as an object of $\mathrm{Galm}(G)$, and we obviously have

$$\mathrm{Hom}_G(B, A) = \mathrm{Hom}_{G/N}(B, A^N)$$

because the image of B by a G-homomorphism is automatically contained in A^N. Considering these Hom as functors of objects B in $\mathrm{Galm}(G/N)$, we see at once that the functor on the right of the equality is exact if and only if the functor on the left is exact.

§2. Cohomology

(a) Existence and uniqueness. One can define the cohomology by means of the standard complex. For $A \in \mathrm{Galm}(G)$, let us put:

$$C^r(G, A) = 0 \quad \text{if} \quad r = 0$$

$$C^0(G, A) = A$$

$$C^r(G, A) = \text{groups of maps} \quad f : G^r \to A \quad (\text{for } r > 0),$$

$$\text{continuous for the discrete topology on } A.$$

We define the coboundary

$$\delta_r : C^r(G, A) \to C^{r+1}(G, A)$$

by the usual formula as in Chapter I, and one sees that $C(G, A)$ is a complex. Furthermore:

Proposition 2.1. *The functor $A \mapsto C(G, A)$ is an exact functor of $\mathrm{Galm}(G)$ into the category of complexes of abelian groups.*

Proof. Let

$$0 \to A' \to A \to A'' \to 0$$

be an exact sequence in $\mathrm{Galm}(G)$. Then the corresponding sequence of standard complexes is also exact, the surjectivity on the right being due to the fact that modules have the discrete topology, and that every continuous map $f : G^r \to A''$ can therefore be lifted to a continuous map of G^r into A.

By Proposition 1.5, we therefore obtain a δ-functor defined in all degrees $r \in \mathbf{Z}$ and 0 for $r < 0$, such that in dimension 0 this functor is $A \mapsto A^G$. We are going to see that this functor vanishes on injectives for $r > 0$, and hence by the uniqueness theory, that this δ-functor is isomorphic to the derived functor of $A \mapsto A^G$, which we denoted by H_G.

Theorem 2.2. *Let G be a group of Galois type. Then the cohomological functor H_G on $\mathrm{Galm}(G)$ is such that:*

$$H^r(G, A) = 0 \quad for \quad r > 0.$$

$$H^0(G, A) = A^G.$$

$$H^r(G, A) = 0 \quad if \ A \ is \ injective, \quad r > 0.$$

Proof. Let $f(\sigma_1, \ldots, \sigma_r)$ be a standard cocycle with $r \geqq 1$. There exists a normal open subgroup U such that f depends only on cosets of U. Let A be injective in $\mathrm{Galm}(G)$. There exists an open normal subgroup V of G such that all the values of f are in A^V because f takes on only a finite number of values. Let $W = U \cap V$. Then f is the inflation of a cocycle $\bar{f}$ of G/W in A^W. By Proposition 1.5, we know that A^W is injective in $\mathrm{Mod}(G/W)$. Hence $\bar{f} = \delta \bar{g}$ with a cochain $\bar{g}$ of G/W in A^W, and so $f = \delta g$ if g is the inflation of $\bar{G}$ to G. Moreover, g is a continuous cochain, and so we have shown that f is a coboundary, hence that $H^r(G, A) = 0$.

In addition, the above argument also shows:

Theorem 2.3. *Let G be a group of Galois type and $A \in \mathrm{Galm}(G)$. Then*

$$H^r(G, A) \approx \mathrm{dir}\ \lim H^r(G/U, A^U),$$

the direct limit dir lim being taken over all open normal subgroups U of G, with respect to inflation. Furthermore, $H^r(G, A)$ is a torsion group for $r > 0$.

Thus we see that we can consider our cohomological functor H_G in three ways: the derived functor, the limit of cohomology groups of finite groups, and the homology of the standard complex.

For the general terminology of direct and inverse limits, cf. *Algebra*, Chapter III, §10, and also Exercises 16 - 26. We return to such limits in (c) below.

Remark. Let G be a group of Galois type, and let $\in \mathrm{Galm}(G)$. If G acts trivially on A, then similar to a previous remark, we have

$$H^1(G, A) = \mathrm{cont}\ \hom(G, A),$$

i.e. $H^1(G, A)$ consists of the continuous homomorphism of G into A. One sees this immediately from the standard cocycles, which are characterized by the condition

$$f(\sigma) + f(\tau) = f(\sigma\tau)$$

in the case of trivial action. In particular, take $A = \mathbf{F}_p$. Then as in the discrete case, we have:

Let G be a p-group of Galois type. If $H^1(G, \mathbf{F}_p) = 0$ then $G = e$, i.e. G is trivial.

Indeed, if $G \neq e$, then one can find an open subgroup U such that G/U is a finite p-group $\neq e$, and then one can find a non-trivial homomorphism $\lambda : G/U \to \mathbf{F}_p$ which, composed with the canonical homomorphism $G \to G/U$ would give rise to a non-trivial element of $H^1(G, \mathbf{F}_p)$.

(b) Changing the group. The theory concerning changes of groups is done as in the discrete case. Let $\lambda : G' \to G$ be a continuous homomorphism of a group of Galois type into another. Then λ gives rise to an exact functor

$$\Phi_\lambda : \mathrm{Galm}(G) \to \mathrm{Galm}(G'),$$

meaning that every object $A \in \mathrm{Galm}(G)$ may be viewed as a Galois module of G'. If

$$\varphi : A \to A'$$

is a morphism in $\mathrm{Galm}(G')$, with $A \in \mathrm{Galm}(G), A' \in \mathrm{Galm}(G')$, with the abuse of notation writing A instead of $\Phi_\lambda(A)$, the pair (λ, φ) determines a homomorphism

$$H^r(\lambda, \varphi) = (\lambda, \varphi)_* : H^r(G, A) \to H^r(G', A'),$$

functorially, exactly as for discrete groups.

One can also see this homomorphism explicitly on the standard complex, because we obtain a morphism of complexes

$$C(\lambda, \phi) : C(G, A) \to C(G', A')$$

which maps a continuous cochain f on the cochain $\varphi \circ f \circ \lambda^r$.

In particular, we have the inflation, lifting, restriction and conjugation:

$$\inf : H^r(G/N, A^N) \to H^r(G, A)$$
$$\mathrm{lif} : H^r(G/N, B) \to H^r(G, B)$$
$$\mathrm{res} : H^r(G, A) \to H^r(S, A)$$
$$\sigma_* : H^r(S, A) \to H^r(S^\sigma, \sigma^{-1}A),$$

for N closed normal in G, S closed in G and $\sigma \in G$.

All the commutativity relations of Chapter II are valid in the present case, and we shall always refer to the corresponding result in Chapter II when we want to apply the result to groups of Galois type.

For U open but not necessarily normal in G, we also have the transfer

$$\text{tr} : H^r(U, A) \to H^r(G, A)$$

with $A \in \text{Galm}(G)$. All the results of Chapter II, §1 for the transfer also apply in the present case, because the proofs rely only on the uniqueness theorem, the determination of the morphism in dimension 0, and the fact that injectives erase the cohomology functor in dimension > 0.

(c) Limits. We have already seen in a naive way that our cohomology functor on $\text{Galm}(G)$ is a limit. We can state a more general result as follows.

Theorem 2.4. *Let (G_i, λ_{ij}) and (A_i, φ_{ij}) be an inverse directed family of groups of Galois type, and a directed system of abelian groups respectively, on the same set of indices. Suppose that for each i, we have $A_i \in \text{Galm}(G_i)$ and that for $i \leqq j$, the homomorphisms*

$$\lambda_{ij} : G_j \to G_i \quad and \quad \varphi_{ij} : A_i \to A_j$$

are compatible. Let $G = \text{inv lim } G_i$ and $A = \text{dir lim } A_i$. Then A has a canonically determined structure as an element of $\text{Galm}(G)$, such that for each i, the maps

$$\lambda_i : G \to G_i \quad and \quad \varphi_i : A_i \to A$$

are compatible. Furthermore, we have an isomorphism of complexes

$$\theta : C(G, A) \xrightarrow{\approx} \text{dir lim } C(G_i, A_i),$$

and consequently isomorphisms

$$\theta_* : H^r(G, A) \to \text{dir lim } H^r(G_i, A_i).$$

Proof. This is a generalization of the argument given for Theorem 2.3. It suffices to observe that each cochain $f : G^r \to A$ is uniformly continuous, and consequently that there exists an open normal subgroup U of G such that f depends only on cosets of U, and takes on only a finite number of values. These values are all represented in some A_i. Hence there exists an open normal subgroup U_i of G_i such that $\lambda_1^{-1}(U_i) \subset U$, and we can construct a cochain $f_i : G_i^r \to A_i$ whose image in $C^r(G, A)$ is f. Similarly, we find that if the image of f_i in $C^r(G, A)$ is 0, then its image in $C^r(G_j, A_j)$ is also 0 for some $j > i$, j sufficiently large. So the theorem follows.

We apply the preceding theorem in various cases, of which the most important are:

(a) When the G_i are all factor groups G/U with U open normal in G, the homomorphisms λ_{ij} then being surjective.

(b) When the G_i range over all open subgroups containing a closed subgroup S, the homomorphisms λ_{ij} then being inclusions.

Both cases are covered by the next lemma.

Lemma 2.5. *Let G be of Galois type, and let (G_i) be a family of closed subgroups, N_i a closed normal subgroup of G_i, indexed by a directed set $\{i\}$, and such that $N_j \subset N_i$ and $G_j \subset G_i$ when $i \leq j$. Then one has*

$$\text{inv } \lim G_i/N_i = (\bigcap G_i)/(\bigcap N_i).$$

Proof. Clear.

Note that Theorem 2.3 is a special case of Theorem 2.4 (taking into account Lemma 2.5). In addition, we get more corollaries.

Corollary 2.6. *Let G be of Galois type and $A \in \text{Galm}(G)$. Let S be a closed subgroup of G. Then*

$$H^r(S, A) = \text{inv } \lim_V H^r(V, A),$$

the inverse limit being taken over all open subgroups V of G containing S.

Corollary 2.7. *Let $G, (G_i), (N_i)$ be as in Lemma 2.5. Let $A \in \text{Galm}(G)$ and let $N = \bigcap N_i$. Then*

$$H^r((\bigcap G_i)/(\bigcap N_i), A^N) \approx \text{dir } \lim H^r(G_i/N_i, A^{N_i}),$$

the limit being taken with respect to the canonical homomorphisms.

Proof. Immediate, because

$$A^N = \bigcup A^{N_i} = \text{dir lim } A^{N_i}$$

because by hypothesis $A \in \text{Galm}(G)$.

Corollary 2.8. *Let G be of Galois type and $A \in \text{Galm}(G)$. Then*

$$H^r(G, A) = \text{dir lim } H^r(G, E)$$

where the limit is taken with respect to the inclusion morphisms $E \subset A$, for all submodules E of A finitely generated over $\mathbf{Z}$.

Proof. By the definition of the continuous operation of G on A, we know that A is the union of G-submodules finitely generated over $\mathbf{Z}$, so we can apply the theorem.

Thus we see that the cohomology group $H^r(G, A)$ are limits of cohomology groups of finite groups, acting on finitely generated modules over $\mathbf{Z}$. We have already seen that these are torsion modules for $r > 0$.

Corollary 2.9. *Let m be an integer > 0, and $A \in \text{Galm}(G)$. Suppose*

$$m_A : A \to A$$

is an automorphism, in other words that A is uniquely divisible by m. Then the period of an element of $H^r(G, A)$ for $r > 0$ is an integer prime to m. If m_A is an automorphism for all positive integers m, then $H^r(G, A) = 0$ for all $r > 0$.

(d) The erasing functor, and induced representations. We are going to define an erasing functor M_G on $\text{Galm}(G)$ similar to the one we defined on $\text{Mod}(G)$ when G is discrete.

Let S be a closed subgroup of G, which we suppose of Galois type. Let $B \in \text{Galm}(S)$ and let $M_G^S(B)$ be the set of all continuous maps $g : G \to B$ (B discrete) satisfying the relation

$$\sigma g(\tau) = g(\sigma\tau) \quad \text{for} \quad \sigma \in S, \tau \in G.$$

Addition is defined in $M_G^S(B)$ as usual, i.e. by adding values in B. We define an action of G by the formula

$$(\tau g)(x) = g(x\tau) \quad \text{for} \quad \tau, x \in G.$$

Because of the uniform continuity, one verifies at once that $M_G^S(B) \in \text{Galm}(G)$.

Taking into account the existence of a continuous section of G/S in G in Proposition 1.1, one sees that:

$M_G^S(B)$ *is isomorphic to the Galois module of all continuous maps* $G/S \to G$.

Thus we find results similar to those of Chapter II, which we summarize in a proposition.

Proposition 2.10. *Notations as above, M_G^S is a covariant, additive exact functor from* $\text{Galm}(S)$ *into* $\text{Galm}(G)$. *The bifunctors*

$$\text{Hom}_G(A, M_G^S(B)) \quad and \quad \text{Hom}_S(A, B)$$

on $\text{Galm}(G) \times \text{Galm}(S)$ *are isomorphic. If B is injective in* $\text{Galm}(S)$, *then $M_G^S(B)$ is injective with* $\text{Galm}(G)$.

The proof is the same as in Chapter II, in light of the condition of uniform continuity and the lemma on the existence of a cross section.

Theorem 2.11. *Let G be of Galois type, and S a closed subgroup. Then the inclusion $S \subset G$ is compatible with the homomorphism*

$$g \mapsto g(e) \quad of \quad M_G^S(B) \to B,$$

giving rise to an isomorphism of functors

$$H_G \circ M_G^S \approx H_S.$$

In particular, if $S = e$, then $H^r(G, M_S(B)) = 0$ for $r > 0$.

Proof. Identical to the proof when G is discrete. For the last assertion, when $S = e$, we put $M_G = M_G^e$.

In particular, we obtain an erasing functor $M_G = M_G^e$ as in the discrete case. For $A \in \text{Galm}(G)$, we have an exact sequence

$$0 \to A \xrightarrow{\varepsilon_A} M_G(A) \to X(A) \to 0,$$

where ε_A is defined by the formula $\varepsilon_A(a) = g_a$ and $g_a(\sigma) = \sigma a$ for $\sigma \in G$.

As in the discrete case, the above exact sequence splits.

Corollary 2.12. *Let G be of Galois type, S a closed subgroup, and $B \in \mathrm{Galm}(S)$. Then $H^r(S, M_G^S(B)) = 0$ for $r > 0$.*

Proof. When $S = G$, this is a special case of the theorem, taking $S = e$. If V ranges over the family of open subgroups containing S, then we use the fact of Corollary 2.6 that

$$H^r(S, A) = \mathrm{dir\ lim}\ H^r(V, A).$$

It will therefore suffice to prove the result when $S = V$ is open. But in this case, $M_S(B)$ is isomorphic in $\mathrm{Galm}(V)$ to a finite product of $M_G^V(B)$, and one can apply the preceding result.

Corollary 2.13. *Let $A \in \mathrm{Galm}(G)$ be injective. Then $H^r(S, A) = 0$ for all closed subgroups S of G and $r > 0$.*

Proof. In the erasing sequence with ε_A, we see that A is a direct factor of $M_G(A)$, so we can apply Corollary 2.12.

(e) Cup products. The theory of cup products can be developed exactly as in the case when G is discrete. Since existence was proved previously with the standard complex, using general theorems on abelian categories, we can do the same thing in the present case. In addition, we observe that

$\mathrm{Galm}(G)$ *is closed under taking the tensor product,*

as one sees immediately, so that tensor products can be used to factorize multilinear maps. Thus $\mathrm{Galm}(G)$ can be defined to be a multilinear category. If $A_1, \ldots, A_n$, B are in $\mathrm{Galm}(G)$, then we define $f : A_1 \times \ldots \times A_N \to B$ to be in $L(A_1, \ldots, A_n, B)$ if f is multilinear in $\mathrm{Mod}(\mathbf{Z})$, and

$$f(\sigma a_1, \ldots, \sigma a_n) = \sigma f(a_1, \ldots, a_n) \quad \text{for all} \quad \sigma \in G,$$

exactly as in the case where G is discrete.

We thus obtain the existence and uniqueness of the cup product, which satisfies the property of the three exact sequences as in the discrete case. Again, we have the same relations of commutativity concerning the transfer, restriction, inflation and conjugation.

(f) Spectral sequence. The results concerning spectral sequences apply without change, taking into account the uniform continuity of cochains. We have a functor $F : \mathrm{Galm}(G) \to \mathrm{Galm}(G/N)$ for a closed normal subgroup N, defined by $A \mapsto A^N$. The group of Galois type G/N acts on $H^r(N, A)$ by conjugation, and one has:

Proposition 2.14. *If N is closed normal in G, then $H^r(N, A)$ is in $\mathrm{Galm}(G/N)$ for $A \in \mathrm{Galm}(G)$.*

Proof. If $\sigma \in N$, from the definition of σ_*, we know that $\sigma_* = \mathrm{id}$. We have to show that for all $\sigma \in H^r(N, A)$ there exists an open subgroup U such that $\sigma_* \alpha = \alpha$ for all $\sigma \in U$. But by shifting dimensions, there exist exact sequences and coboundaries $\delta_1, \ldots \delta_r$ such that

$$\alpha = \delta_1, \ldots, \delta_r \alpha_0 \quad \text{with } \alpha_0 \in H^0(N, B) \text{ for some } B \in \mathrm{Galm}(G).$$

One merely uses the erasing functor r times. We have

$$\sigma_* \alpha = \sigma_* \delta_1 \ldots \delta_r \alpha_0 = \delta_1 \ldots \delta_r \sigma_* \alpha_0$$

and we apply the result in dimension 0, which is clear in this case since σ_* denotes the continuous operation of $\sigma \in S$.

Since the functor $A \mapsto A^N$ transforms an injective module to an injective module, one obtains the spectral sequence of the composite of derived functors. The explicit computations for the restriction, inflation and the edge homomorphisms remain valid in the present case.

(g) Sylow subgroups. As a further application of the fact that the cohomology of Galois type groups is a limit of cohomology of finite groups, we find:

Proposition 2.15. *Let G be of Galois type, and $A \in \mathrm{Galm}(G)$. Let S be a closed subgroup of G. If $(G : S)$ is prime to a prime number p, then the restriction*

$$\mathrm{res} : H^r(G, A) \to H^r(S, A)$$

induces an injection on $H^r(G, A, p)$.

Proof. If S is an open subgroup V in G, then we have the transfer and restriction formula

$$\mathrm{tr} \circ \mathrm{res}(\alpha) = (G : V)\alpha,$$

which proves our assertion. The general case follows, taking into account that

$$H^r(S, A) = \operatorname{dir lim} H^r(V, A)$$

for V open containing S.

§3. Cohomological dimension.

Let G be a group of Galois type. We denote by $\operatorname{Galm_{tor}}(G)$ the abelian category whose objects are the objects A of $\operatorname{Galm}(G)$ which are torsion modules, i.e. for each $a \in A$ there is an integer $n \neq 0$ such that $na = 0$. Given $A \in \operatorname{Galm}(G)$, we denote by A_{tor} the submodule of torsion elements. Similarly for a prime p, we let A_{p^n} denote the kernel of p_A^n in A, and A_{p^∞} is the union of all A_{p^n} for all positive integers n. We call A_{p^∞} the submodule of p-**primary elements**. As usual for an integer m, we let A_m be the kernel of m_A, so

$$A_{tor} = \bigcup A_m \quad \text{and} \quad A_{p^\infty} = \bigcup A_{p^n}$$

the first union being taken for $m \in \mathbf{Z}, m > 0$ and the second for $n > 0$.

The subcategory of elements $A \in \operatorname{Galm}(G)$ such that $A = A_{p^\infty}$ (i.e. A is p-primary) will be denoted by $\operatorname{Galm_p}(G)$.

Let n be an integer > 0. We define the notion of **cohomological dimension**, abbreviated **cd**, and **strict cohomological dimension**, abbreviated **scd**, as follows.

$\operatorname{cd}(G) \leqq n$ if and only if $H^r(G, A) = 0$ for all $r > n$ and $A \in \operatorname{Galm_{tor}}(G)$

$\operatorname{cd}_p(G) \leqq n$ if and only if $H^r(G, A, p) = 0$ for all $r > n$ and $A \in \operatorname{Galm_{tor}}(G)$

$\operatorname{scd}(G) \leqq n$ if and only if $H^r(G, A) = 0$ for all $r > n$ and $A \in \operatorname{Galm}(G)$

$\operatorname{scd}_p(G) \leqq n$ if and only if $H^r(G, A, p) = 0$ for $r > n$ and $A \in \operatorname{Galm}(G)$.

We note that cohomological dimension is defined via torsion modules, and the strict cohomological dimension is defined by means of arbitrary modules (in $\operatorname{Galm}(G)$, of course).

Since

$$H^r(G, A) = \bigoplus_p H^r(G, A, p)$$

one sees that

$$\operatorname{cd}(G) = \sup_p \operatorname{cd}_p(G) \quad \text{and} \quad \operatorname{scd}(G) = \sup_p \operatorname{scd}_p(G).$$

For all $A \in \operatorname{Galm_{tor}}(G)$ we have $A = \bigcup A_{p^\infty}$, the direct sum being taken over all primes p. Hence

$$H^r(G, A) = \bigoplus H^r(G, A_{p^\infty}).$$

To determine $\operatorname{cd}_p(G)$, it will suffice to consider $H^r(G, A_{p^\infty})$, because if we let $A'_{(p)}$ be the p-complementary module

$$A'_{(p)} = \bigcup A_m \quad \text{with} \quad m \quad \text{prime to} \quad p,$$

then $A'_{(p)}$ is uniquely determined by p^n for all integers $n > 0$, so p^n induces an automorphism of $H^r(G, A'_{(p)})$ for $r > 0$, and $H^r(G, A'_{(p)})$ is a torsion group. Hence $H^r(G, A'_{(p)})$ does not contain any element whose torsion is a power of p, and we find:

Proposition 3.1. *Let $A \in \operatorname{Galm_{tor}}(G)$. Then the homomorphism*

$$H^r(G, A_{p^\infty}) \to H^r(G, A, p)$$

induced by the inclusion $A_{p^\infty} \subset A$ is an isomorphism for all r.

Corollary 3.2. *In the definition of $\operatorname{cd}_p(G)$, one can replace the condition $A \in \operatorname{Galm_{tor}}(G)$ by $A \in \operatorname{Galm}_p(G)$.*

We are going to see that the strict dimension can differ only by 1 from the other dimension.

Proposition 3.3. *Let G be of Galois type, and p prime. Then*

$$\operatorname{cd}_p(G) \leqq \operatorname{scd}_p(G) \leqq \operatorname{cd}_p(G) + 1,$$

and the same inequalities hold omitting the index p.

Proof. The first inequality is trivial. For the second, consider the exact sequence

$$0 \to pA \xrightarrow{i} A \to A/pA \to 0$$

$$0 \to A_p \to A \xrightarrow{j} pA \to 0$$

and the corresponding cohomology exact sequences

$$H^{r+1}(pA) \xrightarrow{i_*} H^{r+1}(A) \to H^{r+1}(A/pA)$$

$$H^{r+1}(A_p) \to H^{r+1}(A) \xrightarrow{j_*} H^{r+1}(A/pA).$$

We assume that $\mathrm{cd}_p(G) < n$ and $r > n$. Since $ij = p$, we find $i_* j_* = p_*$. We have $H^{r+1}(A_p) = 0$ by definition, and also $H^{r+1}(A/pA) = 0$. One then sees that j_* is bijective and i_* is surjective. Hence p_* is surjective, i.e. $H^r(A)$ is divisible by p, and hence by an arbitrary power of p. The elements of $H^r(G, A, p)$ being p-primary, it follows that $H^{r+1}(G, A, p) = 0$. This proves the proposition.

For the next result, we need a lemma on the erasing functor M_G^S.

Lemma 3.4. *Let G be of Galois type, and S a closed subgroup. Let $B \in \mathrm{Galm}_{\mathrm{tor}}(S)$ (resp. $\mathrm{Galm}_p(S)$). Then $M_G^S(B)$ is in $\mathrm{Galm}_{\mathrm{tor}}(S)$ (resp. $\mathrm{Galm}_p(G)$). If, in addition B is finitely generated over $\mathbf{Z}$, and S is open, then $M_G^S(B)$ is finitely generated over $\mathbf{Z}$.*

Proof. Immediate from the definitions.

Proposition 3.5. *Let S be a closed subgroup of H. Then*

$$\mathrm{cd}_p \leqq \mathrm{cd}_p(G) \quad \text{and} \quad \mathrm{scd}_p(S) \leqq \mathrm{scd}_p(G),$$

and equality holds if $(G : S)$ is prime to p.

Proof. By Theorem 2.11, we know that $H^r(G, M_G^S(B)) \approx H^r(S, B)$ for all $B \in \mathrm{Galm}(S)$. The assertions are then immediate consequences of the definitions, together with the fact that $(G : S)$ prime to p implies that the restriction is an injection on the p-primary part of cohomology (Proposition 2.15).

As a special case, we find:

Corollary 3.6. *Let G_p be a p-Sylow subgroup of G. Then*

$$\mathrm{cd}_p(G) = \mathrm{cd}_p(G_p) = \mathrm{cd}(G_p),$$

and similarly with scd *instead of* cd. *Furthermore,*

$$\mathrm{cd}(G) = \sup_p \mathrm{cd}(G_p) \quad and \quad \mathrm{scd}(G) = \sup_p \mathrm{scd}(G_p).$$

We now study the cohomological dimension, and leave aside the strict dimension. First, we have a criterion in terms of a category of submodules, easily described.

Proposition 3.7. *We have $\mathrm{cd}_p(G) \leqq n$ if and only if $H^{n+1}(G, E) = 0$ for all elements $E \in \mathrm{Galm}(G)$ such that E is finite of p-power order, and simple as a G-module.*

Proof. Implication in one direction is trivial, taking into account that E is uniquely divisible by every integer m prime to p, and therefore that

$$H^{n+1}(G, E) = H^{n+1}(G, E, p).$$

Conversely, suppose $H^{n+1}(G, E) = 0$ for all E as prescribed. Let $A \in \mathrm{Galm}_{\mathrm{tor}}(G)$ have finite p-power order. If $A \neq 0$, then there is an exact sequence

$$0 \to A' \to A \to A'' \to 0$$

with A' simple. The order of A'' is strictly smaller then the order of A, and the exact cohomology sequence shows by induction that $H^{n+1}(G, A) = 0$. Then let $A \in \mathrm{Galm}_p(G)$. Then A is a direct limit of finite submodules, and we can apply Corollary 2.8. It follows that $H^{n+1}(G, A) = 0$ for $A \in \mathrm{Galm}_p(G)$. Using the erasing functor M_G, one can then proceed by induction, taking into account the fact that M_G maps $\mathrm{Galm}_p(G)$ into $\mathrm{Galm}_p(G)$, and one finds $H^r(G, A) = 0$ for $r > n$ and $A \in \mathrm{Galm}_p(G)$. We can conclude the proof by applying Corollary 3.2.

Lemma 3.8. *Let G be a p-group of Galois type, and let $A \in \mathrm{Galm}_p(G)$. If $A^G = 0$ then $A = 0$. The only simple module $A \in \mathrm{Galm}_p(G)$ is $\mathbf{F}_p$.*

Proof. We already proved this lemma when G is finite, and the general case is an immediate consequence, because G acts continuously on A.

Theorem 3.9. *Let $G = G_p$ be a p-group of Galois type. Then* $\mathrm{cd}(G) \leqq n$ *if and only if* $H^{n+1}(G, \mathbf{F}_p) = 0$.

Proof. This is immediate from Proposition 3.7 and the lemma.

Theorem 3.10. *Let G be a group of Galois type. The following conditions are equivalent:*

$\mathrm{cd}(G) = 0;$
$\mathrm{scd}(G) = 0;$
$G = e.$

If G is a p-group, then $\mathrm{cd}_p(G) = 0$ implies $G = e$.

Proof. It will clearly suffice to prove that if $\mathrm{cd}(G) = 0$ then G is trivial, so suppose $\mathrm{cd}(G) = 0$. For every p-Sylow subgroup G_p of G, we have $\mathrm{cd}(G_p) = 0$ (as one sees from the induced representation), and
$$\mathrm{cd}(G_p) = \mathrm{cd}_p(G_p).$$
Hence $H^1(G_p, \mathbf{F}_p) = 0$. But G_p acts trivially on $\mathbf{F}_p$ so $H^1(G_p, \mathbf{F}_p)$ is just the group of continuous homomorphism cont $\hom(G_p, \mathbf{F}_p)$. If $G \neq e$, then there exists an open normal subgroup U such that G/U is a finite p-group, is equal to e. One could then construct a non-trivial homomorphism of G/U into $\mathbf{F}_p$, contradicting the hypothesis, and concluding the proof.

To show that certain cohomology groups are not 0 in certain dimensions greater than some integer, we have the following criterion.

Lemma 3.11. *Let G be a p-group of Galois type and $\mathrm{cd}(G) = n < \infty$. If $E \in \mathrm{Galm}_p(G)$ has finite order and $E \neq 0$, then $H^n(G, E) \neq 0$.*

Proof. By Lemma 3.8, there is an exact sequence
$$0 \to E' \to E \to E'' \to 0$$
with a maximal submodule E' of E. Since $H^n(G, \mathbf{F}_p) \neq 0$ by hypothesis, one has, again by hypothesis, the exact sequence
$$H^n(G, E) \to H^n(G, \mathbf{F}_p) \to H^{n+1}(G, E') = 0,$$
which shows that $H^n(G, E)$ cannot be trivial.

As an application, we give a refinement of Proposition 3.5.

Proposition 3.12. *Let G be of Galois type, and let S be a closed subgroup of G. If $\operatorname{ord}_p(G : S)$ is finite and $\operatorname{cd}_p(G) < \infty$, then $\operatorname{cd}_p(S) = \operatorname{cd}_p(G)$.*

Proof. Let S_p be a p-Sylow subgroup of S, and similarly G_p a p-Sylow subgroup of G containing S_p. Then

$$\operatorname{ord}_p(G_p : S_p) + \operatorname{ord}_p(G : G_p) = \operatorname{ord}_p(G : S_p)$$
$$= \operatorname{ord}_p(G : S) + \operatorname{ord}_p(S : S_p)$$

Hence $\operatorname{ord}_p(G_p : S_p) = \operatorname{ord}_p(G : S)$. This reduces the proof to the case when G is a p-group, and S is open in G. Suppose

$$n = \operatorname{cd}(G) < \infty.$$

Then by Lemma 3.11,

$$H^n(S, \mathbf{F}_p) = H^n(G, M_G^S(\mathbf{F}_p)) \neq 0,$$

because $M_G^S(\mathbf{F}_p)$ has $p^{(G:S)}$ elements. This concludes the proof.

Corollary 3.13. *If $0 < \operatorname{ord}_p(G : e) < \infty$, then $\operatorname{cd}_p(G) = \infty$. In fact, if G is a finite p-group, then $H^r(G, \mathbf{F}_p) \neq 0$ for all $r > 0$.*

From this corollary, one sees the cohomological dimension is interesting only for infinite groups. We shall give below examples of Galois groups with finite cohomological dimension.

§4. Cohomological dimension ≤ 1.

Let us first remark that if G is a group of Galois type with $\operatorname{scd}_p(G) \leq 1$, then $\operatorname{scd}_p(G) = 0$ and hence every p-Sylow subgroup G_p of G is trivial. Indeed, we have by hypothesis

$$0 = H^2(G_p, \mathbf{Z}) \approx H^1(G_p, \mathbf{Q}/\mathbf{Z}) = \operatorname{cont} \operatorname{hom}(G_p, \mathbf{Q}/\mathbf{Z})$$

from the exact sequence with $\mathbf{Z}, \mathbf{Q}$ and $\mathbf{Q}/\mathbf{Z}$. That $\mathbf{Q}$ is uniquely divisible by every integer $\neq 0$ implies that its cohomology is 0 in dimensions > 0. At the end of the preceding section, we saw that if $G_p \neq e$ then we can find a non-trivial continuous homomorphism of G_p into $\mathbf{F}_p$, which can be naturally imbedded in $\mathbf{Q}/\mathbf{Z}$, and one sees therefore that $G_p = e$, thus proving our assertion.

We then consider the condition $\mathrm{cd}_p(G) \leqq 1$. We shall see that this condition characterizes certain topologically free groups.

We define a group of Galois type to be **p-extensive** if and only if for every finite group F and each abelian p-subgroup E normal in F, and every continuous homomorphism $f : G \to F/E$, there exists a continuous homomorphism $\bar{f} : G \to F$ which makes the following diagram commutative:

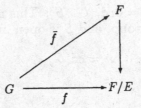

Proposition 4.1. *We have* $\mathrm{cd}_p(G) \leqq 1$ *if and only if* G *is p-extensive.*

Proof. Suppose first that $\mathrm{cd}_p(G) \leqq 1$. We are given F, E, f as above. As usual, we may consider E as an F/E-module, the operation being that of conjugation. Consequently, E is in $\mathrm{Galm}(G)$ via f, namely for $\sigma \in G$ and $x \in E$ we define

$$\sigma x = f(\sigma)x.$$

For each $\sigma \in F/E$, let u_σ be a representative in F. Put

$$e_{\sigma,\tau} = u_\sigma u_\tau u_{\sigma\tau}^{-1}.$$

Then $(e_{\sigma,\tau})$ is a 2-cocycle in $C^2(F/E, E)$, and consequently $(e_{f(\sigma),f(\tau)})$ is a 2-cocycle in $C^2(G, E)$. By hypothesis, there exists a continuous map $\sigma \mapsto a_\sigma$ of G in E such that

$$e_{f(\sigma),f(\tau)} = a_{\sigma\tau}/a_\sigma \sigma a_\tau.$$

We define $\bar{f}(\sigma)$ by

$$\bar{f}(\sigma) = a_\sigma u_{f(\sigma)}.$$

From the definition of the action of G on E, we have

$$\sigma a = u_{f(\sigma)} a u_{f(\sigma)}^{-1}.$$

Thus we find

$$\begin{aligned}
\bar{f}(\sigma)\bar{f}(\sigma) &= a_\sigma u_{f(\sigma)} a_\tau u_{f(\tau)} = a_\sigma a_\tau^\sigma u_{f(\sigma)} u_{f(\tau)} \\
&= a_\sigma a_\tau^\sigma e_{f(\sigma), f(\tau)} u_{f(\sigma\tau)} \\
&= a_{\sigma\tau} u_{f(\sigma\tau)} \\
&= \bar{f}(\sigma\tau),
\end{aligned}$$

which shows that $\bar{f}$ is a homomorphism. It is continuous because (a_σ) is a continuous cochain, and $\sigma \mapsto f(\sigma)$ is continuous. Furthermore, it is clear that $\bar{f}$ is a lifting of f, i.e. that the diagram as in the definition of p-extensive is commutative.

Conversely, let $E \in \mathrm{Galm}_{\mathrm{tor}}(G)$ be of finite order, equal to a p-power, and let $\alpha \in H^2(G, E)$. We have to prove that $\alpha = 0$. Since E is finite, there exists an open normal subgroup U such that U leaves E fixed, i.e. $E = E^U$, and E is therefore a G/U-module. Taking a smaller open subgroup of U if necessary, we can suppose without loss of generality that α comes from the inflation of an element in $H^2(G/U, E)$, i.e. there exists $\alpha_0 \in H^2(G/U, E)$ such that $\alpha = \inf_G^{G/U}(\alpha_0)$. Let F be the group extension of G/U by E corresponding to the class of α_0, so that we have $G/U = F/E$, and let

$$f : G \to F/E$$

be the corresponding homomorphism. We are then in the same situation as in the first part of the proof, and $(e_{f(\sigma), f(\tau)})$ is a 2-cocycle representing α. Since $\bar{f}$ now exists by hypothesis, we define $a_\sigma = \bar{f}(\sigma) u_{f(\sigma)}^{-1}$. The same computation as before shows that

$$a_{\sigma,\tau} = a_\sigma a_\tau^\sigma e_{f(\sigma), f(\tau)},$$

and since (a_σ) is clearly a continuous cochain, one sees that $(e_{f(\sigma), f(\tau)})$ is a coboundary, in other words $\alpha = 0$. This concludes the proof.

Remark. In the definition of p-extensive, without loss of generality, we may assume that f is surjective (it suffices to replace F

by the inverse image of $f(G)$ in F/E). However, we cannot require that $\bar{f}$ is surjective. For instance, let G be the Galois group of the separable closure of a field k. Then F/E is the Galois group of a finite extension K/k, and the problem of finding $\bar{f}$ surjective amounts to finding a finite Galois extension $L \supset K \supset k$ such that F is its Galois group, a problem considered for example by Iwasawa, *Annls of Math.* 1953.

We shall now extend the extension property to the situation when we can take F, E to be of Galois type.

Proposition 4.2. *Let G be of Galois type and p-extensive. Then the p-extension property concerning $(G, f, F, F/E)$ is valid when F is of Galois type (rather than finite), and E is a closed normal p-subgroup.*

Proof. We suppose first that E is finite abelian normal in F. There exists an open normal subgroup U such that $U \cap E = e$. Let $f_1 : G \to F/EU$ be the composite of $f : G \to F/E$ with the canonical homomorphism $F/E \to F/EU$. We can lift f_1 to a continuous homomorphism $\bar{f}_1 : G \to F/U$ by p-extensivity for $F_1 = F/U$ and $E_1 = EU/(U \cap E)$, and f_1. We have a homomorphism

$$(f, \bar{f}_1) : G \to (F/E) \times (F/U),$$

and the canonical map $i : F \to (F/E) \times (F/U)$ is an injection since $U \cap E = e$. The image of G under $(f, \bar{f}_1)$ is contained in the image of i because f and $\bar{f}_1$ lift f_1. Hence $\bar{f} = (f, \bar{f}_1) : G \to F$ solves the extension problem in the present case.

We can now deal with the general case. We want to lift $f : G \to F/E$. We consider all pairs (E', f') where E' is a closed subgroup of E normal in F, and $f' : G \to F/E'$ lifts f. By Zorn's lemma, there is a maximal pair, which we denote also by (E, f). We have to show that $E = e$. If $E \neq e$, then there exists a non trivial element $\theta \in H^1(E, \mathbf{F}_p)$. This character vanishes on an open subgroup V, and has therefore only a finite number of conjugates by elements of F, i.e. it is a Galois module of F/E. Let E_1 be the intersection of the kernel of θ and all its conjugates. Then E_1 is a closed subgroup of E, normal in F, and by the first part of the proof we can lift f to $f_1 : G \to F/E_1$, which contradicts the hypothesis that (E, f) is maximal, and concludes the proof of Proposition 4.2.

Next, we connect cohomological dimension with free groups. We fix a prime p.

Let X be a set and $F_0(X)$ the free group generated by X in the ordinary meaning of the word (cf. *Algebra*, Chapter I, §12). We consider the family of normal subgroups $U \subset X$ such that:

(i) U contains all but a finite number of elements of X.

(ii) U has index a power of p in $F_0(X)$.

We let $F_p(X)$ be the inverse limit

$$F_p(X) = \text{inv lim } F_0/U$$

taken over all such subgroups U. We call $F_p(X)$ the **profinite free p-group generated by** X. Thus $F_p(X)$ is a group of Galois type.

Let G be a group of Galois type and G^0 the intersection of all the kernels of continuous homomorphisms $\theta : G \to \mathbf{F}_p$, i.e. $\theta \in H^1(G, \mathbf{F}_p)$. Then $H^1(G, \mathbf{F}_p)$ is the character group of G/G^0. The converse is also true by Pontrjagin duality.

By definition, if P is a finite p-group, then the continuous homomorphisms $f : F_p(X) \to P$ are in bijection with the maps $f_0 : X \to P$ such that $f_0(x) = e$ for all but a finite number of $x \in X$. Hence $H^1(F_p(X), \mathbf{F}_p)$ is a vector space over $\mathbf{F}_p$, of finite dimension equal to the cardinality of X, and having a basis which can be identified with the elements of X.

Furthermore, we see that $F_p(X)$ is p-extensive, and that cd $F_p(X) \leq 1$. Indeed, we can take f surjective in the definition of p-extensive, and F is then a finite p-group. One uses the freeness of F_0 to see immediately that $F_p(X)$ is p-extensive. We shall prove the converse.

Lemma 4.3. *Let G be a p-group of Galois type, S a closed subgroup. Then $SG^0 = G$ implies $S = G$.*

Proof. This is essentially an analogue of Nakayama's lemma in commutative algebra. Actually, one can prove the lemma first for finite groups, and then extend it to the infinite case, to be left to the reader.

148

Theorem 4.4. *Let G be a p-group of Galois type. Then there exists a profinite free p-group, $F_p(X)$ and a continuous homomorphism*

$$\bar{g} : F_p(X) \to G$$

such that the induced homomorphism

$$H^1(G, \mathbf{F}_p) \to H^1(F_p(X), \mathbf{F}_p)$$

is an isomorphism. The map $\bar{g}$ is then surjective. If $\mathrm{cd}(G) \leqq 1$, then $\bar{g}$ is an isomorphism.

Proof. From the preceding discussion, to obtain an isomorphism

$$H^1(G, \mathbf{F}_p) \xrightarrow{\approx} H^1(F_p(X), \mathbf{F}_p)$$

it suffices to take for X a basis of $H^1(G, \mathbf{F}_p)$ and to form $F_p(X)$. By duality, we obtain an isomorphism

$$F_p(X)/F_p(X)^0 \approx G/G^0,$$

whence a homomorphism

$$g : F_p(X) \to G/G^0.$$

Since $F_p(X)$ is p-extensive, we can lift g to G, to get the commutative diagram

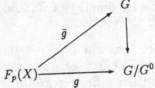

and $\bar{g}$ is surjective by the lemma. Suppose finally that $\mathrm{cd}(G) \leqq 1$, and let N be the kernel of $\bar{g}$. Then we obtain an exact sequence

$$0 \longrightarrow H^1(G, \mathbf{F}_p) \xrightarrow{\mathrm{inf}} H^1(F_p(X), \mathbf{F}_p) \xrightarrow{\mathrm{res}} H^1(N, \mathbf{F}_p)^G \longrightarrow H^2(G, \mathbf{F}_p) = 0.$$

We have $H^2(G, \mathbf{F}_p) = 0$ by the assumption $\mathrm{cd}(G) \leqq 1$. The inflation is an isomorphism, and hence $H^1(N, \mathbf{F}_p)^G = 0$. By Lemma 3.8, we find $H^1(N, \mathbf{F}_p) = 0$, i.e. N has only the trivial character, whence $N = e$, thus proving the theorem.

Corollary 4.5. *Let G be a p-group of Galois type. Then the following conditions are equivalent:*

G *is profinite free;*

G *is p-extensive;*

$\mathrm{cd}(G) \leqq 1$.

We end this section with a discussion of the condition $\mathrm{cd}(G) \leqq 1$ for factor groups. Let G be of Galois type. Let T be the intersection of all subgroups of G which are the kernels of continuous homomorphisms of G into p-groups of Galois type. Then G/T is a p-group which we denote by $G(p)$, and which we call the **maximal p-quotient** of G. One can also characterize T by the condition that it is a closed normal subgroup satisfying:

(a) $(G : T)$ is a p-power.

(b) $H^1(T, \mathbf{F}_p) = 0$.

The characterization is immediate.

Proposition 4.6. *Let G be a group of Galois type. Then $\mathrm{cd}_p(G) \leqq 1$ implies $\mathrm{cd}_p G(p) \leqq 1$.*

Proof. Consider the exact sequence

$$0 \longrightarrow H^1(G/T, \mathbf{F}_p) \longrightarrow H^1(G, \mathbf{F}_p) \longrightarrow H^1(T, \mathbf{F}_p)^{G/T} \longrightarrow H^2(G/T, \mathbf{F}_p) \longrightarrow 0,$$

with a 0 on the right because of the assumption $\mathrm{cd}_p(G) \leqq 1$. By the characterization of T we have $H^1(T, \mathbf{F}_p) = 0$, whence $H^2(G/T, \mathbf{F}_p) = 0$ which suffices to prove the proposition by Theorem 3.9.

In the Galois theory, $G(p)$ is the Galois group of the maximal p-extension of the ground field, and G is the Galois group of the algebraic closure. Cf. §6 below for applications to this context.

§5. The tower theorem

In many cases, one gets information on a group G be considering a normal subgroup N and the factor group G/N. We do this for cohomological dimension, and we shall find

$$\mathrm{cd}(G) \leqq \mathrm{cd}(N) + \mathrm{cd}(G/N),$$

and similar with cd_p instead of cd. We use the spectral sequence with

$$E_2^{r,s} = H^r(G/N, H^s(N, A)) \quad \text{converging to} \quad H(G, A)$$

for $A \in \mathrm{Galm}(G)$. There is a filtration of $H^n(G, A)$ such that the successive quotients are isomorphic to $E_\infty^{r,s}$ for $r + s = n$, and $E_\infty^{r,s}$ is a subgroup of a factor group of $E_2^{r,s}$. Hence $H^n(G, A) = 0$ whenever $H^r(G/N, H^s(N, A)) = 0$, which occurs in the following cases:

$$r > \mathrm{cd}(G/N) \quad \text{and} \quad s > 0 \text{ or } A \in \mathrm{Galm}_{tor}(G);$$
$$r > \mathrm{scd}(G/N) \quad \text{and} \quad s \text{ arbitrary};$$
$$s > \mathrm{cd}(N) \quad \text{and} \quad A \in \mathrm{Galm}_{tor}(G);$$
$$s > \mathrm{scd}(N) \quad \text{and} \quad A \in \mathrm{Galm}(G).$$

From this we find the theorem:

Theorem 5.1. *Let G be of Galois type and N a closed normal subgroup. Then for all primes p,*

$$\mathrm{cd}_p(G) \leqq \mathrm{cd}_p(G/N) + \mathrm{cd}_p(N),$$

and similarly with cd instead of cd_p.

As an application, suppose that G/N is topologically cyclic, and $\mathrm{cd}_p(N) \leqq 1$. Then $\mathrm{cd}_p(G) \leqq 2$. This happens in the following cases: G is the Galois group of the algebraic closure of a totally imaginary number field, or a p-adic field. Indeed, in each case, one can construct a cyclic extension (maximal unramified in the local case, cyclotomic in the global case), which decomposes G into a subgroup N and factor G/N as above. In the next sections, we shall give a criterion with the Brauer group to show that $\mathrm{cd}(N) \leqq 1$ for suitable N.

§6. Galois groups over a field

Let k be a field and k_s its separable closure. Let

$$G_k = \mathrm{Gal}(K_s/k)$$

be the Galois group. If K is a Galois extension of k, we let $G_{K/k}$ be its Galois group. Then G_K is normal in G_k and the factor group G_k/G_K is $G_{K/k}$. All these groups are of Galois type, with the Krull topology.

We shall use constantly Hilbert's Theorem 90, that for the multiplicative group K^*, we have

$$H^1(G_{K/k}, K^*) = 0.$$

Note that $K^* \in \text{Galm}(G_{K/k})$. In the additive case, with the additive group K^+,

$$H^r(G_{K/k}, K^+) = 0 \quad \text{for all} \quad r > 0.$$

One sees this reduction to the case when K is finite Galois over k, so there is a normal basis showing that K^+ is semilocal with local group reduced to e, whence the cohomology is trivial in dimension > 0.

Next, we give a result in characteristic p.

Theorem 6.1. *Let k have characteristic $p > 0$ and let $k(p)$ be the maximal p-extension with Galois group $G(p)$ over k. Then cd $G(p) \leqq 1$, and so $G(p)$ is profinite free. The number of generators is equal to $\dim_{\mathbf{F}_p}(k^+/\wp k^+)$, where $\wp x = x^p - x$.*

Proof. We recall the Kummer theory exact sequence

$$0 \to F_p \to k_s^+ \xrightarrow{\wp} k_s^+ \to 0,$$

whence the cohomology sequence

$$0 \to \mathbf{F}_p \to k^+ \to k^+ \to H^1(G_k, \mathbf{F}_p) \to H^1(G_k, k_s^+) = 0.$$

Consequently,

$$k^+/\wp k^+ \approx H^1(G_k, \mathbf{F}_p) = \text{cont hom}(G_k, \mathbf{F}_p).$$

Since $k(p)$ is the maximal Galois p-extension of k, it has no Galois extension of p-power degree, and hence we have an exact sequence

$$0 \to \mathbf{F}_p \to k(p)^+ \to k(p)^+ \to 0.$$

By the remarks made at the beginning of this section, we get from the exact cohomology sequence that $H^2(G(p), \mathbf{F}_p) = 0$, and hence by the criterion of Theorem 3.9 that cd $G(p) \leqq 1$. Moreover, the beginning of this same exact sequence yields

$$k^+ \xrightarrow{\wp} k^+ \to H^1(G(p), \mathbf{F}_p) \to H^1(G(p), k(p)^+) = 0,$$

whence an isomorphism

$$k^+/\wp k^+ \approx H^1(G(p), \mathbf{F}_p)),$$

which gives us the desired number of generators.

We now go to characteristic $\neq p$ using the multiplicative Kummer sequence instead of the additive one. Cohomological dimension will be studied via Galois cohomology in k_s^*.

Theorem 6.2. *Let k be a field of characteristic $\neq p$ and containing a p-th root of unity. Let $k(p)$ be the maximal p-extension, with Galois group $G(p)$. Then $\mathrm{cd}(G(p)) \leqq n$ if and only if:*

(i) $H^n(G(p), k(p)^*)$ *is divisible by p.*

(ii) $H^{n+1}(G(p), k(p)^*) = 0.$

Proof. We consider the exact sequence

$$0 \to \mathbf{F}_p \to k(p)^* \xrightarrow{P} k(p)^* \to 0,$$

where $\mathbf{F}_p$ gets embedded on the group of p-th roots of unity, and the map on the right is taking p-th powers. We obtain the cohomology exact sequence

$$H^n(k(p)^*) \to H^n(k(p)^*) \to H^{n+1}(\mathbf{F}_p) \to H^{n+1}(k(p)^*) \to H^{n+1}(k(p)^*)$$

with acting group $G(p)$. The theorem follows at once from this exact sequence.

Corollary 6.3. (Kawada) *The Galois group $G(p) = G_{k(p)/k}$ is a profinite p-group if $p = \mathrm{char}\ k$. If $p \neq \mathrm{char}\ k$ and the p-th roots of unity are in k, then it is profinite free if and only if $H^2(G(p), k(p)) = 0$.*

Proof. One always has $H^1(k(p)^*) = 0$, and the rest follows from the preceding two theorems.

Theorem 6.1 can be translated in terms of cohomology with values in k_s^*.

Theorem 6.4. *Let k be a field and p prime $\neq$ char k. Let n be an integer > 0. Then $\operatorname{cd}_p(G_k) \leqq n$ if and only if:*

(i) $H^n(G_E, k_s^*) = 0$ *is divisible by p*

(ii) $H^{n+1}(G_E, k_s^*, p) = 0$

for all finite separable extensions E of k of degree prime to p.

Proof. The Kummer sequence

$$0 \to \mathbf{F}_p \to k_s^* \xrightarrow{p} k_s^* \to 0$$

yields the cohomology sequence with groups G_k:

$$H^n(k_s^*) \xrightarrow{p} H^n(k_s^*) \to H^{n+1}(\mathbf{F}_p) \to H^{n+1}(k_s^*) \xrightarrow{p} H^{n+1}(k_s^*) \to H^{n+2}(\mathbf{F}_p).$$

Suppose $\operatorname{cd}_p(G_k) \leqq n$. Then $H^{n+1}(\mathbf{F}_p) = H^{n+2}(\mathbf{F}_p) = 0$ by Proposition 3.7. Conditions (i) and (ii) are then clear, taking Proposition 3.5 into account. The converse can be proved in a similar way, from the fact that if G_p is a p-Sylow subgroup of G_k, then

$$H^r(G_p, k_s^*) = \operatorname{inv} \lim H^r(G_E, k_s^*),$$

the limit being taken over all finite separable extensions E of k of degree prime to p. The Galois groups G_E constitute precisely the set of open subgroups of G_k containing G_p, or its conjugates, which amounts to the same thing.

Corollary 6.5. *If $H^2(G_E, k_s^*) = 0$ for all finite separable extensions E of k, then $\operatorname{cd}(G_k) \leqq 1$.*

Proof. We have $H^1(G_e, k_s^*) = 0$ automatically, and we apply the theorem for the p-component when $p \neq$ char k. If $p =$ char k, then we saw in Theorem 6.1 that the cohomological dimension is $\leqq 1$.

The above corollary provides the announced criterion in terms of the Brauer group, because that is what $H^2(k_s^*)$ amounts to.

Theorem 6.6. *Let K be an extension of k. Then*

$$\operatorname{cd}_p(G_K) \leqq \operatorname{tr}\ \deg K/k + \operatorname{cd}_p(G_k).$$

(By definition, tr deg *is the transcendence degree.)*

Proof. If in a tower $K \supset K_1 \supset k$ the assertion is true for K/K_1 and for K_1/k, then it is true for K/k. We are therefore reduced to the cases when either K/k is algebraic, in which case G_K is a closed subgroup of G_k, and the assertion is trivial; or when K is a pure transcendental extension $K = k(x)$, in which case we have a field diagram as follows.

$$k(x) \xrightarrow{\ G_k\ } k^a(x) \xrightarrow{\ G_{k^a(x)}\ } k^a(x)_s$$
$$\Big\uparrow \qquad\qquad \Big\uparrow$$
$$k \xrightarrow[\ G_k\ ]{} k^a$$

By Tsen's theorem and the corollary of Theorem 6.3, we know that $\mathrm{cd}(G_{k^a(x)}) \leqq 1$. The tower theorem shows that

$$\mathrm{cd}(G_{k(x)}) \leqq \mathrm{cd}(G_k) + 1,$$

thus proving the theorem.

Theorem 6.7. *In the preceding theorem, there is equality if $\mathrm{cd}_p(G_k) < \infty$ $(p \neq \mathrm{char}\ k)$ and K is finitely generated over k.*

Proof. The assertion is again transitive in towers, and we are reduced either to the case of a finite algebraic extension, when we can apply Proposition 3.12, or to $K = k(x)$ purely transcendental. For this last case, we need a lemma.

Lemma 6.8. *Let G be of Galois type, T a closed normal subgroup such that $\mathrm{cd}_p(T) \leqq 1$. If $\mathrm{cd}_p(G/T) \leqq n$, then there is an isomorphism*

$$H^{n+1}(G, A) \approx H^n(G/T, H^1(T, A))$$

for all $A \in \mathrm{Galm}_{\mathrm{tor}}(G)$.

Proof. We have $H^r(T, A) = 0$ for $r > 1$ and the spectral sequence becomes an exact sequence

$$0 \to H^{n+1}(G/T, A^T) \to H^{n+1}(G, A) \to H^n(G/T, H^1(T, A))$$
$$\to H^{n+2}(G/T, A^T) \to 0,$$

whence the lemma follows.

Coming back to the theorem, put

$$G = G_{k(x)} \text{ and } T = G_{k(x)_s/k_s(x)}.$$

We refer to the diagram for Theorem 6.6. We may replace k be its extension corresponding to a Sylow subgroup of G_k, that is we may suppose that G_k is a p-group. We have $G_k = G/T$. Let us now take $A = \mathbf{F}_p$ in the lemma. Suppose

$$n = \mathrm{cd}_p(G/T) < \infty.$$

We must show that $H^{n+1}(G, A) \neq 0$. By the lemma, this amounts to showing that $H^n(G/T, H^1(T, \mathbf{F}_p)) \neq 0$. Since the p-th roots of unity are in k ($p \neq \mathrm{char}\ k$), Kummer theory shows that

$$H^1(T, F_p) = \mathrm{cont}\ \hom(T, \mathbf{F}_p)$$

is G/T-isomorphic to $k_s(x)^*/k_s(x)^{*p}$. The unique factorization in $k_s(x)$ shows that this group contains a subgroup G-isomorphic to $\mathbf{F}_p$. On the other hand, this group is a direct sum of its orbits under G/T, and one of these orbits is $\mathbf{F}_p$. Hence $H^{n+1}(G, \mathbf{F}_p) \neq 0$ as was to be shown.

The theorem we have just proved, and which occurs here at the end of the theory, historically arose at its beginning. Its conjecture and the sketch of its proof are due to Grothendieck.

CHAPTER VIII

Group Extensions

§1. Morphisms of extensions

Let G be a group and A an abelian group, both written multiplicatively. An **extension** of A by G is an exact sequence of groups

$$0 \to A \xrightarrow{i} U \xrightarrow{j} G \to 0.$$

We can then define an action of G on A. If we identify A as a subgroup of U, then U acts on A by conjugation. Since A is assumed commutative, it follows that elements of A act trivially, so $U/A = G$ acts on A.

For each $\sigma \in G$ and $a \in A$ we select an element $u_\sigma \in U$ such that $j u_\sigma = \sigma$, and we put

$$^\sigma a = [u_\sigma]a = u_\sigma a u_\sigma^{-1}.$$

Each element of U can be written uniquely in the form

$$u = a u_\sigma \text{ with } \sigma \in G \text{ and } a \in A.$$

Then there exist elements $a_{\sigma,\tau} \in A$ such that

$$u_\sigma u_\tau = a_{\sigma,\tau} u_{\sigma\tau},$$

and $(a_{\sigma,\tau})$ is a 2-cocycle of G in A. A different choice of u_σ would give rise to another cocycle, differing from the first one by a coboundary. Hence the cohomology class α of these cocycles is a well defined element of $H^2(G, A)$, determined by the extension, i.e. by the exact sequence.

Conversely, suppose given an element $\alpha \in H^2(G, A)$ with G given, and A abelian in $\mathrm{Mod}(G)$. Let $(a_{\sigma,r})$ be a cocycle representing α. We can then define an extension of A by G as follows. We let U be the set of pairs (a, σ) with $a \in A$ and $\sigma \in G$. We define multiplication in U by

$$(a, \sigma)(b, \tau) = (a\sigma b a_{\sigma, r}, \sigma\tau).$$

One verifies that U is a group, whose unit element is $(a_{e,e}^{-1}, e)$. The existence of the inverse of (a, σ) is determined at once from the definition of multiplication. Defining $j(a, \sigma) = \sigma$ gives a homomorphism of U onto G, whose kernel is isomorphic to A, under the correspondence

$$a \mapsto (aa_{e,e}^{-1}, e).$$

Thus we get a group extension of A by G.

Extensions of groups form a category, the morphisms being triplets of homomorphisms (f, F, φ) which make the following diagram commutative:

$$
\begin{array}{ccccccccc}
0 & \longrightarrow & A & \longrightarrow & U & \longrightarrow & G & \longrightarrow & 0 \\
 & & \downarrow{\scriptstyle f} & & \downarrow{\scriptstyle F} & & \downarrow{\scriptstyle \varphi} & & \\
0 & \longrightarrow & B & \longrightarrow & V & \longrightarrow & H & \longrightarrow & 0
\end{array}
$$

We have the general notion of isomorphism in this category, but we look at the restricted notion of extensions U, U' of A by G (so the same A and G). Two such extensions will be said to be **isomorphic** if there exists an isomorphism $F : U \to U'$ making the following diagram commutative:

$$
\begin{array}{ccccc}
A & \longrightarrow & U & \longrightarrow & G \\
\downarrow{\scriptstyle \mathrm{id}} & & \downarrow{\scriptstyle F} & & \downarrow{\scriptstyle \mathrm{id}} \\
A & \longrightarrow & U' & \longrightarrow & G
\end{array}
$$

Isomorphism classes of extensions thus form a category, the morphisms being given by isomorphisms F as above.

Let (G, A) be a pair consisting of a group G and a G-module A. We denote by $E(G, A)$ the isomorphism classes of extensions of A by G. For G fixed, $A \mapsto E(G, A)$ is a functor on $\text{Mod}(G)$. We may summarize the discussion

Theorem 1.1. *On the category* $\text{Mod}(G)$, *the functors* $H^2(G, A)$ *and* $E(G, A)$ *are isomorphic, by the bijection established at the beginning of the section.*

Next, we state a general result providing the existence of the homomorphism F when pairs of homomorphisms (φ, f) are given, from a pair (G, A) to a pair (G', A').

Theorem 1.2. *Let* $G \approx U/A$ *and* $G' \approx U'/A'$ *be two extensions. Suppose given two homomorphisms*

$$\varphi : G \to G' \quad and \quad f : A \to A'.$$

There exists a homomorphism $F : U \to U'$ *making the diagram commutative:*

$$
\begin{array}{ccccc}
A & \xrightarrow{\ i\ } & U & \xrightarrow{\ j\ } & G \\
{\scriptstyle f}\downarrow & & {\scriptstyle F}\downarrow & & {\scriptstyle \varphi}\downarrow \\
A' & \xrightarrow[\ i'\ ]{} & U' & \xrightarrow[\ j'\ ]{} & G'
\end{array}
$$

if and only if:

(1) f *is a* G-*homomorphism, with* G *acting on* A' *via* φ.

(2) $f_* \alpha = \varphi^* \alpha$, *where* α, α' *are the cohomology classes of the two extensons respectively, and* f_*, φ^* *are the morphisms induced by the morphisms of pairs*

$$(\text{id}, f) : (G, A) \to (G, A') \quad and \quad (\varphi, \text{id}) : (G', A') \to (G, A').$$

Proof. We begin by showing that the conditions are necessary. Without loss of generality, we let i be an inclusion. Let (u_σ) and $(u'_{\sigma'})$ be representatives of σ and σ' respectively in G and G'. For $u = a u_\sigma$ in U we find

$$F(u) = F(a u_\sigma) = F(a)F(u_\sigma) = f(a)F(u_\sigma).$$

One sees that F is uniquely determined by the data $F(u_\sigma)$. We have

$$jF(u_\sigma) = \varphi j u_\sigma = \varphi\sigma = j' u'_{\varphi\sigma}.$$

Hence there exist elements $c_\sigma \in A'$ such that

$$F(u_\sigma) = c_\sigma u'_{\varphi\sigma}.$$

It follows that F is uniquely determined by the data (c_σ), which is a cochain of G in A'. Since F is a homomorphism, we must have

$$F(u_\sigma a u_\sigma^{-1}) = F(u_\sigma)f(a)f(u_\sigma)^{-1}$$
$$F(u_\sigma u_\tau) = F(u_\sigma)F(u_\tau).$$

These conditions imply:

(I) $f(\sigma a) = \varphi\sigma(fa)$ for $a \in A$ and $\sigma \in G$.

(II) $fa_{\sigma,\tau} = b_{\varphi\sigma,\varphi\tau}((\varphi\sigma)c_\tau c_{\sigma\tau}^{-1} c_\sigma)$,

for the cocycles $(a_{\sigma,\tau})$ and $(b_{\sigma',\tau'})$ associated to the representatives (u_σ) and (u'_σ). By definition, these two conditions express precisely the conditions (1) and (2) of the theorem. Conversely, one verifies that these conditions are sufficient by defining

$$F(a u_\sigma) = f(a)c_\sigma u'_{\varphi\sigma}.$$

This concludes the proof.

We also want to describe more precisely the possible F in an isomorphism class of extensions of A by G. We work more generally with the situation of Theorem 1.2. Let f, φ be fixed and let

$$F_1, \ F_2 : U \to U'$$

be homomorphisms which make the diagram of Theorem 1.2 commutative. We say that F_1 is **equivalent** to F_2 if they differ by an inner automorphism of U' coming from an element of A', that is there exists $a' \in A'$ such that

$$F_1(u) = a' F_2(u) a'^{-1} \quad \text{for all} \quad u \in U.$$

This equivalence is the weakest one can hope for.

Theorem 1.3. *Let f, φ be given as in Theorem 1.2. Then the equivalence classes of homomorphisms F as in this theorem form a principal homogeneous space of $H^1(G, A')$. The action of $H^1(G, A')$ on this space is defined as follows. Let (u_σ) be representatives of G in U, and (z_σ) a 1-cocycle of G in A'. Then*

$$(zF)(au_\sigma) = f(a)z_\sigma F(u_\sigma).$$

Proof. The straightforward proof is left to the reader.

Corollary 1.4. *If $H^1(G, A') = 0$, then two homomorphisms $F_1, F_2 : U \to U'$ which make the diagram of Theorem 1.2 commutative are equivalent.*

§2. Commutators and transfer in an extension.

Let G be a finite group and $A \in \text{Mod}(G)$. We shall write A multiplicatively, and so we replace the trace by the norm $\mathbf{N} = \mathbf{N}_G$. We consider an extension of A by G,

$$0 \to A \xrightarrow{i} E \xrightarrow{j} G \to 0,$$

and we suppose without loss of generality that i is an inclusion. We fix a family of representatives (u_σ) of G in E, giving rise to the cocycle $(a_{\sigma,\tau})$ as in the preceding section. Its class is denoted by α. We let E^c be the commutator subgroup of E. The notations will remain fixed throughout this section.

Proposition 2.1. *The image of the transfer*

$$\text{Tr} : E/E^c \to A$$

is contained in A^G, and one has:
(1) $\text{Tr}(aE^c) = \prod_{\sigma \in G} u_\sigma a u_\sigma^{-1} = \mathbf{N}_G(a)$ *for $a \in A$.*
(2) $\text{Tr}(u_\tau E^c) = \prod_{\sigma \in G} u_\sigma u_\tau u_{\sigma\tau}^{-1} = \prod_{\sigma \in G} a_{\sigma,\tau}$ *(the Nakayama map).*

Proof. These formulas are immediate consequences of the definition of the transfer.

Proposition 2.2. *One has $I_G A \subset E^c \cap A \subset A_N$. For the cup product relative to the pairing $\mathbf{Z} \times A \to A$, we have*

$$\alpha \cup \mathbf{H}^{-3}(G, \mathbf{Z}) = \varkappa_G((E^c \cap A)/I_G A).$$

Proof. We have at once $\sigma a/a = u_\sigma a u_\sigma^{-1} a^{-1} \in E^c \cap A$. The other stated inclusion can be seen from the fact that tr is trivial on E^c, and applying Proposition 2.1. Now for the statement about the cup product, recall that a subgroup of an abelian group is determined by the group of characters $f : A \to \mathbf{Q}/\mathbf{Z}$ vanishing on the subgroup. A character $f : A \to \mathbf{Q}/\mathbf{Z}$ vanishes on $E^c \cap A$ if and only if we can extend f to a character of E/E^c, because

$$A/(E^c \cap A) \subset E/E^c.$$

The extension of a character can be formulated in terms of a commutative diagram such as those we considered previously, and of the existence of a map F, namely:

$$
\begin{array}{ccccc}
A & \longrightarrow & E & \longrightarrow & G \\
\downarrow{\scriptstyle f} & & \downarrow{\scriptstyle F} & & \downarrow \\
\mathbf{Q}/\mathbf{Z} & \longrightarrow & \mathbf{Q}/\mathbf{Z} & \longrightarrow & 0
\end{array}
$$

The existence of F is equivalent to the conditions:

(a) f is a G-homomorphism.
(b) $f_* \alpha = 0$.

From the definition of the cup product, we have a commutative diagram:

$$
\begin{array}{ccc}
\mathbf{H}^{-3}(\mathbf{Z}) \times \mathbf{H}^2(\mathbf{Q}/\mathbf{Z}) & \longrightarrow & \mathbf{H}^{-1}(\mathbf{Q}/\mathbf{Z}) = (\mathbf{Q}/\mathbf{Z})_N \\
\uparrow{\scriptstyle \mathrm{id}} \quad \uparrow{\scriptstyle f_*} & & \uparrow{\scriptstyle f_*} \\
\mathbf{H}^{-3}(\mathbf{Z}) \times \mathbf{H}^2(A) & \longrightarrow & \mathbf{H}^{-1}(A) = A_N/I_G A.
\end{array}
$$

The duality theorem asserts that $\mathbf{H}^{-3}(\mathbf{Z})$ is dual to $\mathbf{H}^2(\mathbf{Q}/\mathbf{Z})$. In addition, the effect of f_* on $\mathbf{H}^{-1}(A)$ is induced by f on $A_N/I_G A$.

Suppose that f is a character of A vanishing on A_N. Then

$$f_*(\alpha \cup \mathbf{H}^{-3}(A)) = 0, \text{ and so } f_*(\alpha) \cup \mathbf{H}^{-3}(\mathbf{Z}) = 0.$$

Since $\mathbf{H}^{-3}(Z)$ is the character group of $\mathbf{H}^2(\mathbf{Q}/\mathbf{Z})$, we conclude that $f_*(\alpha) = 0$. The converse is proved in a similar way. This concludes the proof of Proposition 2.2.

In addition, Proposition 1.1 also gives:

Theorem 2.3. *Let*

$$0 \to A \xrightarrow{i} E \xrightarrow{j} G \to 0$$

be an extension, and $\alpha \in H^2(G, A)$ its cohomology class. Then the following diagram is commutative:

$$
\begin{array}{ccccccccc}
0 & \longrightarrow & A/E^c \cap A & \xrightarrow{\ \bar{i}\ } & E/E^c & \xrightarrow{\ \bar{j}\ } & G/G^c = \mathbf{H}^{-2}(G,\mathbf{Z}) & \longrightarrow & 0 \\
& & \downarrow \bar{\mathrm{N}} & & \downarrow \mathrm{Tr} & & \downarrow \alpha_{-2} & & \\
0 & \longrightarrow & NA & \longrightarrow & A^G & \longrightarrow & \mathbf{H}^0(G,A) & \longrightarrow & 0
\end{array}
$$

where $\bar{\mathrm{N}}, \bar{i}, \bar{j}$ are the homomorphisms induced by the norm, the inclusion, and j respectively; and α_{-2} denotes the cup product with α on $H^{-2}(G, \mathbf{Z})$.

Proof. The left square is commutative because of the formula for the norm in Proposition 2.1(1). The transfer maps E/E^c into A^G by Proposition 2.1. The right square is commutative because the Nakayama map is an explicit determination of the cup product, and we can apply Proposition 2.1(2).

The next two corollaries are especially important in the application to class modules and class formations as in Chapter IX. They give conditions under which the transfer is an isomorphism.

Corollary 2.4. *Let $E/A = G$ be an extension with corresponding cohomology class $\alpha \in H^2(G, A)$. With the three homomorphisms*

$$\mathrm{Tr} : E/E^c \to A^G$$
$$\alpha_{-3} : \mathbf{H}^{-3}(G, \mathbf{Z}) \to \mathbf{H}^{-1}(G, A)$$
$$\alpha_{-2} : \mathbf{H}^{-2}(G, \mathbf{Z}) \to \mathbf{H}^0(G, A)$$

we get an exact sequence

$$0 \to \mathbf{H}^{-1}(G, A)/\mathrm{Im}\ \alpha_{-3} \to \ \mathrm{Ker}\ \mathrm{Tr} \to \ \mathrm{Ker}\ \alpha_{-2} \to 0$$

and an isomorphism

$$0 \to A^G/\operatorname{Im} \operatorname{Tr} \to \mathbf{H}^0(G, A)/\operatorname{Im} \alpha_{-2} \to 0.$$

Proof. Chasing around diagrams.

Corollary 2.5. *If α_{-2} and α_{-3} are isomorphisms, then the transfer on A^G/NA is an isomorphism in Theorem 2.3.*

The situation of Corollary 2.5 is realized for class modules or class formations in Chapter IX.

§3. The deflation

Let G be a group and $A \in \operatorname{Mod}(G)$, written multiplicatively. Let E_G be an extension of A by G. Let N be a normal subgroup of G and $E_N = j^{-1}(N)$, so we have two exact sequences:

$$0 \to A \to E_G \xrightarrow{j} G \to 0$$

$$0 \to A \to E_N \to N \to 0.$$

Then E_N is an extension of A by N, and if $\alpha \in H^2(G, A)$ is the cohomology class of E_G then $\operatorname{res}^G_N(\alpha)$ is the cohomology class of E_N.

One sees that E_N is normal in E_G, and in fact $E_G/E_N \approx G/N$. We obtain an exact sequence

$$0 \to E_N \to E_G \to G/N \to 0.$$

Since E_N is not necessaily commutative, we factor by E_N^c to get the exact sequence

$$0 \to E_N/E_N^c \to E_G/E_N^c \to G/N \to 0$$

giving an extension of E_N/E_N^c by G/N, called the **factor extension** corresponding to the normal subgroup N of G. The group lattice is as follows.

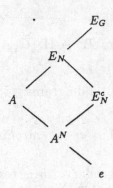

This factor extension corresponds to a cohomology class β in $H^2(G/N, E_N/E_N^c)$. We can take the transfer

$$\mathrm{Tr} : E_N/E_N^c \rightarrow A^N,$$

which is a G/N-homomorphism, the operation of G/N on E_N/E_N^c being compatible with that of E_G/E_N^c. Consequently, there is a induced homomorphism

$$\mathrm{Tr}_* : H^r(G/N, E^N/E_N^c) \rightarrow H^r(G/N, A^N).$$

The image of $\mathrm{Tr}_*(\beta)$ depends only on α. Hence we get a map

$$\mathrm{def} : H^2(G, A) \rightarrow H^2(G/N, A^N) \text{ such that } \alpha \mapsto \mathrm{Tr}_*(\beta).$$

We call this map the **deflation**. It may not be a homomorphism, but we shall see that for G finite, it is. First:

Theorem 3.1. *Let S be a subgroup of a finite group G. Fix right coset representatives of S in G, and for $\sigma \in G$ let $\bar{\sigma}$ be the representative of $S\sigma$. Let $A \in \mathrm{Mod}(G)$ and let $(a_{\sigma, \tau})$ be a 2-cocycle of G in A. Let E_G be the extension of A by S obtained from the restriction of this cocycle to S. Let (u_σ) be representatives of G in E_G. Let $\gamma(\sigma, \tau) = \bar{\sigma}\bar{\tau}\overline{\sigma\tau}^{-1}$. Then*

$$\mathrm{Tr}_A^{E_S}(u_{\bar{\sigma}} u_{\bar{\tau}} u_{\overline{\sigma\tau}}^{-1}) = \prod_{\rho \in S} a_{\rho, \gamma}.$$

Proof. This comes directly from the formulas of Theorem 2.1.

Corollary 3.2. *Let G be a finite group, N normal in G. Let $A \in \mathrm{Mod}(G)$. Then on $H^2(G, A)$, we have*

$$\mathrm{inf}_G^{G/N} \circ \mathrm{def}_{G/N}^G = (N : e).$$

Proof. One computes with the explicit formulas on cocycles. Note that the group A being written multiplicatively, the expression $(N : e)$ on the right is really the map $\alpha \mapsto \alpha^{(N:e)}$ for $\alpha \in H^2(G, A)$.

Theorem 3.3. *Let G be a finite group and N a normal subgroup. Then the deflation is a homomorphism. If $\alpha \in H^2(G, A)$ is represented by the cocycle $(a_{\sigma,\tau})$, then $\mathrm{def}(\alpha)$ is represented by the cocycle*

$$\prod_{\rho \in N} \rho(a_{\bar{\sigma}, \bar{\tau}} a_{\gamma, \overline{\sigma\tau}}^{-1}) \prod_{\rho \in N} a_{\rho, \gamma} = \prod_{\rho \in N} a_{\bar{\sigma}\rho, \bar{\tau}} a_{\rho, \bar{\sigma}} a_{\rho, \overline{\sigma\tau}}^{-1}.$$

As in Theorem 3.1, $\bar{\sigma}$ denotes a fixed coset representative of the coset N^σ, and $\gamma = \bar{\sigma}\bar{\tau}\overline{\sigma\tau}^{-1} = \gamma(\sigma, \tau)$.

Proof. The proof is done by an explicit computation, using the explicit formula for the transfer in Theorem 3.1. The fact that the deflation is a homomorphism is then apparent from the expression on the right side of the equality. One also sees from this right expression that the expression on the left is well defined. The details are left to the reader.

CHAPTER IX

Class Formations

§1. Definitions

Let G be a group of Galois type, with a fundamental system of open neighborhoods of e consisting of open subgroups of finite index $U, V, \ldots$. Let $A \in \mathrm{Galm}(G)$ be a Galois module. We then say that the pair (G, A) is a **class formation** if it satisfies the following two axioms:

CF 1. For each open subgroup V of G one has $H^1(V, A) = 0$.

Because of the inflation-restriction exact sequence in dimension 1, this axiom is equivalent to the condition that for all open subgroups U, V with U normal in V, we have

$$H^1(V/U, A^U) = 0.$$

Example. If k is a field and K is a Galois extension of k with Galois group G, then (G, K^*) satisfies the axiom **CF 1**.

By **CF 1**, it follows that the inflation-restriction sequence is exact in dimension 2, and hence that the inflations

$$\inf : H^2(V/U, A^U) \to H^2(V, A)$$

are monomorphisms for V open, U open and normal in V. We may therefore consider $H^2(V, A)$ as the union of the subgroups

$H^2(V/U, A^U)$. It is by definition the **Brauer group** in the preceding example. The second axiom reads:

CF 2. For each open subgroup V of G we are given an embedding

$$\mathrm{inv}_V : H^2(V, A) \to \mathbf{Q}/\mathbf{Z} \text{ denoted } \alpha \mapsto \mathrm{inv}_V(\alpha),$$

called the **invariant**, satisfying two conditions:

(a) If $U \subset V$ are open and U is normal in V, of index n in V, then inv_V maps $H^2(V/U, A^U)$ onto the subgroup $(\mathbf{Q}/\mathbf{Z})_n$ consisting of the elements of order n in $\mathbf{Q}/\mathbf{Z}$.

(b) If $U \subset V$ are open subgroups with U of index n in V, then

$$\mathrm{inv}_V \circ \mathrm{res}_U^V = n.\mathrm{inv}_V.$$

We note that if $(G : e)$ is divisible by every positive integer m, then inv_G maps $H^2(G, A)$ onto $\mathbf{Q}/\mathbf{Z}$, i.e.

$$\mathrm{inv}_G : H^2(G, A) \to \mathbf{Q}/\mathbf{Z}$$

is an isomorphism. This is the case in both local and global class field theory over number fields:

In the local case, A is the multiplicative group of the algebraic closure of a p-adic field, and G is the Galois group.

In the global case, A is the direct limit of the groups of idele classes. On the other hand, if G is finite, then of course inv_G maps $H^2(G, A)$ only on $(\mathbf{Q}/\mathbf{Z})_n$, with $n = (G : e)$.

Let G be finite, and (G, A) a class formation. Then A is a class module. But for a class formation, we are given an additional structure, namely the specific fundamental elements $\alpha \in H^2(G, A)$ whose invariant is $1/n \pmod{\mathbf{Z}}$.

Let (G, A) be a class formation and $U \subset V$ open subgroups with U normal in V. The element $\alpha \in H^2(V/U, A^U) \subset H^2(V, A)$, whose V-invariant $\mathrm{inv}_V(\alpha)$ is $1/(V : U)$, will be called **the fundamental class** of $H^2(V/U, A^U)$, or by abuse of language, of V/U.

Proposition 1.1. *Let $U \subset V \subset W$ be three open subgroups of G, with U normal in W. If α is the fundamental class of W/U then $\mathrm{res}_V^W(\alpha)$ is the fundamental class of V/U.*

Proof. This is immediate from **CF 2(b)**.

Corollary 1.2. *Let (G, A) be a class formation.*

(i) *Let V be an open subgroup of G. Then (V, A) is a class formation, and the restriction*
$$\mathrm{res} : H^2(G, A) \to H^2(V, A)$$
is surjective.

(ii) *Let N be a closed normal subgroup. Then $(G/N, A^N)$ is a class formation, if we define the invariant of an element in $H^2(VN/N, A^N)$ to be the invariant of its inflation in $H^2(VN, A)$.*

Proof. Immediate.

Proposition 1.3. *Let (G, A) be a class formation and let V be an open subgroup of G. Then:*

(i) *The transfer preserves invariants, that is for $\alpha \in H^2(V, A)$ we have*
$$\mathrm{inv}_G \, \mathrm{tr}_G^V(\alpha) = \mathrm{inv}_V(\alpha).$$

(ii) *Conjugation preserves invariants, that is for $\alpha \in H^2(V, A)$ we have*
$$\mathrm{inv}_{V[\sigma]}(\sigma_* \alpha) = \mathrm{inv}_V(\alpha)$$

Proof. Since the restriction is surjective, the first assertion follows at once from **CF 2** and the formula
$$\mathrm{tr} \circ \mathrm{res} = (G : V).$$
As for the second, we recall that σ_* is the identity on $H^2(G, A)$. Hence
$$\mathrm{inv}_{V[\sigma]} \circ \sigma_* \circ \mathrm{res}_V^G = \mathrm{inv}_{V[\sigma]} \mathrm{res}_{V[\sigma]}^G \circ \sigma_*$$
$$= (G : V[\sigma]) \mathrm{inv}_G \circ \sigma_*$$
$$= (G : V) \mathrm{inv}_G$$
$$= \mathrm{inv}_V \mathrm{res}_V^G.$$

Since the restriction is surjective, the proposition follows.

Theorem 1.4. *Let G be a finite group and (G, A) a class formation. Let α be the fundamental element of $H^2(G, A)$. Then the cup product*

$$\alpha_r : \mathbf{H}^r(G, \mathbf{Z}) \longrightarrow \mathbf{H}^{r+2}(G, A)$$

is an isomorphism for all $r \in \mathbf{Z}$.

Proof. For each subgroup G' of G let α' be the retriction to G', and let α'_r the cup product taken on the G'-cohomology. By the triplets theorem, it will suffice to prove that α'_r satisfies the hypotheses of this theorem in three successive dimensions, which we choose to be dimensions $-1, 0$, and $+1$.

For $r = -1$, we have $H^1(G, A) = 0$ so α'_{-1} is surjective.

For $r = 0$, we note that $\mathbf{H}^0(G', \mathbf{Z})$ has order $(G' : e)$ which is the same order as $H^2(G', A)$. We have trivially

$$\alpha'_0(\varkappa(1)) = \alpha',$$

which shows that α'_0 is an isomorphism.

For $r = 1$, we simply note that $H^1(G, \mathbf{Z}) = 0$ since G is finite and the action on $\mathbf{Z}$ is trivial. This concludes the proof of the theorem.

Next we make explicit some commutativity relations for restriction, transfer, inflation and conjugation relative to the natural isomorphism of $\mathbf{H}^r(G, A)$ with $\mathbf{H}^{r-2}(G, \mathbf{Z})$, cupping with α.

Proposition 1.5. *Let G be a finite group and (G, A) a class formation. let $\alpha \in H^2(G, A)$ be a fundamental element and α' its restriction to G' for a subgroup G' of G. Then for each pair of vertical arrows pointing in the same direction, the following diagram is commutative.*

$$
\begin{array}{ccc}
\mathbf{H}^r(G, \mathbf{Z}) & \xrightarrow{\ \alpha_r\ } & \mathbf{H}^{r+2}(G, A) \\[2pt]
{\scriptstyle\text{res}}\big\downarrow\big\uparrow{\scriptstyle\text{tr}} & & {\scriptstyle\text{res}}\big\downarrow\big\uparrow{\scriptstyle\text{tr}} \\[2pt]
\mathbf{H}^r(G', \mathbf{Z}) & \xrightarrow[\ \alpha'_r\ ]{} & \mathbf{H}^{r+2}(G', A)
\end{array}
$$

Proof. This is just a special case of the general commutativity relations.

Proposition 1.6. *Let G be a finite group and (G, A) a class formation. Let U be normal in G. Let $\alpha \in H^2(G, A)$ be the fundamental element, and $\bar{\alpha}$ the fundamental element for G/U. Then the following diagram is commutative for $r \geq 0$.*

$$
\begin{array}{ccc}
H^r(G/U, \mathbf{Z}) & \xrightarrow{\ \bar{\alpha}_r\ } & H^{r+2}(G/U, A^U) \\[2mm]
{\scriptstyle (U:e)\mathrm{inf}}\big\uparrow & & \big\uparrow {\scriptstyle \mathrm{inf}} \\[2mm]
H^r(G, \mathbf{Z}) & \xrightarrow[\ \alpha_r\ ]{} & H^{r+2}(G, A)
\end{array}
$$

Proof. This is just a special case of the rule

$$
\mathrm{inf}(\alpha \cup \beta) = \mathrm{inf}(\alpha) \cup \mathrm{inf}(\beta).
$$

We have to observe that we deal with the ordinary functor H in dimension $r \geq 0$, differing from the special one only in dimension 0, because the inflation is defined only in this case. The left homomorphism is $(U : e)\mathrm{inf}$ for the inflation, given by the inclusion. Indeed, we have

$$
(G : e) = (G : U)(U : e),
$$

so $\mathrm{inf}(\bar{\alpha}) = (U : e)\alpha$, and we can apply the above rule.

Finally, we consider some isomorphisms of class formations. Let (G, A) and (G', A') be class formations. An **isomorphism**

$$
(\lambda, f) : (G', A') \to (G, A)
$$

consists of a pair isomorphism $\lambda : G \to G'$ and $f : A' \to A$ such that

$$
\mathrm{inv}_G(\lambda, f)_*(\alpha') = \mathrm{inv}_{G'}(\alpha') \text{ for } \alpha' \in H^2(G', A').
$$

From such an isomorphism, we obtain a commutative diagram for U normal in V (subgroups of G):

$$
\begin{array}{ccc}
\mathbf{H}^r(V/U, \mathbf{Z}) & \xrightarrow{\ \alpha_r\ } & \mathbf{H}^{r+2}(V/U, A^U) \\[2mm]
{\scriptstyle (\lambda, 1)_*}\big\downarrow & & \big\downarrow {\scriptstyle (\lambda, f)_*} \\[2mm]
\mathbf{H}^r(\lambda V/\lambda U, \mathbf{Z}) & \xrightarrow[\ \alpha'_r\ ]{} & \mathbf{H}^{r+2}(\lambda V/\lambda U, A'^{\lambda U})
\end{array}
$$

where α, α' denote the fundamental elements in their respective H^2.

Conjugation is a special case, made explicit in the next proposition.

Proposition 1.7. *Let (G, A) be a class formation, and $U \subset V$ two open subgroups with U normal in V. Let $\tau \in G$ and α the fundamental element in $H^2(V/U, A^U)$. Then the following diagram is commutative.*

$$
\begin{array}{ccc}
H^r(V/U, \mathbf{Z}) & \xrightarrow{\ \alpha_r\ } & H^{r+2}(V/U, A^U) \\
\tau_* \downarrow & & \downarrow \tau_* \\
H^r(V[\tau]/U[\tau], \mathbf{Z}) & \xrightarrow[\tau_* \alpha_r]{} & H^{r+2}(V[\tau]/U[\tau], A^{U[\tau]}).
\end{array}
$$

§2. The reciprocity homomorphism

We return to Theorem 8.7 of Chapter IV, but with the additional structure of the class formation. From that theorem, we know that if (G, A) is a class formation and G is finite, then G/G^c is isomorphic to $A^G/S_G A = \mathbf{H}^0(G, A)$. The isomorphism can be realized in two ways. First, directly, and second by duality. Here we start with the duality. We have a bilinear map

$$
A^G \times \hat{G} \to H^2(G, A)
$$

given by

$$
(a, \chi) \mapsto \varkappa(a) \cup \delta\chi.
$$

Following this with the invariant, we obtain a bilinear map

$$
(a, \chi) \mapsto \operatorname{inv}_G(\varkappa(a) \cup \delta\chi) \text{ of } A^G \times \hat{G} \to \mathbf{Q}/\mathbf{Z},
$$

whose kernel on the left is $S_G A$ and whose kernel on the right is trivial. Hence $A^G/S_G A \approx G/G^c$, both groups being dual to $\hat{G}$. We recall the commutative diagram:

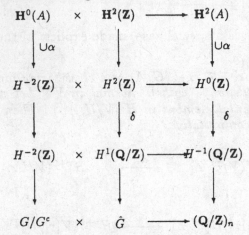

which we apply to the fundamental cocycle $\alpha \in H^2(G, A)$, with $n = (G : e)$ and $\mathrm{inv}_G(\alpha) = 1/n$. We have

$$\varkappa(1) \cup \alpha = \alpha \text{ and } \mathrm{inv}_G(\varkappa(1) \cup \alpha) = \mathrm{inv}_G(\alpha) = 1/n.$$

Thus using the invariant from a class formation, at a finite level, we obtain the following fundamental result.

Theorem 2.1. *Let G be a finite group and (G, A) a class formation. For $a \in A^G$ let σ_a be the element of G/G^c corresponding to a under the above isomorphism. Then for all characters χ of G we have*
$$\chi(\sigma_a) = \mathrm{inv}_G(\varkappa(a) \cup \delta\chi).$$

An element $\sigma \in G$ is equal to σ_a if and only if for all characters χ,
$$\chi(\sigma) = \mathrm{inv}_G(\varkappa(a) \cup \delta\chi).$$

The map $a \mapsto \sigma_a$ induces an isomorphism $A^G/S_G A \approx G/G^c$.

The element σ_a in the theorem will also be denoted by (a, G).

Let now G be of Galois type and let (G, A) be a class formation. Then we have a bilinear map

$$H^0(G, A) \times H^1(G, \mathbf{Q}/\mathbf{Z}) \to H^2(G, A), \text{i.e.} A^G \times \hat{G} \to H^2(G, A)$$

with the ordinary functor H^0, by the formula

$$(a, \chi) \mapsto a \cup \delta\chi,$$

where we identify a character χ with the corresponding element of $H^1(G, \mathbf{Q}/\mathbf{Z})$, and we identify $H^0(G, A)$ with A^G. Since inflation commutes with the cup product, we see that if U is normal open in G, then the following diagram is commutative:

$$
\begin{array}{ccccc}
H^0(G, A) & \times & H^1(G, \mathbf{Q}/\mathbf{Z}) & \longrightarrow & H^2(G, A) \\
\uparrow{\scriptstyle\text{inf}} & & \uparrow{\scriptstyle\text{inf}} & & \uparrow{\scriptstyle\text{inf}} \\
H^0(G/U, A^U) & \times & H^1(G/U, \mathbf{Q}/\mathbf{Z}) & \longrightarrow & H^2(G/U, A^U)
\end{array}
$$

The inflation on the far left is simple the inclusion $A^U \subset A$, and the inflation in the middle is that of characters.

In particular, to each element $a \in A^G$ we obtain a character of $H^1(G, \mathbf{Q}/\mathbf{Z})$ given by

$$\chi \mapsto \operatorname{inv}_G(a \cup \delta\chi).$$

We consider $H^1(G, \mathbf{Q}/\mathbf{Z})$ as a discrete group. Its character group is G/G^c, according to Pontrjagin duality between discrete and compact groups, but G^c now denotes the closure of the commutator group. Thus we obtain a homomorphism

$$\operatorname{rec}_G : A^G \to G/G^c$$

which we call the **reciprocity homomorphism**, characterized by the property that for U open normal in G, and $a \in A^G$ we have

$$\operatorname{rec}_{G/U}(a) = (a, G/U) = (a, G/G^c U).$$

Similarly, we may replace U be any normal closed subgroup of G. This is called the **consistency** of the reciprocity mapping. As when G is finite, we denote

$$\operatorname{rec}_G(a) = (a, G).$$

The next theorem is merely a formal summary of what precedes for finite factor groups, and the consistency.

Theorem 2.2. *Let G be a group of Galois type and (G, A) a class formation. Then there exists a unique homomorphism*

$$\mathrm{rec}_G : A^G \to G/G^c \quad \text{denoted} \quad a \mapsto (a, G)$$

satisfying the property

$$\mathrm{inv}_G(a \cup \delta\chi) = \chi(a, G)$$

for all characters χ of G.

Recall that if $\lambda : G_1 \to G_2$ is a group homomorphism, then λ induces a homomorphism

$$\lambda^c : G_1/G_1^c \to G_2/G_2^c.$$

This also holds for a continuous homomorphism of groups of Galois type, where G^c denotes the closure of the commutator group.

The next theorem summarizes the formalism of class formation theory and the reciprocity mapping.

Theorem 2.3. *Let G be a group of Galois type and (G, A) a class formation.*

(i) *If $a \in A^G$ and S is a closed normal subgroup with factor group $\lambda : G \to G/S$, then $\mathrm{rec}_{G/S} = \lambda^c \circ \mathrm{rec}_G$, that is*

$$(a, G/T) = \lambda^c(a, G).$$

(ii) *Let V be an open subgroup of G. Then $\mathrm{rec}_V = \mathrm{Tr}_V^G \circ \mathrm{rec}_G$, that is for $a \in A^G$,*

$$(a, V) = \mathrm{Tr}_V^G(a, G).$$

(iii) *Again let V be an open subgroup of G and let $\lambda : V \to G$ be the inclusion. Then $\mathrm{rec}_G \circ \mathbf{S}_G^V = \lambda^c \circ \mathrm{rec}_V$, that is for $a \in A^V$,*

$$(\mathbf{S}_G^V(a), G) = \lambda^c(a, V).$$

(iv) *Let V be an open subgroup of G and $a \in A^V$. Let $\tau \in G$. Then*

$$(\tau a, V^\tau) = (a, V)^\tau.$$

These properties are called respectively **consistency, transfer, translation**, and **conjugation** for the reciprocity mapping.

Proof. The consistency property is just the commutativity of inflation and cup product. We already used it when we defined the symbol (a, G) for G of Galois type. The other properties are proved by reducing them to the case when G is finite. For instance, let us consider (ii). To show that two elements of V/V^c are equal, it suffices to prove that for every character $\chi : V \to \mathbf{Q}/\mathbf{Z}$ the values of χ on these two elements are equal. To do this, there exists an open normal subgroup U of G with $U \subset V$ such that $\chi(U) = 0$. Let $\bar{G} = G/U$. Then the following diagram is commutative:

$$
\begin{array}{ccccc}
G/G^c & \xrightarrow{\mathrm{Tr}} & V/V^c & \xrightarrow{\chi} & \mathbf{Q}/\mathbf{Z} \\
\downarrow & & \downarrow & & \downarrow \\
\bar{G}/\bar{G}^c & \xrightarrow{\mathrm{Tr}} & \bar{V}/\bar{V}^c & \xrightarrow{\chi} & \mathbf{Q}/\mathbf{Z}
\end{array}
$$

the vertical maps being canonical. Furthermore, by consistency, we have
$$(a, G)U = (a, G/U) = (a, \bar{G}).$$
This reduces the property to the finite case $\bar{G}$.

But when G is finite, then we can also write
$$(a, G) = \sigma \Leftrightarrow \varkappa_G(a) = \zeta_\sigma \cup \alpha,$$
where ζ_σ is the element of $\mathbf{H}^{-2}(G, \mathbf{Z}) \approx G/G^c$ corresponding to σ, and α is the fundamental class. The restriction $\mathrm{res}^G_V(\alpha) = \alpha'$ is the fundamental class of $H^2(V, A)$, and we know that
$$\zeta_{\mathrm{Tr}(\sigma)} \cup \alpha' = \mathrm{res}^G_V(\zeta_\sigma \cup \alpha).$$
Since
$$\mathrm{res}^G_V(\varkappa_G(a)) = \varkappa_V(a),$$
one sees that $\mathrm{Tr}(a, G) = (a, V)$.

For (iii), note that the diagram is commutative,

$$
\begin{array}{ccccc}
V/V^c & \longrightarrow & G/G^c & \xrightarrow{\chi} & \mathbf{Q}/\mathbf{Z} \\
\downarrow & & \downarrow & & \downarrow \\
\bar{V}/\bar{V}^c & \longrightarrow & \bar{G}/\bar{G}^c & \xrightarrow{\chi} & \mathbf{Q}\mathbf{Z}
\end{array}
$$

where as previously $U \subset V$ is normal in G and $\bar{G} = G/U, \bar{V} = V/U$. This reduces the property to the case when G is finite. In this case, let

$$\lambda^c : V/V^c \to G/G^c$$

be the homomorphism induced by inclusion. Let α be the fundamental class of (G, A). Then $\operatorname{res}_V^G(\alpha) = \alpha'$ is the fundamental class of (V, A). The transfer and cup product are related by the formula

$$\operatorname{tr}(\zeta_\tau \cup \alpha') \doteq \zeta_{\lambda\tau} \cup \alpha.$$

But the transfer amounts to the trace on $H^0(V, A) = A^V$, so the assertion is proved.

The fourth property is just a transport of structure for algebraically defined notions and relations.

We state one more property somewhat different from the others.

Theorem 2.4. Limitation Theorem. *Let G be of Galois type, V an open subgroup, and (G, A) a class formation. Then the image of $S_G^V(A^V)$ by the reciprocity mapping rec_G is contained in VG^c/G^c, and we have an isomorphism induced by rec_G, namely*

$$\operatorname{rec}_G : A^G/S_G^V A^V \xrightarrow{\approx} G/VG^c.$$

Proof. The first assertion is Property (iii) of Theorem 2.3. Conversely, since V is open, we may assume without loss of generality that G is finite. In this case, there exists $b \in A^V$ such that $\lambda^c(b, V) = (a, G)$. By this same Property (iii), this is equal to $(S_G^V(b), G)$. But we know that the kernel of rec_G is equal to $S_G A$. Hence a and $S_G^V(b)$ are congruent mod $S_G(A)$. Since

$$S_G(A) \subset S_G^V(A),$$

we have proved the theorem.

Corollary 2.5. *Let G be finite and (G, A) a class formation. Let $G' = G/G^c$ and $A' = A^{G^c}$. Then $S_G(A) = S_{G'}(A')$ and $\operatorname{rec}_G, \operatorname{rec}_{G'}$ are equal, their kernels being $S_G(A)$.*

Note that $G' = G/G^c$ can be written G^{ab}, and can be viewed as the maximal abelian quotient of G in Corollary 2.5. The corollary shows that the information in the reciprocity mapping is entirely concerned with this maximal abelian quotient.

Theorem 2.6. *Let G be abelian of Galois type. Let (G, A) be a class formation. Then the open subgroups V of G are in bijection with the subgroups of A of the form $\mathbf{S}_G^V(A^V)$, called the **trace group**. If we denote this subgroup by B_V, and U is an open subgroup of G, then $U \subset V$ if and only if $B_V \subset B_U$, and $B_{UV} = B_U \cap B_V$. If in addition B is a subgroup of A^G such that $B \supset B_V$ for some open subgroup V of G, then there exists U open subgroup of G such that $B = B_U$.*

Proof. All the assertions are special cases of what has previously been proved, except possibly for the last one. But for this one, one may suppose G finite and consider $(G/V, A^V)$ instead of (G, A). We let $U = \mathrm{rec}_G(B)$, and we find an isomorphism $B/\mathbf{S}_G(A) \approx U$, to which we apply Theorem 2.4 to conclude the proof.

A subgroup B of A^G will be called **admissible** if there exists V open subgroup of G such that $B = \mathbf{S}_B^V(A^G)$. We then write $B = B_V$. The next reuslt is an immediate consequence of Theorem 2.6 and the basic properties of the reciprocity map.

Corollary 2.7. *Let G be a group of Galois type and (G, A) a class formation. Let $B \subset A^G$ be admissible, $B = B_U$, and suppose U normal, G/U abelian. Let V be an open subgroup of G and put*

$$C = (\mathbf{S}_G^V)^{-1}(B),$$

so C is a subgroup of A^V. Then C is admissible for the class formation (V, A), and C corresponds to the subgroup $U \cap V$ of V.

In the next section, we discuss in greater detail the relations between class formations and group extensions. However, we can already formulate the theorem of Shafarevich-Weil. Note that if G is of Galois type and U is open normal in G, then U/U^c is a Galois module for G, or in other words, U^c is normal in G. Consequently, G/U acts on U/U^c, and we obtain a group extension

(1) $0 \to U/U^c \to G/U^c \to G/U \to 0.$

If in addition (G, A) is a class formation, then the reciprocity mapping

$$\mathrm{rec}_U : A^U \to U/U^c$$

is a G/U-homomorphism.

Theorem 2.8 (Shafarevich-Weil). *Let G be of Galois type, U open normal in G, and (G, A) a class formation. Then*

$$\text{rec}_{U^*} : H^2(G/U, A^U) \to H^2(G/U, U/U^c)$$

maps the fundamental class on the class of the group extension (1). There exists a family of coset representatives $(\bar{\sigma})_{\sigma \in G}$ of U in G such that if $a_{\bar{\sigma}, \bar{\tau}}$ is a cocycle representing α, then

$$(a_{\bar{\sigma}, \bar{\tau}}, U) = \bar{\sigma} \bar{\tau} \overline{\sigma\tau}^{-1} U^c.$$

Proof. Let $V \subset U$ be open normal in G. Ultimately, we let V tend to e. By the deflation operation of Chapter VIII, Theorem 3.2, there exists a cocycle $b_{\bar{\sigma}, \bar{\tau}}$ representing the fundamental class of $H^2(G/V, A^V)$ and representatives $\bar{\sigma}$ of U/V such that

$$(2) \qquad a_{\bar{\sigma}, \bar{\tau}} = \mathbf{S}_{U/V}(b_{\bar{\sigma}, \bar{\tau}} / b_{\gamma(\sigma, \tau), \overline{\sigma\tau}}) \prod_{\rho \in U/V} b_{\rho, \gamma(\sigma, \tau)},$$

where $\gamma(\sigma, \tau) = \bar{\sigma} \bar{\tau} \overline{\sigma\tau}^{\ -1}$. Therefore, we find

$$(a_{\bar{\sigma}, \bar{\tau}} U/V) = \bar{\sigma} \bar{\tau} \overline{\sigma\tau}^{\ -1} V U^c.$$

We take a limit over V as follows. let $C_1, \dots, C_m$ be the cosets of U in G. They are closed and compact. The product space $(C_1, \dots, C_m)$ is compact. If $\bar{\sigma}_1, \dots, \bar{\sigma}_m$ are representatives of U/V satisfying (2), then any representatives of the cosets $\bar{\sigma}_1 V, \dots, \bar{\sigma}_m V$ will also satisfy (2). The subset $(\bar{\sigma}_1 V, \dots \bar{\sigma}_m V)$ is closed in $(C_1, \dots, C_m)$. From the consistency of the reciprocity map, these subsets have the finite intersection property. Hence their intersection taken over all V is not empty, and there exists representatives $\bar{\sigma}$ of the cosets of U in G which all give the same $a_{\bar{\sigma}, \bar{\tau}}$. The theorem is now clear.

§3. Weil groups

Let G be a group of Galois type and (G, A) a class formation. At the end of the preceding section, we saw the exact sequence

$$0 \to U/U^c \to G/U^c \to G/U \to 0$$

for every open subgroup U normal in G. Furthermore, U/U^c is isomorphic to the factor group $A^U/S_G^U(A)$. We now seek an extension X of A^U by G/U and a commutative diagram

$$
\begin{array}{ccccccccc}
0 & \longrightarrow & A^u & \longrightarrow & X & \longrightarrow & G/U & \longrightarrow & 0 \\
& & \mathrm{rec}_U \downarrow & & \downarrow & & \downarrow \mathrm{id} & & \\
0 & \longrightarrow & U/U^c & \longrightarrow & G/U^c & \longrightarrow & G/U & \longrightarrow & 0
\end{array}
$$

satisfying various properties made explicit below. The problem will be solved in the following discussion.

We start first with the finite case, so let G be finite. By a **Weil group** for (G, A) we mean a triple $(E, g, \{f_U\})$, consisting of a group E and a surjective homomorphism

$$
E \xrightarrow{g} G \rightarrow 0
$$

(so a group extension) such that, if we put $E_U = g^{-1}(U)$ for U an open subgroup of G (and so $E = E_G$), then f_U is an isomorphism

$$
f_U : A^U \rightarrow E_U/E_U^c.
$$

These data are assumed to satisfy four axioms:

W 1. For each pair of open subgroups $U \subset V$ of G, the following diagram is commutative:

$$
\begin{array}{ccc}
A^U & \xrightarrow{\ f_U\ } & E_U/E_U^c \\
\mathrm{inc} \uparrow & & \uparrow \mathrm{Tr} \\
A^V & \xrightarrow[\ f_V\ ]{} & E_V/E_V^c
\end{array}
$$

W 2. For $x \in E_G$ and every open subgroup U of G, the diagram is commutative:

$$
\begin{array}{ccc}
A^U & \xrightarrow{\ f_U\ } & E_U/E_U^c \\
\downarrow & & \downarrow \\
A^{U[x]} & \xrightarrow[\ f_{U[x]}\ ]{} & E_{U[x]}/E_{U[x]}^c
\end{array}
$$

The vertical isomorphisms are the natural ones arising from x.

Let $U \subset V$ be open subgroups of G, and U normal in V. Then we have a canonical isomorphism

$$E_V/E_U \approx V/U$$

and an exact sequence

(3) $$0 \to E_U/E_U^c \to E_V/E_U^c \to V/U \to 0.$$

Then V/U acts on E_U/E_U^c and **W 2** guarantees that f_U is a G/U-isomorphism. This being the case we can formulate the third axiom.

W 3. Let $f_{U_*} : H^2(V/U, A^U) \to H^2(V/U, E_U/E_U^c)$ be the induced homomorphism. Then the image of the fundamental class of $(V/U, A^U)$ is the class corresponding to the group extension defined by the exact sequence (3).

Finally we have a separation condition.

W 4. One has $E_e^c = e$, in other words the map $f_e : A \to E_e$ is an isomorphism.

Theorem 3.1. *Let G be a finite group and (G, A) a class formation. Then there exists a Weil group for (G, A). Its uniqueness will be described in the subsequent theorem.*

Proof. Let E_G be an extension of A by G,

$$0 \to A \to E_G \xrightarrow{g} G \to 0$$

corresponding to the fundamental class in $H^2(G, A)$. This extension is uniquely determined up to inner automorphisms by elements of A, because $H^1(G, A)$ is trivial (Corollary 1.4 of Chapter VIII), and we have an isomorphism

$$f_e : A \to E_e,$$

so **W 4** is satisfied.

For each $U \subset G$, we let $E_U = g^{-1}(U)$, the extreme cases being given by A and E_G. Thus we have an exact sequence

$$0 \to A \to E_U \to U \to 0$$

of subextension, and its class in $H^2(U, A)$ is the restriction of the fundamental class, i.e. it is a fundamental class for (U, A).

Consequently, if U is normal in G, we may form the factor extension

$$0 \to E_U/E_U^c \to E_G/E_U^c \to G/U \to 0.$$

By Corollary 2.5 of Chapter VIII, we know that the transfer

$$\text{Tr} : E_U/E_U^c \to A^U$$

is an isomorphism, and one sees at once that it is a G/U-isomorphism. Its inverse gives us the desired map

$$f_U : A^U \to E_U/E_U^c.$$

It is now easy to verify that the objects $(E_G, g, \{f_U\})$ as defined above form a Weil group. The Axioms **W 1, W 2, W 4** are immediate, taking into account the transitivity of the transfer and its functoriality. For **W 3**, we have to consider the deflation. In light of the "functorial" definition of E_G, one may suppose that $V = G$ in axiom **W 3**. If α is the fundamental class in $H^2(G, A)$, then $(U : e)\alpha$ is the inflation of the fundamental class in $(G/U, A^U)$. By Corollary 3.2 of Chapter VIII, one sees that the deflation of the fundamental class of (G, A) to $(G/U, A^U)$ is the fundamental class of $(G/U, A^U)$. Since f_U is the inverse of the transfer, one sees from the definition of the deflation that axiom **W 3** is satisfied. This concludes the proof of existence.

We now consider the uniqueness of a Weil group. Suppose G finite, and let (G, A) be a class formation. let $(E, g, \{f_U'\})$ be two Weil groups. An **isomorphism** φ of the first on the second is a group isomorphism

$$\varphi : E \to E'$$

satisfying the following conditions:

ISOW 1. The diagram is commutative:

$$\begin{array}{ccc} E & \xrightarrow{\ g\ } & G \\ {\scriptstyle \varphi}\big\downarrow & & \big\downarrow{\scriptstyle \text{id}} \\ E' & \longrightarrow & G. \end{array}$$

From **ISOW 1** we see that $\varphi(E_U) = E'_U$ for all open subgroups U, whence an isomorphism

$$\varphi^c_U : E_U/E^c_U \longrightarrow E'_U/E'^c_U.$$

The second condition then reads:

ISOW 2. The diagram

$$
\begin{array}{ccc}
A^U & \xrightarrow{\ f_U\ } & E_U/E^c_U \\
{\scriptstyle\text{id}}\downarrow & & \downarrow{\scriptstyle\varphi^c_U} \\
A^U & \xrightarrow[\ f'_U\ ]{} & E'_U/E'^c_U
\end{array}
$$

is commutative for all open subgroups U of G.

Theorem 3.2. *Let G be a finite group. Two Weil groups associated to a class formation (G, A) are isomorphic. Such an isomorphism is uniquely determined up to an inner automorphism of E' by elements of E'_e.*

Proof. Let φ be an isomorphism. The following diagram is commutative by definition.

$$
\begin{array}{ccccccccc}
0 & \longrightarrow & A & \xrightarrow[\approx]{f_e} & E_e & \longrightarrow & E & \longrightarrow & G & \longrightarrow & 0 \\
& & {\scriptstyle\text{id}}\downarrow & & \downarrow & & \downarrow{\scriptstyle\varphi} & & \downarrow & & \\
0 & \longrightarrow & A & \xrightarrow[f'_e]{\approx} & E'_e & \longrightarrow & E' & \longrightarrow & G & \longrightarrow & 0.
\end{array}
$$

Conversely, we claim that any homomorphism φ which makes this diagram commutative is an isomorphism of Weil groups. Indeed, the exactness of the sequences shows that φ is a group isomorphism of E on E', and $\varphi(E_U) = E'_U$ for all subgroups U of G. Hence φ induces an isomorphism

$$\varphi^c_U : E_U/E^c_U \longrightarrow E'_U/E'^c_U.$$

We consider the cube:

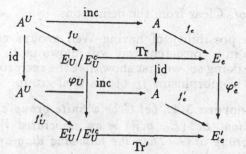

The top and bottom squares are commutative by **ISOW 1**. The back square is clearly commutative. The front face is commutative because the transfer is functorial. The square on the right is commutative because of the commutative diagram in Theorem 3.2. Hence the left square is commutative because the horizontal morphisms are injective.

Thus the study of a Weil isomorphism is reduced to the study of φ in the diagram. Such φ always exists since the group extensions have the same cohomology class. Uniqueness follows from the fact that $H^1(G, A) = 0$, using Theorem 1.3 of Chapter VIII, which was put there for the present purpose.

We already know that a class formation gives rise to others by restriction or deflation with respect to a normal subgroup. Similarly, a Weil group for (G, A) gives rise to Weil groups at intermediate levels as follows.

Theorem 3.3. *Let (G, A) be a class formation, and suppose G finite. Let (E_G, g_G, F_G) be the corresponding Weil group, where $\mathfrak{F}_G$ is the family of isomorphisms $\{f_U\}$ for subgroups U of G. Let V be a subgroup of G. Let*

$$E_V = g_G^{-1}(V) \text{ and } g_V = \text{restriction of } g_G \text{ to } V;$$
$$\mathfrak{F}_V = \text{subfamily of } \mathfrak{F}_G \text{ consisting of those } f_U \text{ such that}$$
$$U \subset V.$$

Then:

(i) *$(E_V, g_V, \mathfrak{F}_V)$ is a Weil group associated to (V, A).*
(ii) *If V is normal in G, then $(E_G/E_V^c, \bar{g}_G, \bar{\mathfrak{F}}_G)$ is a Weil group associated with $(G/V, A^V)$, the family $\bar{\mathfrak{F}}_G$ consisting of the isomorphism*

$$f_U : A^U \to E_U/E_U^c \approx (E_U/E_V^c)/(E_U^c/E_V^c),$$

where U ranges over the subgroups of G containing V.

184

Proof. Clear from the definitions.

The possibility of having Weil groups associated with factor groups in a consistent way will allows us to take an inverse limit. Before doing so, we first show that the reciprocity maps are induced by the isomorphisms f_U of the Weil group.

Theorem 3.4. *Let G be a finite group and (G, A) a class formation. Let $(E_G, g, \mathfrak{F})$ be an associated Weil group. Let V be a subgroup of G. Then the following diagram is commutative.*

$$\begin{array}{ccc} A^V & \xrightarrow{f_V} & E_V/E_V^c \\ s_G^V \downarrow & & \downarrow \text{inc} \\ A^G & \xrightarrow{f_G} & E_G/E_G^c \end{array}$$

Proof. Since we have not assumed that V is normal in G, we have to reduce the proof to this special case by means of a cube:

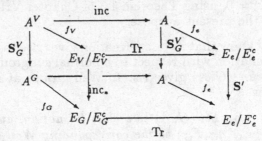

The vertical arrow $\mathbf{S}_G^V$ on the back face is defined by means of representatives of cosets of V in G. The front vertical arrow $\mathbf{S}'$ is defined to make the right face commutative. In other words, we lift these representatives in E_G be means of g^{-1}. Thus if $G = \bigcup \sigma_i V$ we choose $u_i \in E_G$ such that

$$g(\sigma_i) = u_i$$

and we define

$$g'(x) = \prod_i x^{u_i} (\mathrm{mod}\ E_e^c).$$

We note that $E_G = \bigcup u_i E_V$, in other words that the u_i represent the cosets of E_V in E_G. Then the front face is commutative, that is

$$\mathbf{S}'(\mathrm{Tr}'(u)) = \mathrm{Tr}(\mathrm{inc}_*(u)) \text{ for } u \in E_V/E_V^c,$$

immediately from the definition of the transfer. It then follows that the left face is commutative, thus finishing the proof.

Corollary 3.5. *Let G be finite and (G, A) a class formation. Let $(E_G, g, \mathfrak{F})$ be an associated Weil group. If $U \subset V$ are subgroups of G, then f_U and f_V induce isomorphisms:*

$$A^V / S_V^U(A^U) \approx E_V / E_U E_V^c \quad \text{and} \quad (S_V^U)^{-1}(e) \approx (E_U \cap E_V^c) E_U^c.$$

If U is normal in V, then the first isomorphism is the reciprocity mapping, taking into account the isomorphism $E_V / E_U \approx V / U$.

Note that Corollary 3.5 is essentially the same result as Theorem 2.8. The proof of Corollary 3.5 is done by expliciting the transfer in terms of the Nakayama map, and the details are left to the reader.

In practice, in the context of class field theory, the group A has a topology (idele classes globally or multiplicative group of a local field locally). We shall now sketch the procedure which axiomatizes this topology, and allows us to take an inverse limit of Weil groups.

Let G be a group of Galois type and $A \in \text{Galm}(G)$. We say that A is a **topological Galois module** if the following conditions are satisfied:

TOP 1. Each A^U (for U open subgoup of G) is a topological group, and if $U \subset V$, the topology of A^V is induced by the topology of A^U.

TOP 2. The group G acts continuously on A and for each $\sigma \in G$, the natural map $A^U \to A^{U[\sigma]}$ is a topological isomorphism.

Note that if $U \subset V$, it follows that the trace $S_V^U : A^U \to A^V$ is continuous.

Let G be of Galois type and $A \in \text{Galm}(G)$ topological. If (G, A) is a class formation, we then say it is a **topological class formation**. By a **Weil group associated to such a topological class formation**, we mean a triplet $(E_G, g, \mathfrak{F})$ consisting of a topological group E_G, a morphism $g : E_G \to G$ in the category of topological groups (i.e. a continuous homomorphism) whose image is dense in G (so that for each open subgroups $V \supset U$ with U normal in G we have an isomorphism $E_V / E_U \approx V / U$), and a family of topological isomorphisms

$$f_U : A^U \to E_U / E_U^c$$

(where E_U^c is the closure of the commutator group), satisfying the following four axioms.

WT 1. For each pair of open subgroups $U \subset V$ of G, the following diagram is commutative:

$$
\begin{array}{ccc}
A^U & \xrightarrow{\;f_U\;} & E_U/E_U^c \\
{\scriptstyle \text{inc}}\downarrow & & \downarrow{\scriptstyle \text{Tr}} \\
A^V & \xrightarrow[\;f_V\;]{} & E_V/E_V^c
\end{array}
$$

Note that the transfer on the right makes sense, because it extends continuously to the closure of the commutator subgroups.

WT 2. Let $x \in E_G$ be such that $\sigma = g(x)$. Then for all open subgroups U of G the following diagram is commutative:

$$
\begin{array}{ccc}
A^U & \xrightarrow{\;f_U\;} & E_U/E_U^c \\
{\scriptstyle [\sigma]}\downarrow & & \downarrow{\scriptstyle [x]} \\
A^{U[\sigma]} & \xrightarrow[\;f_{U[\sigma]}\;]{} & E_{U[x]}/E_{U[x]}^c
\end{array}
$$

WT 3. If $U \subset V$ are open subgroups of G with U normal in V, then the class of the extension

$$0 \to A^U \approx E_U/E_U^c \to E_V/E_U^c \to E_V/E_U \approx V/U \to 0$$

is the fundamental class of $H^2(V/U, A^U)$.

WT 4. The intersection $\bigcap E_U^c$ taken over all open subgroups U of G is the unit element e of G.

To prove the existence of a topological Weil group, we shall need two sufficient conditions as follows.

WT 5. The trace $\mathbf{S}_V^U : A^U \to A^V$ is an open morphism for each pair of open subgroups $U \subset V$ of G.

WT 6. The factor group A^U/A^V is compact.

Then there exists a topological Weil group associated to the formation.

Theorem 3.6. *Let G be a group of Galois type, $A \in \mathrm{Galm}(G)$, and (G, A) a topological class formation satisfying* **WT 5** *and* **WT 6**.

Proof. The proof is essentialy routine, except for the following remarks. In the uniqueness theorem for Weil groups when G is finite, we know that the isomorphism φ is determined only up to an inner automorphism by an element of $A = E_e$. When we want to take an inverse limit, we need to find a compatible system of Weil groups for each open U, so the topological A^U intervene at this point. The compactness hypothesis is sufficient to allow us to find a coherent system of Weil groups for pairs $(G/U, A^U)$ when U ranges over the family of open normal subgroups of G. The details are now left to the reader.

CHAPTER X

Applications of Galois Cohomology in Algebraic Geometry

by
John Tate

Notes by
Serge Lang

1959

Let k be a field and G_k the Galois group of its algebraic closure (or separable closure). It is compact, totally disconnected, and inverse limit of its factor groups by normal open subgroups which are of finite index, and are the Galois groups of finite extension.

We use the category of Galois modules (discrete topology on A, continuous operation by G) and a cohomological functor H_G such that $H(G, A)$ is the limit of $H(G/U, A^U)$ where U is open normal in G.

The Galois modules $\mathrm{Galm}(G)$ contains the subcategory of the torsion (for $\mathbf{Z}$) modules $\mathrm{Galm}_{\mathrm{tor}}(G)$. Recall that G has **cohomological dimension** $\leq n$ if $H^r(G, A) = 0$ for all $r > n$ and all $A \in \mathrm{Galm}_{\mathrm{tor}}(G)$, and that G has **strict cohomological dimen-**

sion $\leq n$ if $H^r(G, A) = 0$ for $r > n$ and all $A \in \text{Galm}(G)$.

We shall use the **tower theorem** that if N is a closed normal subgroup of G, then $\text{cd}(G) \leq \text{cd}(G/N) + \text{cd}(N)$, as in Chapter VII, Theorem 5.1. If a field k has trivial Brauer group, i.e.

$$H^2(G_E, \Omega^*) = 0 \qquad \text{(all } E/k \text{ finite)}$$

where $\Omega = k_s$ (separable closure) then $\text{cd}(G_k) \leq 1$. By definition, a $\mathfrak{p}$-adic field is a finite extension of $\mathbf{Q}_p$. The maximal unramified extension of a $\mathfrak{p}$-adic field is cyclic and thus of $\text{cd} \leq 1$. Hence:

If k is a $\mathfrak{p}$-adic field, then $\text{cd}(G_k) \leq 2$.

This will be strengthened later to $\text{scd}(G_k) \leq 2$ (Theorem 2.3).

§1. Torsion-free modules

We use principally the dual of Nakayama, namely: Let G be a finite group, (G, A) a class formation, and M finitely generated torsion free (over $\mathbf{Z}$, and so $\mathbf{Z}$-free). Then

$$\mathbf{H}^r(G, \text{Hom}(M, A)) \times \mathbf{H}^{2-r}(G, M) \to H^2(G, A)$$

is a dual pairing, (i.e. puts the two groups in exact duality).

We suppose k is $\mathfrak{p}$-adic, and let Ω be its algebraic closure. We know G_k has $\text{cd}(G_k) \leq 2$ by the tower theorem. We shall eventually show $\text{scd}(G_k) \leq 2$.

Let $X = \text{Hom}(M, \Omega^*)$. Then X is isomorphic to a product of Ω^* as $\mathbf{Z}$-modules, their number being the rank of M, and we can define the operation of G_k on X in the natural manner, so that $X \in \text{Galm}(G_k)$.

By the existence theorem of local class field theory for L ranging over the finite extensions of k, the groups

$$N_{L/k} X_L$$

are cofinal with the groups $n X_k$, if we write $X_L = X^{G_L}$ and similarly $M_L = M^{G_L}$. This is clear since there exists a finite extension K of k which we may take Galois, such that $M^{G_K} = M$. Then for $L \supset K$, our statement is merely local class field theory's existence theorem, and then we use the norm $N_{L/K}$ to conclude the proof (transitivity of the norm).

We shall keep K fixed with the property that $M_K = M$. We wish to analyze the cohomology of X and M with respect to G_k.

Proposition 1.1. $H^r(G_k, X) = 0$ for $r > 2$.

Proof. X is divisible and we use cd ≤ 2, with the exact sequence

$$0 \to X_{\text{tor}} \to X \to X/X_{\text{tor}} \to 0,$$

where X_{tor} is the torsion part of X.

Theorem 1.2. *Induced by the pairing*

$$X \times M \to \Omega^*$$

we have the pairings

P0. $H^2(G_k, X) \times H^0(G_k, M) \to H^2(G_k, \Omega^*) = \mathbf{Q}/\mathbf{Z}$

P1. $H^1(G_k, X) \times H^1(G_k, M) \to \mathbf{Q}/\mathbf{Z}$

P2. $H^0(G_k, X) \times H^2(G_k, M) \to \mathbf{Q}/\mathbf{Z}$

*In **P0**, $H^2(G_k, X) = M_k^\wedge = $ by definition $\mathrm{Hom}(M_k, \mathbf{Q}/\mathbf{Z})$.*
*In **P1**, the two groups are finite, and the pairing is dual.*
*In **P2**, $H^2(G_k, M)$ is the torsion submodule of $\mathrm{Hom}(X_k, \mathbf{Q}/\mathbf{Z})$,*
i.e.
$$H^2(G_k, M) = (X_k^\wedge)_{\text{tor}}.$$

Proof. The pairing in each case is induced by inflation in a finite layer $L \supset K \supset k$. In **P0**, the right hand kernel wil be the intersection of $N_{L/k} M_L$ taken for all $L \supset K$, and this is merely $[L : K] N_{K/k} M_K$ which shrinks to 0. The kernel on the left is obviously 0.

In **P1**, the inflation-restriction sequence together with Hilbert's Theorem 90 shows that

$$\inf : H^1(G_{K/k}, X_K) \to H^1(G_{L/k}, X_L)$$

is an isomorphism, and the trivial action of $G_{L/K}$ on M shows similarly that

$$\inf : H^1(G_{K/k}, M) \to H^1(G_{L/k}, M)$$

is an ismorphism. It follows that both groups are finite, and the duality in the limit is merely the duality in any finite layer $L \supset K$.

In **P3**, we dualize the argument of **P0**, and note that $H^2(G_k, M)$ will produce characters on X_k which are of finite period, i.e., which are trivial on some nX_k for some integer n. Otherwise, nothing is changed from the formalism of **P0**.

§2. Finite modules

The field k is again p-adic and we let A be a finite abelian group in $\text{Galm}(G_k)$. Let $B = \text{Hom}(A, \Omega^*)$. Then A and B have the same order, and $B \in \text{Galm}(G_k)$. Since Ω^* contains all roots of unity, $B = \hat{A}$, and $A = \hat{B}$ once an identification between these roots of unity and $\mathbf{Q}/\mathbf{Z}$ has been made.

Let M be finitely generated torsion free and in $\text{Galm}(G_k)$ such that we have an exact sequence in $\text{Galm}(G_k)$:

$$0 \to N \to M \to A \to 0.$$

Since Ω^* is divisible, i.e. injective, we have an exact sequence

$$0 \leftarrow Y \leftarrow X \leftarrow B \leftarrow 0$$

where $X = \text{Hom}(M, \Omega^*)$ and $Y = \text{Hom}(N, \Omega^*)$.

By the theory of cup products, we shall have two dual sequences

$$(\mathbf{B}) \quad H^1(B) \to H^1(X) \to H^1(Y) \to H^2(B) \to H^2(X) \to H^2(Y)$$

$$(\mathbf{A}) \quad H^1(A) \leftarrow H^1(M) \leftarrow H^1(N) \leftarrow H^0(A) \leftarrow H^0(M) \leftarrow H^0(N)$$

with $H = H_{G_k}$. If one applies $\text{Hom}(\cdot, \mathbf{Q}/\mathbf{Z})$ to sequence $(\mathbf{A})$ we obtain a morphism of sequence $(\mathbf{B})$ into $\text{Hom}((\mathbf{A}), \mathbf{Q}/\mathbf{Z})$. The 5-lemma gives:

Theorem 2.1. *The cup product induced by $A \times B \to \Omega^*$ gives an exact duality*

$$H^2(G_k, B) \times H^0(G_k, A) \to H^2(G, \Omega^*).$$

Theorem 2.2. *With A again finite in $\mathrm{Galm}(G_k)$ and $B = \mathrm{Hom}(A, \Omega^*)$ the cup product*

$$H^1(G_k, A) \times H^1(G_k, B) \to H^2(G_k, \Omega^*)$$

gives an exact duality between the H^1, both of which are finite groups.

Proof. Let us first show they are finite groups. By the inflation-restriction sequence, it suffices to show that for L finite over k and suitably large, $H^1(G_L, A)$ is finite. Take L such that G_L operates trivially on A. Then $H^1(G_L, A) = \mathrm{cont}\ \mathrm{Hom}(G_L, A)$ and it is known from local class field theory or otherwise, that $G_L/G_L^n G_L^c$ is finite.

Now we have two sequences

(B) $\quad H^0(B) \to H^0(X) \to H^0(Y) \to H^1(B) \to H^1(X) \to H^1(Y)$

(A) $\quad H^2(A) \leftarrow H^2(M) \leftarrow H^2(N) \leftarrow H^1(A) \leftarrow H^1(M) \leftarrow H^1(N)$.

We get a morphism from sequence **(A)** into the torsion part of $\mathrm{Hom}((\mathbf{B}), \mathbf{Q}/\mathbf{Z}))$, and the 5-lemma gives the desired result.

Next let n be a large integer, and let us look at the sequences above with $A = M/nM$ and $N = M$. We have the left part of our sequences

$$0 \to H^0(B) \to H^0(X)$$

$$H^3(M) \leftarrow H^3(M) \leftarrow H^2(A) \leftarrow H^2(M)$$

I contend that $H^2(M) \to H^2(A)$ is surjective, because every character of $H^0(B)$ is the restriction of some character of $H^0(X)$ since $B_k \cap nX_k = 0$ for n large. Hence the map

$$n : H^3(M) \to H^3(M)$$

is injective for all n large, and since we deal with torsion groups, they must be 0. This is true for every M torsion free finitely generated, in $\mathrm{Galm}(G_k)$. Looking at the exact sequence factoring out the torsion part, and using $\mathrm{cd} \leq 2$, we see that in fact:

Theorem 2.3. *We have* $\mathrm{scd}(G_k) \leq 2$, *i.e.* $H^r_{G_k} = 0$ *for* $r \geq 3$ *and any object in* $\mathrm{Galm}(G_k)$.

We let χ be the (multiplicative) Euler characteristic. cf. *Algebra*, Chapter XX, §3.

Theorem 2.4. *Let* k *be* $\mathfrak{p}$-*adic, and* A *finite* G_k *Galois module. Let* $B = \mathrm{Hom}(A, \Omega^*)$. *Then* $\chi(G_k, A) = \|A\|_k$.

Proof. The Euler characteristic χ is multiplicative, so can assume A simple, and thus a vector space over $\mathbf{Z}/\ell\mathbf{Z}$ for some prime ℓ. We let $A_K = A^{G_K}$. For each Galois K/k, either $A_k = 0$ or $A_k = A$ by simplicity.

Case 1. $A_k = A$, so G_k operates trivially, so order of A is prime ℓ. Then

$$h^0 = \ell, \quad h^1 = (k^* : k^{*\ell}), \quad h^2(A) = h^0(B) = (k^*_\ell : 1).$$

So the formula checks.

Case 2. $B_k = B$. The situation is dual, and checks also.

Case 3. $A_k = 0$ and $B_k = 0$. Then

$$\chi(G_k, A) = 1/h^1(G_k, A).$$

Let K be maximal tamely ramified over k. Then G_K is a p-group. If $A_K = 0 = A^{G_K}$ then $\ell \neq p$ (otherwise G_K must operate trivially). Hence $H^1(G_K, A) = 0$. The inflation restriction sequence of G_k, G_K and $G_{K/k}$ shows $H^1(G_k, A) = 0$, so we are done.

Assume now $A_K \neq 0$, so $A_K = A$. Let L_0 be the smallest field containing k such that $A_{L_0} = A$. Then L_0 is normal over k, and cannot contain a subgroup of ℓ-power order, otherwise stuff left fixed would be a submodule $\neq 0$, so all of A, contradicting L_0 smallest. In particular, the ramification index of L_0/k is prime to ℓ.

Adjoin ℓ-th roots of unity to L_0 to get L. Then L has the same properties, and in particular, the ramification index of L/k is prime to ℓ.

194

Let T be the inertia field. Then the order of $G_{L/T}$ is prime to ℓ.

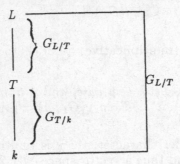

Hence $H^r(G_{L/T}, A) = 0$ for all $r > 0$. By spectral sequence, we conclude $H^r(G_{L/k}, A_L) = H^r(G_{T/k}, A_T)$ all $r > 0$. But $G_{T/k}$ is cyclic, A_T is finite, hence $H^1(G_{L/k}, A_L)$ and $H^2(G_{L/k}, A_L)$ have the same number of elements. In the exact sequence

$$0 \longrightarrow H^1(G_{L/k}, A_L) \longrightarrow H^1(G_k, A) \longrightarrow H^1(G_L, A)^{G_{L/k}} \longrightarrow H^2(G_{L/k}, A) \longrightarrow 0$$

we get 0 on the right, because $H^2(G_k, A)$ is dual to $H^0(G_k, B) = B_k = 0$. We can replace $H^2(G_{L/k}, A)$ by $H^1(G_{L/k}, A)$ as far as the number of elements is concerned, and then the hexagon theorem of the Herbrand quotient shows

$$h^1(G_k, A) = \text{ order } H^1(G_L, A)^{G_{L/k}}.$$

Since G_L operates trivially on A, the H^1 is simply the homs of G_L into A, and thus we have to compute the order of

$$\text{Hom}_{G_{L/k}}(G_L, A).$$

Such homs have to vanish on G_L^ℓ and on the commutator group, so if we let G'_L be the abelianized group, then $G_L^*/G_L^{*\ell}$. But this is $G_{L/k}$-isomorphic to $L^*/L^{*\ell}$, by local class field theory. So we have to compute the order of

$$\text{Hom}_{G_{L/k}}(L^*/L^{*\ell}, A).$$

If $\ell \neq p$, then $L^*/L^{*\ell}$ is $G_{L/k}$-isomorphic to $\mathbf{Z}/\ell\mathbf{Z} \times \mu_\ell$ where μ_ℓ is the group of ℓ-th roots of unity. Also, $G_{L/k}$ has trivial action on $\mathbf{Z}/\ell\mathbf{Z}$. So no hom can come from that since $A_k = 0$. As for $\mathrm{Hom}_{G_{L/k}}$ of μ_ℓ, if f is such, then for all $\sigma \in G_{L/k}$,

$$f(\sigma\zeta) = \sigma f(\zeta).$$

But $\sigma\zeta = \zeta^\nu$ for some ν, so $a = f(\zeta)$ generates a submodule of order ℓ, which must be all of A, so its inverse gives an element of B_k, contradicting $B_k = 0$. Hence all $G_{L/k}$-homs are 0, so Q.E.D.

If $\ell = p$, we must show the number of such homs is $1/\|A\|_k$. But according to Iwasawa,

$$L^*/L^{*p} = \mathbf{Z}/p\mathbf{Z} \times \mu_p \times \mathbf{Z}_p(G_{L/k})^m.$$

Using some standard facts of modular representations, we are done.

§3. The Tate pairing

Let V be a complete normal variety defined over a field k such that any finite set of points can be represented on an affine k-open subset of V. We denote by $A = A(V)$ its Albanese variety, defined over k, and by $B = B(V)$ its Picard variety also defined over k. Let $D_a(V)$ and $D_\ell(V)$ be the groups of divisors algebraically equivalent to 0, resp. linearly equivalent to 0. We have the Picard group $D_a(V)/D_\ell(V)$ and an isomorphism between this group and B, induced by a Poincaré divisor D on the product $V \times B$, and rational over k.

For each finite set of simple points S on V we denote by $\mathrm{Pic}_S(V)$ the factor group $D_{a,S}/D_{\ell,S}^{(1)}$ where $D_{a,S}$ consists of divisors algebraically equivalent to 0 whose support does not meet S, and $D_{\ell,S}^{(1)}$ is the subgroup of $D_{a,S}$ consisting of the divisors of functions f such that $f(P) = 1$ for all points P in S. Then there are canonical surjective homomorphisms

$$\mathrm{Pic}_{S'}(V) \to \mathrm{Pic}_S(V) \to \mathrm{Pic}(V)$$

whenever $S' \supset S$.

Actually we may work rationally over a finite extension K of k which in the applications will be Galois, and with obvious definitions, we form

$$\mathrm{Pic}_{S,K}(V) = D_{a,S,K}/D_{\ell,S,K}^{(1)}$$

the index K indicating rationality over K.

We may form the inverse limit inv $\lim_S \mathrm{Pic}_{S,K}(V)$. For our purposes we assume merely that we have a group $C_{a,K}$ together with a coherent set of surjective homomorphisms

$$\varphi_S : C_{a,K} \to \mathrm{Pic}_{S,K}(V)$$

thus defining a homomorphism φ (their limit) whose kernel is denoted by U_K. We have therefore the exact sequence

$$(1) \qquad 0 \to U_K \to C_{a,K} \xrightarrow{j} B(K) \to 0.$$

We assume throughout that a divisor class (for all our equivalences) which is fixed under all elements of G_K contains a divisor rational over K. Similarly, we shall assume throughout that the sequence

$$(2) \qquad 0 \to Z_{\alpha,K} \to Z_{0,K} \to A(K) \to 0$$

is exact, (where Z_0 are the 0-cycles of degree 0, and Z_α is the kernel of Albanese), for each finite extension K of k.

Relative to our exact sequences, we shall now define a Tate pairing.

Since $C_{a,K}$ is essentially a projective limit, we shall use the exact sequence

$$0 \to D^1_{\ell,S,K}/D^{(1)}_{S,K} \to \mathrm{Pic}_{S,K} \xrightarrow{j} B(k) \to 0$$

because if $S' \supset S$, then we have a commutative and exact diagram

$$
\begin{array}{ccccccccc}
0 & \longrightarrow & D_{\ell,S',K}/D^{(1)}_{\ell,S',K} & \longrightarrow & \mathrm{Pic}_{S',K} & \longrightarrow & B(k) & \longrightarrow & 0 \\
& & \downarrow & & \downarrow & & \downarrow \mathrm{id} & & \\
0 & \longrightarrow & D_{\ell,S,K}/D^{(1)}_{\ell,S,K} & \longrightarrow & \mathrm{Pic}_{S,K} & \longrightarrow & B(k) & \longrightarrow & 0 \\
& & & & \downarrow & & & & \\
& & & & 0 & & & &
\end{array}
$$

Now we wish to define a pairing

$$Z_{0,K} \times U_K \to K^*.$$

Let $u \in U_K$, and $\mathfrak{a} \in Z_{0,K}$. Write

$$\mathfrak{a} = \sum n_Q Q$$

where the Q are distinct algebraic points. Let S be a finite set of points containing all those of $\mathfrak{a}$, and rational over K. Then u has a representative in $\mathrm{Pic}_{S,K}$ and a further representative function f_S defined over K, and defined at all points of $\mathfrak{a}$. We define

$$\langle \mathfrak{a}, u \rangle^{-1} = f_S(\mathfrak{a}) = \prod f_S(Q)^{n_Q}.$$

It is easily seen that $\langle \mathfrak{a}, u \rangle^{-1}$ does not depend on the choice of S and f_S subject to the above conditions. It is then clear that this is a bilinear pairing.

We define a further pairing

$$Z_{a,K} \times C_a, K \to K^*$$

as follows. Let $\gamma \in C_{a,K}$, so $\gamma = \lim \gamma_S, \gamma_S \in D_{a,S,K}/D_{\ell,S,K}^{(1)}$. Let $\mathfrak{a} \in Z_{\alpha,K}$ and let S contain $\mathrm{supp}(\mathfrak{a})$. Let $\mathfrak{b}$ be a 0-cycle on B, rational over K, corresponding to the point $b = j\gamma$. Let X_S be a divisor on V rational over K, representing γ_S. Let D be a Poincaré divisor whose support does not meet that of $S \times \mathfrak{b}$. We define:

$$\langle \mathfrak{a}, \gamma \rangle = \frac{[^tD(\mathfrak{b}) - X_S](\mathfrak{a})}{D(\mathfrak{a}, \mathfrak{b})}$$

observing that $^tD(\mathfrak{b}) - X_S$ is the divisor of a function whose support does not meet S and is thus defined at $\mathfrak{a}$.

Using the reciprocity law of [La 57], see also [La 59], Chapter VI, §4, Theorem 10, one verifies that this is independent of the successive choice of S, $\mathfrak{b}$, X_S, and D subject to the above conditions.

One verifies finally that our pairings agree on $Z_\alpha \times U$.

Thus to summarize: we have exact sequences

$$0 \to Z_{\alpha,K}(V) \to Z_{0,K}(V) \to A(K) \to 0$$
$$0 \to U_K \to C_{a,K}(V) \to B(K) \to 0$$

and we have a Tate pairing:

$$\langle \mathfrak{a}, \gamma \rangle = \frac{[{}^t D(\mathfrak{b}) - X_S](\mathfrak{a})}{D(\mathfrak{a}, \mathfrak{b})} \in K^*$$

where $\mathfrak{b} \in Z_0(B)$ maps on the point $b \in B$, the same point as $\gamma \in C_a(V)$, S is a finite set of points containing $\operatorname{supp}(\mathfrak{a}), X_S$ represents γ_S, and

$$\langle \mathfrak{a}, u \rangle = f_S(\mathfrak{a})^{-1}$$

where f_S is a function representing u. We take S so large that everything is defined.

Proposition 3.1. *The induced bilinear map on (A_m, B_m) coincides with $e_m(a, b)$, i.e. with ${}^t D(m\mathfrak{b}, \mathfrak{a})/D(m\mathfrak{a}, \mathfrak{b})$.*

Proof. Clear. We are using [La 57] and [La 59], Chapter VI.

The above statements refer to the Tate augmented product of Chapter V. The augmented product exists whenever one is given two exact sequences

$$0 \to A' \to A \xrightarrow{j} A'' \to 0$$

$$0 \to B' \to B \xrightarrow{j} B'' \to 0$$

an object C, two pairings $A \times B' \to C$ and $A' \times B \to C$ which agree on $A' \times B'$. Such an abstract situation induces an augmented product

$$H^r(A'') \times H^s(B'') \xrightarrow{\cup_a} H^{r+s+1}(C)$$

which may be defined in terms of cocycles as follows. If f'' and g'' are cocycles in A'' and B'' respectively, their augmented cup is represented by the cocycle

$$\delta f \cup g + (-1)^{\dim f} f \cup \delta g$$

where $jf = f''$ and $jg = g''$, i.e. f and g are cochains of A and B respectively pulled back from f'' and g''.

In dimensions $(0, 1)$, the most important for what follows, we may make the Tate pairing explicit in the following manner. Let (b_σ) be a 1-cocycle representing an element $\beta \in H^1(G_{K/k}, B(K))$

and let $a \in A(K)$ represent $\alpha \in H^0(G_{K/k}, A(K))$. Let $\mathfrak{a} \in Z_{0,k}(V)$ belong to a, and let $\mathfrak{b}_\sigma$ be in $Z_{0,K}(B)$ and such that $S(\mathfrak{b}_\sigma) = b_\sigma$. Let D be a Poincaré divisor on $V \times B$ whose support does not meet $\mathfrak{a} \times \mathfrak{b}_\sigma$ for any σ. Then it is easily verified that putting $\mathfrak{b} = \mathfrak{b}_\sigma + \sigma\mathfrak{b}_\tau - \mathfrak{b}_{\sigma\tau}$, the cocycle

$$^t D(\mathfrak{b}, \mathfrak{a})$$

represents $\alpha \cup_{\mathrm{aug}} \beta$.

§4. The $(0,1)$ duality for abelian varieties

We assume for the rest of this section that k is $\mathfrak{p}$-adic.

Theorem 4.1. *The augmented product of the Tate pairing described in Section 3 induces a duality between $H^0(G_k, A)$ and $H^1(G_k, B)$, with values in $H^2(G_k, \Omega) = \mathbf{Q}/\mathbf{Z}$.*

Proof. According to the general theory of the augmented cupping, we have for each integer $m > 0$,

$$
\begin{array}{ccccccccc}
0 & \to & A(k)/mA(k) & \to & H^1(G_k, A_m) & \to & H^1(G_k, A)_m & \to & 0 \\
 & & \downarrow & & \downarrow & & \downarrow & & \\
0 & \to & (H^1(G_k, B)_m)^\wedge & \to & H^1(G_k, B_m)^\wedge & \to & (B(k)/mB(k))^\wedge & \to & 0
\end{array}
$$

a morphism of the first sequence into the second. Since the pairing between A_m and B_m is an exact duality, so is the pairing between their H^1 by Theorem 2.1. We wish to prove the end vertical arrows are isomorphisms, and for this we count. We have:

$$(A(k) : mA(k)) \le h^1(B)_m \qquad (B(k) : mB(k)) \le h^1(A)_m$$

$$h^1(A)_m(A(k) : mA(k)) = h^1(A_m)$$

$$h^1(B)_m(B(k) : mB(k)) = h^1(B_m)$$

$$\frac{(A(k) : mA(k))}{(A(k)_m : 0)} = \|m\|_k^{-r} = \frac{(B(k) : mB(k))}{(B(k)_m : 0)}$$

(by cyclic cohomology, trivial action)

$$\chi(A_m) = \|m^{2r}\|_k = \chi(B_m) \text{ and } h^0(B_m) = h^2(A_m)$$

by duality. Putting everything together, we get equality in the first inequalities; this proves the desired isomorphism.

Theorem 4.2. *We have $H^2(G_k, B) = 0$. (This is special for abelian varieties, better than $\mathrm{scd} \leq 2$.)*

Proof. We have an exact sequence

$$0 \to H^1(B)/mH^1(B) \to H^2(B_m) \to H^2(B_m) \to H^2(B)_m \to 0.$$

But $H^2(B_m)$ is dual to $H^0(A_m)$ and in particular has the same number of elements. Also, $H^1(B)$ being dual to $H^0(A)$, we see that $H^1(B)/mH^1(B)$ is dual to $H^0(A)_m = A(k)_m$, which is also $H^0(A_m)$. Hence the two terms on the left have the same number of elements, since $H^2(B)_m = 0$ for all m, so 0 since it is torsion group.

Now we have the duality for H^1, H^0 in finite layers.

Theorem 4.3. *Let K/k be finite Galois with group $G = G_{K/k}$. Then the pairing*

$$H^0(G, A(K)) \times H^1(G, B(K)) \to H^2(G, K^*)$$

is a duality.

Proof. This follows from the abstract fact that restriction is dual to the transfer valid for any Tate pairing and the induced augmented cupping.

If one uses the inflation-restriction sequence, together with the commutativity derived abstractly for d_2, and Theorem 4.2, we get the following $\pm$ commutative diagram, putting $U = G_K$, and $G = G_{K/k}$,

$$
\begin{array}{ccccc}
H^1(U,A)^{G/U} \times (B^U)_{G/U} & \xrightarrow{\ \cup_{\mathrm{aug}}\ } & H^2(U,\Omega^*)_{G/U} & \xrightarrow{\ \mathrm{tr}\ } & H^2(G,\Omega^*) \\
\Big\downarrow{\scriptstyle d_2} \qquad \Big\uparrow{\scriptstyle \mathrm{inc}} & & & & \Big\downarrow{\scriptstyle \mathrm{id.}} \\
H^2(G/U,A^U) \times H^1(G/U,B^U) & \xrightarrow{\ \cup_{\mathrm{aug}}\ } & H^2(G/U,\Omega^{*U}) & \xrightarrow{\ \mathrm{inf}\ } & H^2(G,\Omega^*)
\end{array}
$$

Identifying $H^2(G, \Omega^*)$ with $\mathbf{Q}/\mathbf{Z}$ we get the duality between H^2 and $\mathbf{H}^{-1}$:

Identifying $H^2(G, \Omega^*)$ with $\mathbf{Q}/\mathbf{Z}$ we get the duality between H^2 and $\mathbf{H}^{-1}$:

Theorem 4.4. *If K/k is a finite Galois extension with group G, then the augmented cupping*

$$\mathbf{H}^2(G, A(K)) \times \mathbf{H}^{-1}(G, B(K)) \to \mathbf{H}^2(G, K^*)$$

is a perfect duality.

§5. The full duality

We wish to show how the following theorem essentially follows from the $(0, 1)$ duality without any further use of arithmetic, only from abstract commutative diagrams.

Theorem 5.1. *Let k be a $\mathfrak{p}$-adic field, A and B an abelian variety and its Picard variety defined over k, and consider the Tate pairing described in §3. Then the augmented cupping*

$$\mathbf{H}^{1-r}(G_{K/k}, A(K)) \times \mathbf{H}^r(G_{K/k}, B(K)) \to \mathbf{H}^2(G_{K/k}, K^*)$$

puts the two groups (which are finite) in exact duality.

(Of course, the right hand H^2 is $(\mathbf{Q}/\mathbf{Z})_n$, where $n = (G_{K/k} : e)$.)

As usual, the $\wedge$ means Hom into $\mathbf{Q}/\mathbf{Z}$.

Put $G_K = U$ and $G = G_k$. We have a compact discrete duality

$$A^U \times H^1(U, B) \to H^2(U, \Omega^*) = \mathbf{Q}/\mathbf{Z}$$

and we know from this that A^U is isomorphic to $H^1(U, B)^\wedge$ as a G/U-module. Hence the commutative diagram say for $r \geq 3$:

$$
\begin{array}{ccc}
\mathbf{H}^{1-r}(G/U, H^1(U,B)^\wedge) \times \mathbf{H}^{r-2}(G/U, H^1(U,B)) & \overset{\cup}{\longrightarrow} & \mathbf{H}^{-1}(G/U, \mathbf{Q}/\mathbf{Z}) \\
\downarrow \qquad\qquad\qquad \downarrow{\scriptstyle\text{id}} & & \downarrow{\scriptstyle\text{id}} \\
\mathbf{H}^{1-r}(G/U, A^U) \quad \times \mathbf{H}^{r-2}(G/U, H^1(U,B)) & \underset{\cup}{\longrightarrow} & \mathbf{H}^{-1}(G/U, \mathbf{Q}/\mathbf{Z}).
\end{array}
$$

The top line comes from the cup product duality theorem, and the arrow on the left is an ismorphism, as described above.

Of course, we have $H^{-1}(G/U, \mathbf{Q}/\mathbf{Z}) = (\mathbf{Q}/\mathbf{Z})_n$ if n is the order of G/U. Note also that the $\mathbf{Q}/\mathbf{Z}$ in the lower right stands for $H^2(U, \Omega^*)$, because of the invariant isomorphism.

Now from the spectral sequence and Theorem 4.2 to the effect that H^2 of an abelian variety is trivial over a $\mathfrak{p}$-adic field, we get an isomorphism

$$d_2 : H^{r-2}(G/U, H^1(U, B)) \to H^r(G/U, H^0(U, B))$$

and we use another abstract diagram:

$$
\begin{array}{ccc}
\mathbf{H}^{1-r}(G/U, A^U) \times \mathbf{H}^{r-2}(G/U, H^1(U,B)) & \xrightarrow{\;\cup\;} & \mathbf{H}^{-1}(G/U, H^2(U,\Omega^*)) \\
\downarrow{\scriptstyle\text{id}} \qquad\qquad \downarrow{\scriptstyle d_2} & & \\
\mathbf{H}^{1-r}(G/U, A^U) \times \ \mathbf{H}^{r}(G/U, H^0(U,B))) & \xrightarrow[\cup_{\text{aug}}]{} & \mathbf{H}^2(G/U, \Omega^{*U})
\end{array}
$$

In order to complete it to a commutative one, we complete the top line and the bottom one respectively as follows:

$$
\begin{array}{ccc}
\mathbf{H}^{-1}(G/U, H^2(U,\Omega^*)) \xrightarrow[\text{inc}]{} H^2(U,\Omega^*)_{G/U} & \xrightarrow{\;\text{tr}\;} & H^2(G, \Omega^*) \\
& & \downarrow{\scriptstyle\text{id.}} \\
\mathbf{H}^2(G/U, \Omega^{*U}) \xrightarrow[\qquad\text{inf}\qquad]{} & & H^2(G, \Omega^*)
\end{array}
$$

and since the transfer and inflation perserve invariants, we see that our duality has been reduced as advertised.

We observe that we have the ordinary cup on the top line and the augmented cup on the bottom. The top one is relative to the $A^U, H^1(U, B)$ duality, derived previously.

§6. The Brauer group

We continue to work with a variety V defined over a $\mathfrak{p}$-adic field k. We assume V complete, non-singular in codimension 1, and such that any finite set of points can be represented on an affine k-open subset of V.

We let $G = G_k$ and all cohomology groups in this section will be taken relative to G. We observe that the function field $\Omega(V)$ has group G over $k(V)$, and we wish to look at its cohomology.

By Hilbert's Theorem 90 it is trivial in dimension 1, and hence we look at it in dimension 2: It is nothing but that part of the Brauer group over $k(V)$ which is split by a constant field extension. We make the following assumptions.

Assumption 1. *There exists a 0-cycle on V rational over k and of degree 1.*

Assumption 2. *Let NS denote the Néron-Severi group of V (it is finitely generated). Then the natural map*

$$\mathrm{Div}(V)^{G_k} \to NS^{G_k} = NS_k$$

of divisors rational over k into that part of Néron-Severi which is fixed under G_k is surjective, i.e. every class rational over k has a representative divisor rational over k.

These assumptions can be translated into cohomology, and it is actually in this latter form that we shall use them. This is done as follows.

To begin with, note that Assumption 1 guarantees that there is a canonical map of V into its Albanese variety defined over k (use the cycle to get an origin on the principal homogeneous space of Albanese). Hence by pull-back from Albanese, given a rational point b on the Picard variety, there is a divisor $X \in D_a(V)$ rational over k such that $Cl(X) = b$. In other words, the map

$$D_a(V)^{G_k} \to B(k)$$

is surjective. Now consider the exact sequence

$$H^0(\mathrm{Div}(V)) \to H^0(NS) \to H^1(D_a) \to H^1(\mathrm{Div}(V)).$$

Then $\mathrm{Div}(V)^{G_k}$ is a direct sum over $\mathbf{Z}$ of groups generated by the irreducible divisors, and putting together a divisor and its conjugates, we get

$$\mathrm{Div}(V)^{G_k} = \bigoplus_{\xi} \bigoplus_{X \in \xi} \mathbf{Z} \cdot X$$

where ξ ranges over the prime rational divisors of V over k and X ranges over its algebraic components. Now the inside sum is semilocal, and by semilocal theory we get $H^1(G_K, \mathbf{Z})$ where $K = k_X$ is the smallest field of definition of X. This is 0 because G_K is of Galois type and the cohomology comes from finite things. Thus:

Proposition 6.1. $H^1(\mathrm{Div}(V)) = 0$.

From this, one sees that our Assumption 1 is equivalent with the condition $H^1(D_a) = 0$.

Now looking at the other sequence

$$H^0(D_a) \to H^0(B) \to H^1(D_\ell) \to H^1(D_a)$$

we see that Assumption 2 is equivalent to $H^1(D_\ell) = 0$. Thus:

Assumptions 1 and 2 are equivalent with

$$H^1(D_a) = 0 \quad and \quad H^1(D_\ell) = 0.$$

Now we have two exact sequences

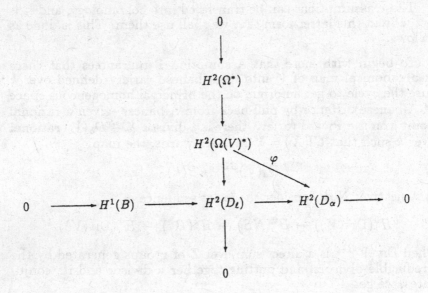

and from them we get a surjective map

$$\varphi : H^2(\Omega(V)^*) \to H^2(D_a) \to 0.$$

We *define* $H^2_u(\Omega(V)^*)$ to be its kernel, and call it the **unramified** part of the Brauer group $H^2(\Omega(V)^*)$. In view of the exact cross we get a map

$$H^2_u(\Omega(V)^*) \to H^1(B).$$

Thus

$$H^2(\Omega(V)^*)/H_u^2(\Omega(V)^*) \approx H^2(D_a)$$

$$H_u^2(\Omega(V)^*)/H^2(\Omega) \approx H^1(B) \approx \mathrm{Char}(A(k))$$

$$H^2(\Omega^*) \approx \mathbf{Q}/\mathbf{Z} = \mathrm{Char}(\mathbf{Z})$$

where Char means continuous character (or here equivalently character of finite order, or torsion part of $\hat{A}(k) = \mathrm{Hom}(A(k), \mathbf{Q}/\mathbf{Z})$). This gives us a good description of our Brauer group, relative to the filtration

$$H^2(\Omega(V)^*) \supset H_u^2(\Omega(V)^*) \supset H^2(\Omega^*) \supset 0.$$

We wish to give a more concrete description of H_u^2 above, making explicit its connection with the Tate pairing. Relative to the sequence

$$0 \to A(k) \to Z_k/Z_{\alpha,k} \to \mathbf{Z} \to 0$$

and taking characters Char, we shall get a commutatiove exact diagram as follows:

$$
\begin{array}{ccccccccc}
0 & \longrightarrow & \mathrm{Char}(\mathbf{Z}) & \longrightarrow & \mathrm{Char}(Z_k/Z_{\alpha,k}) & \longrightarrow & \mathrm{Char}(A(k)) & \longrightarrow & 0 \\
& & \uparrow & & \uparrow & & \uparrow & & \\
0 & \longrightarrow & H^2(\Omega^*) & \longrightarrow & H^2(\Omega(V)^*) & \longrightarrow & H^1(B) & \longrightarrow & 0
\end{array}
$$

The two end arrows are as we have just described them, and are isomorphisms. We must now define the middle arrow and prove commutativity.

Let $v \in H_u^2(\Omega(V)^*)$. For each prime rational 0-cycle $\mathfrak{p}$ of V over k we shall define its reduction mod $\mathfrak{p}$, $v_\mathfrak{p} \in H^2(\Omega^*)$ and then a character $Z_k/Z_{\alpha,k}$ by the formula

$$\theta_V(\mathfrak{a}) = \sum_\mathfrak{p} \nu_\mathfrak{p} \, \mathrm{inv}(v_\mathfrak{p})$$

whenever $\mathfrak{a}$ is a rational 0-cycle,

$$\mathfrak{a} = \sum \nu_\mathfrak{p} \cdot \mathfrak{p}, \ \nu_\mathfrak{p} \in \mathbf{Z}.$$

We shall prove that θ_V vanished on the kernel of Albanese, whence the character, and then we shall prove commutativity.

Definition of $v_{\mathfrak{p}}$. Let $(f_{\sigma,\tau})$ be a representative cocycle. By definition it splits in D_a, so that there is a divisor $X_\sigma \in D_a$ such that

$$(f_{\sigma,\tau}) = X_\sigma + \sigma X_\tau - X_{\sigma\tau}.$$

For each σ, choose a function g_σ such that $X_\sigma = (g_\sigma)$ at $\mathfrak{p}$. Put

$$f'_{\sigma,\tau} = f_{\sigma,\tau}/(\delta g)_{\sigma,\tau}$$

then $(f'_{\sigma,\tau}) = 0$ at $\mathfrak{p}$, i.e. $f'_{\sigma,\tau}$ is a unit at $\mathfrak{p}$. We now put

$$a_{\sigma,\tau} = \prod_{P \in \mathfrak{p}} f'_{\sigma,\tau}(P),$$

where $\{P\}$ ranges over the algebraic points into which $\mathfrak{p}$ splits. We contend that $(a_{\sigma,\tau})$ is a cocycle, and that its class does not depend on the choices made during its construction.

Let $(f^*_{\sigma,\tau})$ be another representative cocycle which is a unit at $\mathfrak{p}$, and obtained by the same process. Then

$$f^* = f' \cdot \delta g$$

with $(\delta g)_{\sigma,\tau} = g_\sigma g_\tau^\sigma/g_{\sigma\tau}$ a unit at $\mathfrak{p}$. Hence taking divisors,

$$(g_\sigma) + \sigma(g_\tau) - (g_{\sigma\tau}) = 0$$

at all points in supp($\mathfrak{p}$). Let this support be S, and let Div^S be the group of divisors passing through some point of $\mathfrak{p}$. Then $H^1(\mathrm{Div}^S) = 0$ by the same argument as in Proposition 6.1 (semilocal and $H^1(\mathbf{Z}) = 0$) and hence taking the image of (g_σ) in Div^S we conclude that there exists a divisor $X \in \mathrm{Div}^S$ such that $(g_\sigma) = \sigma X - X$. Let h be a function such that $X = (h)$ at $\mathfrak{p}$. Replace g_σ by $g_\sigma h^{1-\sigma}$. Then still $f^* = f' \cdot \delta g$ and now g_σ is a unit at $\mathfrak{p}$, for all σ. From this we get

$$\prod_{P \in \mathfrak{p}} (f^*/f')_{\sigma,\tau}(P) = \prod_{P \in \mathfrak{p}} (\delta g)_{\sigma,\tau}(P)$$

from which we see that it is the boundary of the 1-cochain

$$\prod_{P \in \mathfrak{p}} g_\sigma(P).$$

Thus we have proved our reduction mapping

$$v \mapsto v_{\mathfrak{p}}$$

well defined.

Now the image of v in $H^1(B)$ is by definition reprsented by the cocycle $Cl(X_\sigma) = b_\sigma$ (notation as in the above paragraph) and if $\mathfrak{a} \in Z_{0,k}$ then

$$\theta_v(\mathfrak{a}) = \pm S(\mathfrak{a}) \cup_{\mathrm{aug}} \beta$$

where β is represented by the cocycle (b_σ).

Thus θ_v vanishes on the kernel of Albanese, and the right side of our diagam is commutative.

As for the left side, given $\omega \in H^2(\Omega^*)$, represented by a cocycle $(c_{\sigma,\tau})$, then $\omega_{\mathfrak{p}}$ is represented by

$$\prod_{P \in \mathfrak{p}} c_{\sigma,\tau} = c_{\sigma,\tau}^m$$

where $m = \deg(\mathfrak{p})$. Then

$$\theta_\omega(\mathfrak{a}) = \sum_{\mathfrak{p}} (\deg \mathfrak{p}) \nu_{\mathfrak{p}} \mathrm{inv}(\omega)$$

$$= \deg(\mathfrak{a}) \cdot \mathrm{inv}(\omega)$$

whence commutativity. This concludes the proof of the following theorem.

Theorem 6.2. *There is an isomorphism*

$$H_u^2(\Omega(V)^*) \approx \mathrm{Char}(Z/Z_\alpha)_k$$

under the mapping θ_v and the diagram

$$
\begin{array}{ccccccccc}
0 & \longrightarrow & \mathrm{Char}(\mathbf{Z}) & \longrightarrow & \mathrm{Char}(Z/Z_\alpha)_k & \longrightarrow & \mathrm{Char}(A(k)) & \longrightarrow & 0 \\
& & \uparrow & & \uparrow & & \uparrow & & \\
0 & \longrightarrow & H^2(\Omega^*) & \longrightarrow & H^2(\Omega(V)^*) & \longrightarrow & H^1(B) & \longrightarrow & 0
\end{array}
$$

which is exact and commutative.

We conclude this section with a description of $H^2(D_a)$ also in terms of characters.

We have the exact sequence

$$0 \to D_a \to \mathrm{Div} \to NS \to 0$$

and hence

$$0 \to H^1(NS) \to H^2(D_a) \to H^2(\mathrm{Div}) \to H^2(NS) \to 0$$

the last 0 by scd ≤ 2.

Now $H^2(\mathrm{Div})$ is easy to describe since Div is essentially a direct sum. In fact,

$$\mathrm{Div}_k = \bigoplus_{\xi} \bigoplus_{X \in \xi} \mathbf{Z} \cdot X$$

where the sum is taken over all prime rational divisors ξ over k and all algebraic components X in ξ. Using the semilocal theory, we get

$$H^2(\mathrm{Div}) = \bigoplus_{\xi} H^2(G_{k_\xi}, \mathbf{Z}) = \bigoplus_{\xi} H^1(G_{k_\xi}, \mathbf{Q}/\mathbf{Z})$$

$$= \bigoplus_{\xi} \mathrm{Char}(k_\xi^*)$$

where k_ξ is the smallest field of definition of an algebraic point X in ξ and G_{k_ξ} is the Galois group over k_ξ. Thus we see that our H^2 is a direct sum of character groups.

The Case of Curves. If we assume that V has dimension 1, i.e. is a non singular curve, then this result simplifies considerably since $NS = \mathbf{Z}$ is infinite cyclic, and we have also

$$NS = NS_k = (NS)^{G_k}.$$

We have

$$H^1(NS) = 0, \quad H^2(NS) = H^2(\mathbf{Z}) = \mathrm{Char}(k^*)$$

and we get the commutative diagram

$$0 \longrightarrow H^2(D_a) \longrightarrow H^2(\mathrm{Div}) \longrightarrow H^2(\mathbf{Z}) \longrightarrow 0$$

$$\| \qquad\qquad \| \qquad\qquad \|$$

$$0 \longrightarrow \bigoplus{}^0 \mathrm{Char}(k_\xi^*) \longrightarrow \bigoplus \mathrm{Char}(k_\xi^*) \longrightarrow \mathrm{Char}(k^*) \longrightarrow 0$$

where the $\bigoplus^0$ on the left means those elements whose sum gives 0. The morphism on the lower right is given by the restriction of a character from k_ξ^* to k^* and the sum mapping. Thus an element of $H^2(\text{Div})$ is given by a vector of characters $(\chi_{v,\mathfrak{p}})$ where $\mathfrak{p}$ ranges over the prime rational cycles of V, i.e. the ξ since cycles and divisors coincide.

We observe also that by Tsen's theorem, $H^2(\Omega(V)^*)$ is the full Brauer group over $k(V)$ since $\Omega(V)$ does not admit any division algebras of finite rank over itself.

Finally, we have slightly better information on H_u^2:

Proposition 6.3. *If V is a curve, then*

$$H_u^2(\Omega(V)^*) = \bigcap_\mathfrak{p} H_\mathfrak{p}^2(\Omega(V)^*)$$

where $H_\mathfrak{p}^2$ consists of those cohomology classes having a cocycle representative $(f_{\sigma,\tau})$ in the units at $\mathfrak{p}$.

We leave the proof as an exercise to the reader.

We shall discuss ideles for arbitrary varieties in the next section. Here, for curves, we take the usual definition, and we then have the same theorem as in class field theory.

Proposition 6.4. *Let V be a curve, and for each $\mathfrak{p}$ let $k(V)_\mathfrak{p}$ be the completion at the prime rational cycle $\mathfrak{p}$. Let $\mathrm{Br}(k(V))$ be the Brauer group over $k(V)$, i.e. $H^2(G_k, k(V)_s)$ where $k(V)_s$ is the separable ($=$ algebraic) closure of $k(V)$, and similarly for $\mathrm{Br}(k(V)_\mathfrak{p})$. Then the map*

$$\mathrm{Br}(k(V)) \to \prod_\mathfrak{p} \mathrm{Br}(k(V)_\mathfrak{p})$$

is injective.

One can give a proof based on the preceding discussion or by proving that $H^1(C_a) = 0$, just as in class field theory. We leave the details to the reader.

§7. Ideles and idele classes

Let k be a field and V a complete normal variety defined over k and such that any finite set of points can be represented on an affine k-open subset. By a cycle, we shall always mean a 0-cycle.

For each prime rational cycle $\mathfrak{p}$ over k on V we have the integers $\mathfrak{o}_\mathfrak{p}$, the units $U_\mathfrak{p}$ and maximal ideal $\mathfrak{m}_\mathfrak{p}$ in $k(V)$.

There are several candidates to play the role of ideles, and we shall describe here what would be a factor group of the classical ideles in the case of curves. We let

$$F_\mathfrak{p} = k(V)^*/(1 + \mathfrak{m}_\mathfrak{p}).$$

Then we let I_k be the subgroup of the Cartesian product of all the $F_\mathfrak{p}$ consisting of the vectors

$$\mathbf{f} = (\dots, f_\mathfrak{p}, \dots) \qquad f_\mathfrak{p} \in F_\mathfrak{p}$$

such that there exists a divisor X, rational over k, such that

$$X = (f_\mathfrak{p}) \text{ at } \mathfrak{p} \text{ for all } \mathfrak{p}.$$

(In the case of curves, this means unit almost everywhere.) We call this divisor X (obviously unique) the **divisor associated with the idele f**, and write $X = (\mathbf{f})$.

We have two subgroups $I_{a,k}$ and $I_{\ell,k}$ consisting of the ideles whose divisor is algebraically equivalent to 0 and linearly equivalent to 0 respectively.

Since every divisor is linearly equivalent to 0 at a simple point, we have an exact sequence

$$0 \to I_{\ell,k} \to I_{a,k} \to B(k) \to 0$$

where B is the Picard variety of V, defined over k.

As usual, we have an imbedding

$$K(V)^* \subset I_k$$

on the diagonal: if $f \in k(V)^*$, then f maps on $(\dots, f, f, f, \dots)$ (of course in the vector, it is the class of $f \mod 1 + \mathfrak{m}_\mathfrak{p}$).

We recall our Picard groups $\mathrm{Pic}_S(V)$ associated with a finite set of points of V and here we assume that S is a finite set of prime rational cycles. We have $\mathrm{Pic}_{S,k}(V) = D_{a,S,k}(V)/D_{\ell,S,k}^{(1)}$ where $D_{a,S,k}$ consists of the divisors on V algebraically equivalent to 0, not passing through any point of S, and rational over k, and $D_{\ell,S,k}^{(1)}$ consists of those which are linarly equivalent to 0, belonging to a function which takes the value 1 at all points of S, and is defined over k.

We contend that we have a surjective map

$$\varphi_S : I_{a,k} \to \mathrm{Pic}_{S,k}$$

for each S as follows. Given $\mathbf{f}$ in $I_{a,k}$ there exists $f \in k(V)^*$ such that we can write

$$\mathbf{f} = f f_{\mathfrak{p}}' \text{ with } \mathbf{f}' = 1 \in F_{\mathfrak{p}}$$

for all $\mathfrak{p} \in S$. This is easily proved by moving the divisor of $\mathbf{f}$ by a linear equivalence, and then using the Chinese remainder theorem in an affine ring of an affine open subset of V. We then put

$$\varphi_S(\mathbf{f}) = Cl_S((\mathbf{f}'))$$

where Cl_S is the equivalence class mod $D_{\ell,S,k}^{(1)}$. Our collection of maps φ_S is obviously consistent, and thus we can define a mapping

$$\varphi : I_{a,k} \to \lim \mathrm{Pic}_{S,k}(V).$$

For our purposes here, we denote by $C_{a,k}$ the image of φ in the limit and call it the group of **idele classes**. This is all right:

Contention. *The kernel of φ is $k(V)^*$.*

Proof. If $\mathbf{f}$ is in the kernel, then for all S there exists a function f_S such that

$$\mathbf{f} = \mathbf{f}_S f_S$$

where $\mathbf{f}_S$ is 1 in S, and $(f_S) = 0$. All f_S have the same divisor, namely $(\mathbf{f})$. Looking at one prime $\mathfrak{p}$ in S, we see that all f_S are equal to the same function f, and we see that $\mathbf{f}$ is simply the function f.

We have the **unit ideles** $I_{u,k}$ consisting of those ideles whose divisor is 0, the **idele classes** $C_k = I_k/K(V)^*$, and also the obvious subgroups of idele classes:

$$C_{a,k} = I_{a,k}/k(V)^*$$
$$C_{u,k} = k(V)^* I_{u,k}/k(V)^* = I_{u,k}/k^*.$$

We keep working under Assumptions 1 and 2, of course. In that case, if K is a finite Galois extension of k, we have the two fundamental exact sequences

(1) $$0 \to Z_{\alpha,K} \to Z_{0,K} \to A(K) \to 0$$

(2) $$0 \to C_{u,K} \to C_{a,K} \to B(K) \to 0$$

in the category of $G_{K/k}$-mpodules. For the limit, with respect to Ω one will of course take the injectve limit over all K.

From the definition, we see that

$$I_{u,k} = \prod_{\mathfrak{p}/k} k(\mathfrak{p})^*$$

where $k(\mathfrak{p})$ is the residue class field of the prime rational cycle $\mathfrak{p}$ over k, i.e. $k(\mathfrak{p}) = \mathfrak{o}_\mathfrak{p}/\mathfrak{m}_\mathfrak{p}$.

If K/k is finite Galois, then we write

$$I_{u,K} = \prod_{\mathfrak{P}/K} k(\mathfrak{P})^*$$

where $\mathfrak{P}$ ranges over the prime rational cycles over K.

§8. Idele class cohomology

Aside from the fundamental sequences (1) and (2), we have three sequences.

$$0 \to Z_{0,K} \to Z_K \to \mathbf{Z} \to 0$$
$$0 \to K^* \to I_{u,K} \to C_{u,K} \to 0$$
$$0 \to 0 \to K^* \to K^* \to 0$$

and pairings giving rise to cup products:

$$Z_K \times I_{u,K} \to K^*$$

defined in the obvious manner: Given $\mathbf{f} \in I_{u,K}$ and a cycle

$$\mathfrak{a} = \sum_\mathfrak{p} \nu_\mathfrak{p} \cdot \mathfrak{p},$$

the pairing is

$$(\mathfrak{a}, \mathbf{f}) = \prod \mathbf{f}_\mathfrak{p}^{\nu_\mathfrak{p}}.$$

It induces pairings

$$Z_{0,K} \times C_{u,K} \to K^*$$
$$\mathbf{Z} \times K^* \to K^*$$
$$Z_{0,K} \times K^* \to 0$$

and we get an exact commutative diagram from the cup product

$$
\begin{array}{ccccccc}
\mathbf{H}^r(K^*) & \longrightarrow & \mathbf{H}^r(I_u) & \longrightarrow & \mathbf{H}^r(C_u) & \longrightarrow & \mathbf{H}^{r+1}(K^*) \\
\downarrow{\varphi_1} & & \downarrow{\varphi_2} & & \downarrow{\varphi_3} & & \downarrow{\varphi_1} \\
\mathbf{H}^{2-r}(\mathbf{Z})^\wedge & \longrightarrow & \mathbf{H}^{2-r}(\mathbf{Z})^\wedge & \longrightarrow & \mathbf{H}^{2-r}(Z_0)^\wedge & \longrightarrow & \mathbf{H}^{1-r}(\mathbf{Z})^\wedge
\end{array}
$$

taking into account that

$$H^2(G_{K/k}, K^*) = (\mathbf{Q}/\mathbf{Z})_n$$

where $n = (G : e)$ the cup products taking their values in this $\mathbf{H}^2$. Here, as in the next diagram, $\mathbf{H}$ is taken with respect to $G_{K/k}$, we omit the index K on the modules, and $r \in \mathbf{Z}$ so $\mathbf{H} = \mathbf{H}_{G_{K/k}}$ is the special functor.

From the exact sequence in the last section, we get

$$
\begin{array}{ccccccccc}
\mathbf{H}^{r-1}(B) & \longrightarrow & \mathbf{H}^r(C_u) & \longrightarrow & \mathbf{H}^r(C_a) & \longrightarrow & \mathbf{H}^r(B) & \longrightarrow & \mathbf{H}^{r+1}(C_u) \\
\downarrow{\varphi_4} & & \downarrow{\varphi_3} & & \downarrow{\varphi_5} & & \downarrow{\varphi_4} & & \downarrow{\varphi_3} \\
\mathbf{H}^{2-r}(A)^\wedge & \longrightarrow & \mathbf{H}^{2-r}(Z)_0)^\wedge & \longrightarrow & \mathbf{H}^{2-r}(Z_a)^\wedge & \longrightarrow & \mathbf{H}^{1-r}(A)^\wedge & \longrightarrow & \mathbf{H}^{1-r}(Z_0)^\wedge
\end{array}
$$

and φ_4 is induced by the augmented cup, the others by the cup.

Theorem 8.1. *All φ_i are isomorphisms.*

Proof. We proceed stepwise.

φ_1 is an isomorphism by Tate's theorem.
φ_2 by a semilocal analysis and again by Tate's theorem.
φ_3 by the 5-lemma and the result for φ_1 and φ_2.
φ_4 by the augmented cup duality already done.
φ_5 by the 5-lemma and the result for φ_3 and φ_4.

So that's it.

Corollary 8.2. $H^1(G_{K/k}, C_{a,K}) = 0.$

Proof. It is dual to $H^1(Z_\alpha)$ which is 0 since we assumed the existence of a rational cycle of degree 1.

In the case of a curve, if we had worked with the true ideles J_K instead of our truncated ones I_K, we would also have obtained (essentially in the same way) the above corollary. Thus from the sequence

$$0 \to K(V)^* \to J_{a,K} \to C_{a,K} \to 0$$

we would get exactly

$$0 \to H^2(K(V)^*) \to H^2(J_{a,K})$$

thus recovering the fact that an element of the Brauer group which splits locally everywhere splits globally (H^2 is taken with $G_{K/k}$).

Furthermore, the curves exhibit one more duality, a self duality, of our group $F_{\mathfrak{p}}$. This is a local question. We take k a $\mathfrak{p}$-adic field, K a finite extension, Galois with group $G_{K/k}$, and consider the power series $k((t))$ and $K((t))$. We let F be our local group

$$F = K((t))^*/(1 + \mathfrak{m})$$

where $\mathfrak{m}$ is the maximal ideal. Then $1 + \mathfrak{m}$ is uniquely divisible, and so its cohomology is trivial. Hence

$$\mathbf{H}^r(G_{K/k}, K((t))^*) = \mathbf{H}^r(G_{K/k}, F).$$

We have the exact sequence

$$0 \to K^* \to F \to \mathbf{Z} \to 0$$

and a pairing

$$K((t))^* \times K((t))^* \to K^*$$

defined by

$$(f,g) = (-1)^{\operatorname{ord} f \ \operatorname{ord} g} (f^{\operatorname{ord} g}/g^{\operatorname{ord} f})(0),$$

which induces a pairing

$$F \times F \to K^*.$$

Now we get the commutative diagram

$$
\begin{array}{ccccccccc}
0 & \longrightarrow & \mathbf{H}^r(K^*) & \longrightarrow & \mathbf{H}^r(F) & \longrightarrow & \mathbf{H}^r(\mathbf{Z}) & \longrightarrow & 0 \\
& & \downarrow & & \downarrow & & \downarrow & & \\
0 & \longrightarrow & \mathbf{H}^{2-r}(\mathbf{Z}^\wedge) & \longrightarrow & \mathbf{H}^{2-r}(F)^\wedge & \longrightarrow & \mathbf{H}^{2-r}(K^\wedge) & \longrightarrow & 0
\end{array}
$$

and by the five lemma, together with Tate's theorem, we see that the middle arrow is an isomorphism. Hence

$$\mathbf{H}^r(F) \quad \text{is dual to} \quad \mathbf{H}^{2-r}(F)$$

by the cup product.

Bibliography

[ArT 67] E. ARTIN and J. TATE, *Class Field Theory*, Benjamin 1967; Addison Wesley, 1991

[CaE 56] H. CARTAN and S. EILENBERG, *Homological Algebra*, Princeton Univ. Press 1956

[Gr 59] A. GROTHENDIECK, Sur quelques points d'algèbre homologique, *Tohoku Math. J.* **9** (1957) pp. 119-221

[Ho 50a] G. HOCHSCHILD, Local class field thoery, *Ann.Math.* **51** No. 2 (1950) pp. 331-347

[Ho 50b] G. HOCHSCHILD, Note on Artin's reciprocity law, *Ann. Math.* **52** No. 3 (1950) pp. 694-701

[HoN 52] G. HOCHSCHILD and T. NAKAYAMA, Cohomology in class field theory, *Ann.Math.* **55** No. 2 (1952) pp. 348-366

[HoS 53] G. HOCHSCHILD and J.-P. SERRE, Cohomology of group extensions, *Trans. AMS* **74** (1953) pp. 110-134

[Ka 55a] Y. KAWADA, Class formations, *Duke Math. J.* **22** (1955) pp. 165-178

[Ka 55b] Y. KAWADA, Class formations III, *J. Math. Soc. Japan* **7** (1955) pp. 453-490

[Ka 63] Y. KAWADA, Cohomology of group extensions, *J. Fac. Sci. Univ. Tokyo* **9** (1963) pp. 417-431

[**Ka 69**] Y. KAWADA, Class formations, *Proc. Symp. Pure Math.* **20** AMS, 1969

[**KaS 56**] Y. KAWADA and I. SATAKE, Class formations II, *J. Fac. Sci. Univ. Tokyo* **7** (1956) pp. 353-389

[**KaT 55**] Y. KAWADA and J. TATE, On the Galois cohomology of unramified extensions of function fields in one variable, *Am. J. Math.* **77** No. 2 (1955) pp. 197-217

[**La 57**] S. LANG, Divisors and endomorphisms on abelian varieties, *Amer. J. Math.* **80** No. 3 (1958) pp. 761-777

[**La 59**] S. LANG, *Abelian Varieties*, Interscience, 1959; Springer Verlag, 1983

[**La 66**] S. LANG, *Rapport sur la cohomologie des groupes*, Benjamin 1966

[**La 71/93**] S. LANG, *Algebra*, Addison-Wesley 1971, 3rd edn. 1993

[**Mi 86**] J. MILNE, *Arithmetic Duality Theorems*, Academic Press, Boston, 1986

[**Na 36**] T. NAKAYAMA, Über die Beziehungen zwischen den Faktorensystemen und der Normklassengruppe eines galoisschen Erweiterungskörpers, *Math. Ann.* **112** (1936) pp. 85-91

[**Na 43**] T. NAKAYAMA, A theorem on the norm group of a finite extension field, *Jap. J. Math.* **18** (1943) pp. 877-885

[**Na 41**] T. NAKAYAMA, Factor system approach ot the isomorphism and reciprocity theorems, *J. Math. Soc. Japan* **3** No. 1 (1941) pp. 52-58

[**Na 52**] T. NAKAYAMA, Idele class factor sets and class field theory, *Ann. Math.* **55** No. 1 (1952) pp. 73-84

[**Na 53**] T. NAKAYAMA, Note on 3-factor sets, *Kodai Math. Rep.* **3** (1949) pp. 11-14

[**Se 73/94**] J.-P. SERRE, *Cohomologie Galoisienne*, Benjamin 1973, Fifth edition, Lecture Notes in Mathematics No. 5, Springer Verlag 1994

[**Sh 46**] I. SHAFAREVICH, On Galois groups of *p*-adic fields, *Dokl. Akad. Nauk SSSR* **53** No. 1 (1946) pp. 15-16 (see also *Collected Papers*, Springer Verlag 1989, p. 5)

[Ta 52] J. TATE, The higher dimensional cohomology groups of class field theory, *Ann. Math.* **56** No. 2 (1952) pp. 294-27

[Ta 62] J. TATE, Duality theorems in Galois cohomology over number fields, *Proc. Int. Congress Math. Stockholm* (1962) pp. 288-295

[Ta 66] J. TATE, The cohomology groups of tori in finite Galois extensions of number fields, *Nagoya Math. J.* **27** (1966) pp. 709-719

[We 51] A. WEIL, Sur la théorie du corps de classe, *J. Math. Soc. Japan* **3** (1951) pp. 1-35

Complementary References

A. ADEM and R.J. MILGRAM, *Cohomology of Finite Groups*, Springer-Verlag 1994

K. BROWN, *Cohomology of Groups*, Springer-Verlag 1982

S. LANG, *Algebraic Number Theory*, Addison-Wesley 1970; Springer-Verlag 1986, 2nd edn. 1994

S. MAC LANE, *Homology*, Springer-Verlag 1963, 4th printing 1995

Table of Notation

A_{p^∞} : Elements of A annihilated by a power of p

A_φ : If φ is a homomorphism, kernel of φ in A

$\hat{A}$: $\mathrm{Hom}(A, \mathbf{Q}/\mathbf{Z})$

A^G : Elements of A fixed by G

A_m : Kernel of the homomorphism $m_A :: A \to A$ such that $a \mapsto ma$

cd : Cohomological dimension

$\mathbf{F}_p$: $\mathbf{Z}/p\mathbf{Z}$

$\hat{G}$: Character group, $\mathrm{Hom}(G, \mathbf{Q}/\mathbf{Z})$

$\varkappa_G$: Natural homomorphism of A^G onto $H^0(G, A)$ or $\mathbf{H}^0(G, A)$

$\maltese_G$: Natural homomorphism of A_S into $\mathbf{H}^{-1}(G, A)$

$\mathrm{Galm}(G)$: Galois modules

$\mathrm{Galm}_p(G)$: Galois modules whose elements are annihilated by a p-power

$\mathrm{Galm}_{\mathrm{tor}}(G)$: Torsion Galois modules

G^c : Commutator group, or closure of commutator if G is topological

G_p : p-Sylow subgroup of G

Grab : Category of abelian groups

$h_{1/2}$: Herbrand quotient, order of H^2 divided by order of H^1

H_G : Functor such that $H_G(A) = A^G$

$\mathbf{H}_G$: Functor such that $\mathbf{H}_G(A) = A^G/S_G A$

I_G : Augmentation ideal, generated by the elements $\sigma - e, \sigma \in G$

$M_G(A)$: Functions (sometimes continuous) from G into A

$\mathbf{M}_G$: $\mathbf{Z}[G] \otimes A$

M_G^S : Induced functions

$\text{Mod}(G)$: Abelian category of G-modules

$\text{Mod}(\mathbf{Z})$: Abelian category of abelian groups

scd : Strict cohomological dimension

$\mathbf{S}_G$: The relative trace, from a subgroup U of finite index, to G

$\mathbf{S}_G$: The trace, for a finite group G

Tr : Transfer of group theory

tr : Transfer of cohomology

$\mathbf{Z}[G]$: Group ring

Index

Abutment of spectral sequence 117

Admissible subgroup 177

Augmentation 10, 27

Augmented cupping 109, 198

Bilinear map of complexes 84

Brauer group 167, 202, 209

Category of modules 10

Characters 28

Class formation 166

Class module 71

Coerasing functor 5

Cofunctor 4

Cohomological cup functor 76

Cohomological dimension 138

Cohomological period 96

Cohomology ring 89

Complete resolution 23

Conjugation 41, 174

Consistency 173

Cup functor 76

453

Cup product 75

Cyclic groups 32

Deflation, def 164

Delta-functor 3

Double cosets 58

Duality theorems 93, 190, 192, 199

Edge isomorphisms 118

Equivalent extensions 159

Erasable 4

Erasing functor 4, 15, 134

Extension of groups 156

Extreme isomorphisms 118

Factor extension 163

Factor sets 28

Filtered object 116

Filtration 116

Fundamental class 71, 167

G-module 10

G-morphism 11

G-regular 17

Galm(G) 127

Galm$_p(G)$ 138

Galm$_{tor}(G)$ 138

Galois group 151, 195

Galois module 127

Galois type 123

Grab 11

Herbrand lemma 35

Herbrand quotient 35

Hochschild-Serre spectral sequence 118

$\text{Hom}_G(A, B)$ 11

Homogeneous standard complex 27

Idele 210

Idele classes 211

Induced representation 52, 134

Inflation $\text{inf}_G^{G/G'}$ 40

Invariant inv_G 167

Lifting morphism 38

Limitation theoreem 176

Local component 55

Maximal generator 95

Maximal p-quotient 149

$M_G(A)$ 13, 19

$\text{Mod}(G)$ 10

$\text{Mod}(R)$ 10

$M_G^S(B)$ 52

Morphism of pairs 38

Multilinear category 73

Nakayama maps 101

Periodicity 95

p-extensive group 144

p-group 50, 126

Positive spectral sequence 118

Profinite group 124, 147

Projective 17

Reciprocity law 198

Reciprocity mapping 173

Regular 17

Restriction res_U^G 39

Semilocal 71

Shafarevich-Weil theorem 177

Spectral functor 117

Splitting functor 5

Splitting module 70

Standard complex 26, 27

Strict cohomological dimension 138, 193

Supernatural number 125

Sylow group 50, 126, 137

Tate pairing 195

Tate product 109

Tate theorems 23, 70, 98

Tensor product 21

Topological class formation 185

Topological Galois module 185

Trace 12, 15

Transfer of cohomology tr_G 43

Transfer of group theory Tr_G 48, 160, 174

Transgression tg 120

Translation 46, 174

Triplet theorem 68, 88

Triplet theorem for cup products 88

Twin theorem 65

Uniqueness theorems 5, 6

Unramified Brauer group 204

Weil group 179, 185

[1998]

The following article was submitted for publication in the Forum of the AMS *Notices* on 5 January 1998, but was rejected. Cf. my *Response to the Steele Prize* below.

The Kirschner Article and HIV: Scientific and Journalistic (Ir)responsibilities

Serge Lang

Part One: Editorial and Scientific Responsibility

In February 1996, the AMS *Notices* published a 12-page article "Using Mathematics to Understand HIV Immune Dynamics" by Denise Kirschner (pp. 191–202). This paper dealt with the mathematical modeling of HIV infection. Kirschner explicitly thanked "the editors for helpful comments and support in the writing of this article." For six years, I have been involved in gathering information about an extraordinary situation concerning HIV. I have a file more than an inch thick on the subject. The bottom line is that the hypothesis that HIV is a harmless virus is compatible with all the evidence I have studied; that purportedly scientific papers which I have followed up on HIV claiming otherwise are subject to very severe criticisms, pointing to severe faults; and that there is an ongoing phenomenon of mass misinformation, spread by NIH (especially in publications of the Centers for Disease Control—CDC), and spread in the scientific journals such as *Nature* and *Science* as well as in the press at large. I even published two articles on the subject in the *Yale Scientific* (Fall 1994, Winter 1995), reproduced updated in the Kluwer collection (see footnote 1) (and subsequently reproduced in my book *Challenges*, see below). I was therefore shocked to see the *Notices* spreading the orthodoxy uncritically.

In light of what I knew about the HIV situation, I immediately phoned Hugo Rossi, editor in chief of the *Notices*, and I sent him my HIV documentation. The documentation included:

— my *Yale Scientific* articles;

— articles by the mathematician Mark Craddock (School of Mathematics, University of New South Wales, Sydney, Australia), specifically directed at the use or misuse of mathematics in connection with HIV infection;[1]

---

[1]Mark Craddock, in the Kluwer collection *AIDS: Virus or Drug Induced?*, Peter H. Duesberg editor, Kluwer Academic Publishers, 1996:
"Some mathematical considerations on HIV and AIDS" pp. 89–95;

— letters to and from government officials, such as Harold Jaffe, Director of the CDC;

— files containing critical analyses of published articles which received wide attention in the press (scientific and otherwise, including the *New York Times*).

— subsequently these documents were complemented by my *Journalistic Suppression and Manipulation File* (1995–1996), and the File entitled: *Throwing Math and Statistics at People* (Summer 1997).

On the phone I suggested to Rossi that if the publication of Kirschner's article was to be taken seriously, it would involve the *Notices* in a morass for which the AMS was not equipped institutionally. To minimize the time wasted by everybody, I suggested to him that after he processed the documentation, he might write his own editorial statement to the effect that when he arranged for the publication of the Kirschner article, he did not know about the simmering controversy on HIV. Given the additional information, he could ask readers simply to disregard the published article, which readers were not in a position to evaluate without a substantial amount of additional material. Providing this material might result in an open-ended controversy in the *Notices*. Just for a start, the *Notices* might solicit an article by Mark Craddock analyzing the Kirschner article. Rossi answered me by mail without even waiting to receive the material, and he wrote that no matter what this material contained, he would not make a "Stalinesque confession" (his interpretation of what I was asking). I did not read his letter further, and I sent it to Cathleen Morawetz, president of the AMS, together with my resignation from the AMS, because I wanted no part of the responsibility as a member of the AMS to deal with the situation the editors had created with the publication of the Kirschner article, and with Rossi's subsequent position.

Public realtions.   There is some evidence that the Kirschner article was not even meant to be read, but was merely a public relations gesture using HIV combined with math to emphasize the importance of "relevance," "applications," and "social responsibility." Indeed, when the AMS President wrote me back, she suggested that I write a letter to the editors for publication, and she added: "I would recommend a short letter—it takes

"A critical appraisal of the Vancouver men's study *Does it refute the drugs/AIDS hypothesis?* pp. 105–110;

"Science by Press Conference" pp. 127–130.

My articles are:

"HIV and AIDS: Have we been misled? Questions of scientific and journalistic responsibility" pp. 271–295;

"To fund or not to fund, that is the question: proposed experiments on the drug-AIDS hypothesis" pp. 297–307.

less time to write and is much more likely to be read—since limited time and space is a problem for us all." However, the very extensive space occupied by the 12 pages of the Kirschner article, written with the "helpful comments and support" of the editors, did not present a problem to them, nor apparently to the AMS president. Of course, I refused to engage in the superficial dealings the AMS president was suggesting. Barring a possibly short statement by the editor as I was requesting, what I saw as my responsibility as an AMS member would be to insure publication of an extensive documented evaluation of the type Mark Craddock provided in his articles. I was neither able nor did I have the time available to do it myself, but if the AMS higher-ups were serious about informing the readership properly, they could have solicited Mark Craddock as I suggested.

The Landau editorial. Subsequently, in February 1997, the *Notices* published an editorial by Susan Landau (Associate Editor) entitled "Mathematicians and Social Responsibility." The editorial is presumptuous, and Landau subsequently evaded the very responsibilities she invoked in the big-time rhetoric of her editorial. Among other things, Landau asserts: "Our responsibilities extend to preparing the biology students for the work they will actually do (rather than giving them a standard calculus course with the odd population biology example thrown in)." First, I object to her put down of the "standard calculus course with the odd population biology example thrown in." The population biology example is not "odd." Principally, what does her admonition mean in the specific case of HIV and AIDS, in light of the criticisms which have been leveled at the orthodox line on HIV? I sent her my HIV file. What would the *Notices* do about the Kirschner article? What would she do? She wrote me on 12 September 1997: "For the last several months I have been receiving mail from you regarding HIV and AIDS. Despite being an Associate Editor of the *Notices*, I am not really following these issues, and I would like to be removed from your mailing list." So how do "our responsibilities" apply to her, especially since she is an Associate Editor of the *Notices* and the editors provided "helpful comments and support in the writing" of Kirschner's article? Despite having shared the responsibility to publish the Kirschner article, she claims that she is "not really following these issues" and she rejects information about them. Thus *de facto* she is evading her responsibilities in at least two respects: those invoked in her editorial, and those arising *ex officio* as an Associate Editor of the *Notices*. Some letters to the editor paid lip service to her editorial, e.g. in April and May 1997. The authors of these letters were apparently unaware of the HIV pathogeny controversy, the problems with the original Kirschner article, and the post-publication abdication of responsibility by the editors of the *Notices*. I shall return to questions about the Landau editorial at the end of Part Two.

A letter from Arthur Gottlieb, rethinking the problem. Certain events induced me to reconsider the importance of the Kirschner article, and to follow up more actively on the AMS involvement. On 16 May 1997, Arthur Gottlieb M.D., Chair of the Microbiology/Immunology Department at Tulane University, wrote me to ask for my professional opinion concerning the Craddock articles analyzing certain mathematical defects in published and famous articles on HIV/AIDS (see Part Two below, and especially footnotes 4 and 7). I had corresponded previously with Gottlieb, because he had heard of me through the grapevine, and had sent me a letter which he had written to the editors of the *New York Times*, but which was not published. Of course, I circulated his letter to my cc list. I strongly supported Craddock's analysis which concerned especially a "model" by Ho and Shaw, who are two famous HIV researchers. For example, a year ago, Ho was named TIME Man of the Year. Gottlieb wrote me:

I met Mark [Craddock] on a visit in Sydney last year and have been particularly interested in his views of the Ho/Shaw model of HIV pathogenesis which has now acquired the status of a law of nature in the AIDS-HIV community....

I think there is more than a matter of scientific debate here. My experience has been that when models of this type are presented to broad biological-medical audiences, the math is rarely critically analyzed—most people are content with the declaration that a biostatistician has come up with the particular equations that are said to describe the situation. It is the rare individual, indeed, who would raise a meaningful question in such a context. The Ho/Shaw model is now a widely accepted paradigm for HIV pathogenesis. Moreover, it is being used as a basis for therapeutic guidelines in respect of HIV ("treat early and hard"). That, I think, is of concern, if indeed there are serious questions about the validity of the model. It would be good to have your views on this.

Two years ago, at the time of Kirschner's article, it did not seem to me worth while getting further involved setting up the AMS. However, since a person as solidly placed as Gottlieb in the medical establishment has now raised questions which involve joint responsibilities with mathematicians, I revised my estimate of the importance of dealing more thoroughly with an evaluation of various uses of "mathematics" in connection with HIV. I am now dealing with the AMS as an outsider, but the higher-ups at the AMS had, and still have, the responsibility to follow up on Kirschner's article, and they have the responsibility to take into account articles such as those by Craddock, and other articles which are beginning to appear (cf. footnote 7 below). The evidence so far is that they won't do it without some outside intervention. For two years I have kept some higher-ups in the AMS abreast of the situation and my HIV file, with no visible result. In particular, the *Notices* Staff Writer Allyn Jackson did not report the events surrounding my resignation, nor did she report the documentation which I provided on HIV.

My book *Challenges*. In November 1997, my book *Challenges* appeared, published by Springer-Verlag. The book contains an extensive chapter (114 pages) on HIV. Beyond my two articles, the chapter is based on my various files on HIV. The existence of this book now makes it easier to disseminate information about HIV, and thus also contributed to my decision finally to write a piece for publication in the *Notices*. Readers will note that Dr. Gottlieb provided a one-page statement at the end of the HIV chapter, p. 714, where he says about the controversy over HIV pathogeny:

> ...In this chapter, Prof. Serge Lang has well documented the basis of this controversy, and has provided a sobering picture for the reader of the polity of thinking that has characterized this field.... Models of how HIV and cells of the immune system replicate, which have not yet sustained the rigor of thorough scientific discussion and critique at both the biological and mathematical level, are accepted as if they were laws of nature....
>
> A review of the scenarios which Lang has painted should give the thoughtful reader pause as well as some insight into how doctrinaire thinking can develop and be perpetuated.

In a piece addressed to the AMS *Notices*, it is appropriate to go into certain technicalities. In a second part, I shall deal more specifically with mathematical aspects of the HIV problem, and the Kirschner article in particular. Be it noted that I sent my HIV file and various criticisms (by Craddock, Gottlieb, and me) to Kirschner in August 1997, but I have had no reply from her.

Part Two: Specific Mathematical Criticisms

Craddock's articles. I have distributed widely the Craddock article on Ho & Shaw's work: "HIV: Science by Press Conference" (cf. footnote 1). Craddock provides 3 pages of detailed documentation for his conclusions: "...this new work is about as convincing as a giraffe trying to sneak into a polar bears only picnic by wearing sunglasses (as Ben Elton might say)."[2] The mathematics Craddock analyzes here are at the level of freshman calculus. In the other article "Some mathematical considerations on HIV and AIDS," the level is even more elementary. Craddock writes in a very convincing way, by using unpretentious language and making his

---

[2] The work under review is: Ho et al. *Nature* Vol. 373, 12 Jan 1995 pp. 123–126; and Wei et al. *ibid* pp. 117–122.

objections very specific about very concrete items. I have found his articles so well formulated that I have asked various scientists to take them
into consideration, without success. To give an unqualified endorsement of
Craddock I would have to read the original papers by Ho & Shaw, which I
have not done, and am not really competent to do, lacking training in biology. But I don't need any further competence to recognize the legitimacy
of Craddock's criticisms. His specific, documented criticisms include:

— Objections about the mathematical modeling and certain assumptions, not made explicit, and not justified by empirical evidence; un-
   justified assumptions unrelated to the empirical data.

— Questions about the meaning or significance of the data used by Ho
   and Shaw.

— Lack of control groups, in two contexts p. 129:

   (a) "Neither group [Ho and Shaw] compared the rate of T4 cells
       generated in the HIV positive patients with HIV negative con-
       trols!"[3]

   (b) "It must surely be admitted that the system they are trying
       to study, namely the interaction of HIV with T4 cells, might
       behave substantially differently in people who are not being
       pumped full of new drugs, in addition to 'antiretrovirals' like
       Zidovudine?"[4]

— Lack of warning that certain purportedly therapeutic drugs have toxic
   effects.

— Lack of justification for attributing the production (rather than destruction!) of T4 cells to HIV.[5]

Finally, Craddock points out that if one formulates the model correctly,
then what it predicts is not the same as what Ho & Shaw say it predicts.[6]

---

[3]He goes on: "Both groups assert that in HIV infected individuals, up to 5% of the
circulating T4 cells are replaced every 2 days. This information is hardly new, Peter Duesberg says something similar in a paper in the Proceedings of the National
Academy of Sciences from 1989. Except he states that 5% of the bodies T cells will
be replaced every 2 days, in healthy people."

[4]This is similar to the reason Arthur Gottlieb wrote to me in his letter of 16 May
1997: "I might say that I have been skeptical of the validity of the Ho/Shaw model
for several reasons, but principally because it is based on observations in subjects who
were therapeutically perturbed by use of a protease inhibitor."

[5]As Craddock writes: "The logic here is remarkable. It is claimed that HIV sends the
immune system into overdrive as measured by a supposedly accelerated production of
T4 cells. Between 100 million and 2 billion are produced each day in the patients that
were studied."

[6]As he writes: "When correctly formulated (Craddock, Ibid), what emerges is stun-
ning. Ho et al.'s observations combined with their simple model for T cells and virus,

His remarks are in line with the implausibility that it takes ten years for a virus with generation time of 1 to 2 days to achieve effects causing death.

The responsibility for confronting these criticisms lies with the authors he criticizes, and with the relevant scientific journals (such as *Nature*, *Science*, or *The Lancet*) for publishing both the criticisms and whatever replies the authors make. If they make none, scientific and journalistic standards require that readers of these journals be so informed. However, the scientific journals have actually failed in their responsibility. They have skewed and prejudiced scientific discourse, and obstructed the usual self-correcting mechanisms of science. For extensive documentation of these assertions, cf. my book *Challenges*.

I see no reason to deviate from the standards that scientific discourse take place openly in journals, and that the scientists whose works are questioned or criticized be held responsible for answering the questions and criticisms. In particular, it would be entirely appropriate for Ho and Shaw to be confronted directly with the Craddock criticisms, and for them to answer these criticisms, whether to acknowledge their validity, or to counter them if Ho & Shaw are able to do so. Barring specific justified rebuttals to Craddock's specific criticisms, we are entitled to regard these criticisms as valid, and they invalidate the Ho and Shaw papers which Craddock analyzes.[7]

---

predict that the T cell count should reach an equilibrium state quickly. Meaning exponentially fast.... When you add terms to the equation to describe the effects of Virus (inexplicably, they do not include the effects of the virus on the T-cell population in their model. I thought HIV was supposed to be killing these cells somehow), then include the expression for the amount of virus that they give on p. 124, you get a picture of 'HIV disease' that bears no relation to what happens in actual patients. AIDS should develop in days or weeks. There is no possible way it can take ten years. This emerges from Ho et al's own model."

[7](a) Some criticisms of the Ho and Shaw articles already appeared in letters to the editor in *Nature* (**375**, 18 May 1995). One of these letters, by Bukrinsky et al. (pp. 195–196) stated: "A definitive answer awaits accurate estimates of the turnover and half-life of both proliferating and peripheral CD4+ T cells in healthy individuals, normative data for which the immunological community strangely lacks a robust appraisal." In plain English, Bukrinsky et al. make the same point already mentioned, that no control groups were used to compare the behavior of CD4+ T-cells in individuals who are healthy, sick, HIV positive, or HIV negative, in various combinations. Ho and Shaw answered the Bukrinsky et al. comment quoted above as follows: "...we do not understand their logic of comparing our calculated CD4 lymphocyte turnover rates with previous estimates for normal peripheral blood mononuclear cells...." But the logic is clear to me. In plain English, the fact that turnover of T-cells is the same in Ho & Shaw's CD4 lymphocytes as in previous estimates for peripheral blood as in mononuclear cells constitutes clear evidence that HIV is neither the cause of T-cell destruction, nor of harm to the immune system (which has been claimed). I wrote to Bukrinsky on 18 July 1997 to ask him to straighten me out if I misunderstood the situation. He did not answer my letter.

The *Notices* article by Denise Kirschner. The Kirschner article in the *Notices* is an echo of Ho and Shaw. The mathematics in her article are somewhat more involved than the mathematics in the Ho & Shaw articles (her differential equations are more complicated). I have not checked them. But even if correct, to what extent is her use of mathematics useful to understand whether HIV is pathogenic or not, and if so, how? I fully share Craddock's conception of science: "Science is about making observations and trying to fit them into a theoretical framework. Having the theoretical framework allows us to make predictions about phenomena that we can then test. HIV 'science' long ago set off on a different path." Kirschner asserts p. 195: "Clinical data are becoming more available, making it possible to get actual values (or orders of values) directly for the individual parameters in the model." So the paper itself does not contain "actual values." The way the paper is written does not fit the definition of science recalled above, and does not inspire my confidence. I shall give a few concrete reasons why not.

— Kirschner repeats one orthodox line (p. 191) that "HIV is the virus which causes AIDS (Acquired Immune Deficiency Syndrome)" without any acknowledgement that in the Centers for Disease Control list of 29 diseases defining AIDS in the presence of HIV, about 40% of these diseases do not involve immunodeficiency, and that a low T-cell count is only one of the 29 diseases. The assumption that "HIV causes AIDS" is made without justification and without reference to a scientific paper justifying this assumption. After six years of looking into the HIV pathogeny question, I have not learned of the existence of any such paper.

— She repeats the orthodox line (p. 193): "When HIV infects the body, its target is $CD4^+$ T cells. Since $CD4^+$ T cells play the key role in the immune response, this is cause for alarm and a key reason for HIV's devastating impact.... Clearly, there is a necessity for treatment of HIV infection." Here she relies unquestioningly on the orthodox line, which I and a number of other scientists do not automatically accept. There is evidence going against all three assertions: $CD4^+$ T cells being a target of HIV, a devastat-

---

(b) Another letter by Ascher et al. (p. 196) stated: "...But the central paradox of AIDS pathogenesis remains... there is about 100–1,000-fold more cell death than can be accounted for by the observed rate of virus production[5]. It is a murder scene with far more bodies than bullets."

(c) There is a detailed critique of Wei et al. and Ho et al. in an article by Peter Duesberg and Harvey Bialy, "Responding to 'Deusberg and the new view of HIV'," Kluwer collection pp. 115–119.

(d) Further critiques of the mathematical analysis of Ho and Shaw (Wei et al.) have recently begun to appear. See Z. Grossman and R. Herberman, *Nature Medicine* Vol. 3 (1997) pp. 486–490; and G. Pantaleo, *ibid.* pp. 483–486. Cf. also the accompanying editorial: "Two commentaries challenge current thinking in HIV research and treatment."

ing impact being due to HIV, and the necessity for treatment of HIV infection. Aside from the point raised in footnote 7, what about T-cells which live in the presence of HIV? As some scientists including Peter Duesberg have pointed out, HIV is mass produced in immortal T-cells, both by scientists and drug companies. Her only qualification is: "The course of infection with HIV is not clear-cut. Clinicians are still arguing about what causes the eventual collapse of the immune system, resulting in death." However, barring further evidence to the contrary, the way she builds up her proposed model fits Craddock's characterization of "arcane speculations about molecular interactions."

— Several of Craddock's criticisms of the Ho & Shaw article are applicable to her article to the extent the following objections are valid. For example, she writes: ". . . it has been shown that infected $CD4^+$ T cells live less than 1–2 days [10]; therefore, we choose the rate of loss of infected T-cells, mu_T, to be values between .5 and 1.0." How justified is this choice? Her reference [10] is not even an original scientific paper but is partly a laudatory review in *Science* of the Wei et al. and Ho et al. articles, editorializing about what is seen as their implications.[8] Is her model a priori irrelevant because she did not take into account certain essential factors? For instance, she gives no evidence that she took control groups into account. The half-life of T-cells for infected or uninfected people is apparently the same. (Cf. footnote 3.) How did she take into account the presence of drugs or, as Arthur Gottlieb has brought up, protease inhibitors? (Cf. footnote 4.) She does state: "To include AZT chemotherapy in the model, it is necessary to mimic the effects of the drug which serves to reduce viral infectivity. . . ." But there is no evidence that she even considered possible toxic effects of AZT, and she only mentions a parameter which "is multiplied by a function which is 'off' outside the treatment period and 'on' during the treatment period." It's not clear that this kind of "model" represents what actually goes on. As far as I can tell, we are witnessing here a cumulative chain of defective science, uncritically invoking defective results by others, and propagating misinformation combined with an irrelevant mathematical formalism.

— In addition, am I reading correctly that the Kirschner model is in direct contradiction with the Ho & Shaw model, and also with empirical evidence for production rather than destruction of T-cells? Indeed, as we have recalled above, the Ho & Shaw model leads to "accelerated production of T4 cells," and an exponential approach to equilibrium. (Cf. footnotes 5 and 6.) So what's going on?

Funding. I note that the NSF supported Kirschner's work. As far as I am concerned, publication of her article in the *Notices* came at a time

---

[8]Her reference [10] is to J. M. Coffin, "HIV population dynamics in vivo: Implications for genetic variation, pathogenesis and therapy," *Science* **267** (1995) pp. 483–489.

when money is more than tight for mathematics. Higher-ups in the AMS including editors of journals want to make "mathematics" appear relevant to society at large, so that mathematicians get more support from the government. But invoking relevance is not a license for funding and disseminating uncritically certain points of view reinforcing the orthodoxy. To the extent that substantial criticisms of the Kirschner article are valid, including the possibility that it is worthless even as an "arcane speculation," the NSF funding of the 12-page Kirschner article is questionable; and its uncritical publication by the AMS, giving a mathematical aura to HIV and an applied aura to mathematics, is journalistically and scientifically irresponsible without a critical follow-up, which the editors or AMS higher-ups so far have refused to provide.

Math and Medicine. I see no evidence that her paper fits her conclusion p. 201: "Through this simple example, I hope it is also clear that there can and should be a role for mathematics in medicine." Even though her paper might be defective, I am not questioning the big-time generality whether there is a role for mathematics in medicine. However, it is NOT clear to me that her paper is a positive contribution to medicine. This remains to be seen, after competent persons (including Craddock and Gottlieb) have scrutinized it. Furthermore, so far, the "model" she proposes is disjoint from experimental testing or evidence, and from medicine. It is just presented as an independent entity, and I don't see any indication how it might be used clinically, although she writes: "Now that we have a model that we believe mimics a clinical picture, we can use the model to incorporate treatment strategies." Thus she substitutes beliefs ("we believe") for scientific experimental verification. The conclusions she draws are only based on the theoretical model, not actual practice in patients, and her model is biased in favor of the orthodox view. I hope someone such as Craddock or Gottlieb will be willing to give a more extensive analysis, which I am not able to give at the moment.

In summer 1997 I sent a copy of Kirschner's article to Arthur Gottlieb, and he answered: "I have put the Kirschner article on my list of things to do and will read it with a critical eye. Cursory review of same indicates no reference to functional CD4+ cells as a parameter to be considered. That is probably a fatal flaw, as every CD4+ cell is not equal to every other CD4+ cell." However, he also wrote me that he would be very busy with his course last fall, and I have not heard from him since the end of last summer.

Kirschner also states: "The biggest obstacle facing collaboration is the inability of clinicians to understand advanced mathematics, and, on the mathematician's part, the lack of knowledge of the underlying medical problem." With such a sentence, she bypasses the problems raised by Mark

Craddock's criticisms of the Ho & Shaw articles, and the problems which exist with her own article as listed above, as well as the problems with her references. Obstacles to collaboration are not totally ordered, and there may not be a biggest one; but as far as I am concerned, a big obstacle facing collaboration is that criticisms of existing articles are ignored by authors, ignored by editors such as those of the AMS *Notices*, and suppressed by journals such as *Nature*, *Science*, and the *Lancet*. Cf. for instance the exchange between Duesberg and *Nature* editor John Maddox in the Kluwer collection, pp. 111–125. For further documentation, cf. my book *Challenges*.

The existence of various articles on mathematical modeling, especially in connection with HIV, raises further questions about the use of mathematical modeling in biology generally. To what extent has such modeling been used scientifically, resp. medically? To what extent has it just amounted to throwing mathematics and statistics at people, thereby producing "mystification and intimidation" (as Koblitz once characterized this activity by some practitioners of some political science), but making no genuine scientific or medical contribution?

Da Capo

Returning to the issue of responsibility raised in Susan Landau's editorial: when mathematicians teach calculus, or biologists teach the use of mathematical modeling, to what extent do teachers warn students about passing off "mathematical modeling" as science, when a purported "model" is not based on empirical data, and is proposed (let alone accepted) quite independently of empirical verification? How does one document the warnings? Both the Ho & Shaw and Kirschner articles are based on assumptions which are not rooted in empirical evidence. Does one include a warning about making such assumptions explicit when teaching calculus and biology? What are the implications of holding up resp. not holding up in the classroom the Ho & Shaw and Kirschner articles as models of so-called mathematical modeling not justified by empirical conditions? *De facto* can we, do we, shall we engage a calculus class in a discussion of the Ho & Shaw and Kirschner articles (among others), bringing up documentation to the attention of the class to justify the criticisms I and others have made? What would happen if we did so? The social, academic, and practical forces against doing so are multiple, and obviously very strong. For an even broader context in which such questions can be raised, including the context of social sciences, cf. my book *Challenges*.

Additional Comments, 1999

After I wrote the above article for the AMS *Notices*, more fundamental criticisms of the Ho and Shaw papers have arisen, some even coming from members of the orthodoxy. In February 1998, *Nature Medicine* published two technical articles and one commentary by Mario Roederer, a professor at the Stanford Medical School, who wrote (p. 145):[1]

> In this issue of *Nature Medicine*, reports by Pakker *et al.* and Goro-chov *et al.* provide the final nails in the coffin for models of T cell dynamics in which a major reason for changes in T cell numbers is the death of HIV-infected cells.

This sentence of course also applies to the Kirschner article.

On the other hand, Roederer (like other critics of the Ho *et al.* article) accepted uncritically the axiom that HIV destroys the immune system, but he provided no justification for this axiom. At the same time he recognized that he and other medical scientists do not know how HIV destroys the immune system, when he concluded:

> Finally, the facts (1) that HIV uses CD4 as its primary receptor, and (2) that CD4$^+$ T cell numbers decline during AIDS, are only an unfortunate coincidence that have led us astray from understanding the immunopathogenesis of this disease. HIV leads to the progressive destruction of all T cell subsets, irrespective of CD4 expression. Ultimately, AIDS is a disease of perturbed homeostasis. Only when we understand how the body regulates T cell numbers will we be able to find the mechanism(s) by which HIV destroys the immune system.

Roederer's assertion "HIV leads to the progressive destruction of all T cell subsets" was and remains unsupported. What does "lead" mean, and what is the evidence for the assertion if "leads" means "causes in some fashion"? No evidence is given in the Roederer article.

A year later, *Nature Medicine* (January 1999) published further criticisms of the David Ho article, partly reinforcing Roederer's "nail in the coffin," and partly going in other direction.[2] These criticisms came from

---

[1]N. Roederer, Getting to the HAART of T cell dynamics, *Nature Medicine* Vol. 4 No. 2 (1998) pp. 145–146.

Pakker et al., Biphasic kinetics of peripheral blood T cells after triple combination therapy in HIV-1 infection: A composite of redistribution and proliferation, *Nature Medicine* Vol. 4 (1998) pp. 208–214.

Gorochov et al., Perturbation of CD4 and CD8 Tcell repertoires during progression to AIDS and regulation of the CD4 repertoire during anti-viral therapy, *Nature Medicine* Vol. 4 (1998) pp. 215–221.

[2]Hellerstein *et al.*, Directly measured kinetics of circulating T lymphocytes in normal and HIV-1-infected humans, *Nature Medicine* Vol. 5 No. 1 (1999) pp. 83–89.

researchers Hellerstein *et al.* at San Francisco General Hospital, UCSF and UC Berkeley.

Accelerated production or destruction of CD4$^+$ T cells? The Ho and Shaw articles in *Nature* had claimed an original increase of T-cell production following HIV infection, in conjunction with high replication of the HIV virus. The Ho *et al* article concluded: "Taken together, our findings strongly support the view that AIDS is primarily a consequence of continuous, high-level replication of HIV-1, leading to virus- and immune-mediated killing of CD4 lymphocytes." On the other hand, Hellerstein *et al.* write in opposition to these claims:

> **p. 87.** The CD4 lymphopenia of HIV-infection was associated with reduced survival (shorter half-life) of CD4$^+$ T cells in the circulation combined with an inability to increase production of CD4 cells in compensation. Although we cannot identify the reason for the failure to increase CD4$^+$ T-cell production... our results are inconsistent with a highly accelerated destruction of circulating CD4$^+$ T cells that overcomes a higher than normal total production rate. ..."

Of course, it becomes important to determine the reasons for the discrepancy between the Ho-Shaw articles and the Hellerstein article. Do they have to do with differences in the people in their samples? With samples which are not statistically significant? With different techniques? With unrecognized artifacts? Ad lib.

In any case, like Roederer, the authors of the new study in *Nature Medicine* accept HIV pathogenesis as an unquestioned axiom. They interpret the data in this context. But the data do not provide evidence for the axiom. The findings do NOT show that a shorter half-life and inability to increase production of CD4$^+$ cells is due to HIV (according to Hellerstein *et al.*).

One also has to take into account the role of prescription drugs. Are HIV negative but sick people with the same symptoms also treated with AZT and protease inhibitors? What is the effect on T-cells of AZT, or protease inhibitors, or whatever purportedly anti-HIV prescription was administered to the AIDS patients?

The data given in the Hellerstein *et al.* article show that T-cell turnover increases in the group exposed to anti-retroviral drugs.

For a lengthier analysis of the Roederer article, the Hellerstein article and other problems with the HIV-AIDS hypothesis, cf. my third *Yale Scientific* article "The Case of HIV: We Have Been Misled," *The Yale Scientific*, Spring 1999, pp. 9–19.

Notices of the AMS
Volume 46, Number 4 (1999): 458

Response to the Steele Prize

Serge Lang

I thank the Council of the AMS and the Selection Committee for the Steele Prize, which I accept. It is of course rewarding to find one's work appreciated by people such as those on the Selection Committee.

At the same time, I am very uncomfortable with the situation, because I resigned from the AMS in early 1996, after nearly half a century's membership. On the one hand, I am now uncomfortable with spoiling what could have been an unmitigated happy moment, and on the other hand, I do not want this moment to obscure important events which have occurred in the last two to three years, affecting my relationship with the AMS.

Indeed, the *Notices*, February 1996, published a 12-page article "Using Mathematics to Understand HIV Immune Dynamics" by Denise Kirschner, pp. 191–202. Having had occasion to be well informed on the issue of HIV pathogenesis and of strong objections (not only by me) against certain abuses of mathematical modeling in connection with HIV, I communicated an extensive file of documentation to AMS higher-ups at the time concerning the hypothesis that HIV is not pathogenic. This hypothesis of course is incompatible with the official orthodoxy. Readers can evaluate some of my documentation, published in a 114-page chapter of my recent book, *Challenges*.

I resigned from the AMS because of the way my documentation was handled in 1996, principally by the *Notices* editor, Hugo Rossi, in connection with the Kirschner article, and the way official responsibilities were met by those involved. Subsequently, about two years later on 5 January 1998, I submitted a 7-page piece for publication in the "Forum" of the *Notices*. The piece explained:

- encouraging events (see for example p. 714 of *Challenges*) which led me to submit a piece for publication in the *Notices*, rather than disengaging as I had done up to that point;

- my detailed objections to the responses which I got from the AMS officials at the time in 1996;

- direct criticisms of the Kirschner piece per se.

I regard all three as important. Although the "Forum" editor, Susan Friedlander, told me she would have accepted the piece, it was rejected for publication by the 1998 editor-in-chief, Tony Knapp. Thus members of the AMS at large have not been informed through official channels of my resignation, nor of the very serious context of continued problems after the resignation, including the rejection of my "Forum" piece. I tried to inform some members by a direct mailing to 160 chairs of departments in January 1998, but such a mailing can reach only few among the total membership (nearly 30,000).

Torn in various directions, sadly but firmly, I do not want my accepting the Steele Prize to further obscure the history of my recent dealings with the AMS.

ISBN 0-387-98804-1

EAN

9 780387 988047 >